Working Postures and Movements

Tools for Evaluation and Engineering

Working Postures and Movements

Tools for Evaluation and Engineering

Edited by
Nico J. Delleman
Christine M. Haslegrave
Don B. Chaffin

CRC PRESS

Boca Raton   London   New York   Washington, D.C.

Library of Congress Cataloging-in-Publication Data

Working postures and movements : tools for evaluation and engineering
/ edited by Nico J. Delleman, Christine M. Haslegrave, and Don B.
Chaffin.
     p. cm.
Includes bibliographical references and index.
ISBN 0-415-27908-9 (alk. paper)
    1. Human engineering. 2. Posture. 3. Human locomotion. I.
Delleman, N. J. II. Haslegrave, C. M. III. Chaffin, Don B. IV. Title.

TA166.W625 2004
620.8′2—dc22   2004049380

Visit the CRC Press Web site at www.crcpress.com

Foreword

The problem of the growing prevalence of musculoskeletal injuries is of worldwide concern, with an increasing amount of attention from researchers and practitioners alike. The introduction of another book dealing with the subject is not surprising, since the science and the practice of prevention are both developing rapidly.

This book is authored and edited by three leading specialists in the field and has chapters by international experts in their subjects. It gives the reader up-to-date information on recent developments together with practical information on how the knowledge of the nature and causes of musculoskeletal damage may be applied for its prevention. In particular, the last decade has seen a more holistic approach to musculoskeletal problems and this is reflected in the text. Although the physical and physiological behaviour of the body at work is as important as ever, the consequences of the work environment and work organisation are recognised as major influences in injury causation. People move, sit, look, or reach for the purposes of their work, and work is done in a context that imposes its own pressures. Thus, this context must be recognised and taken into account when seeking the causes and consequences of working postures.

While teachers and students will find the book a valuable and well-structured source of knowledge, a major aim is to provide a sourcebook for those who are faced with identifying and improving jobs in those many industries where such workplace injuries arise. All the authors have had experience in dealing with industrial problems, and their wish to see the book in use as a practical manual is influenced by their working knowledge and their desire to reduce the vast scale of human damage and suffering that musculoskeletal injuries cause.

E.N. Corlett
Institute for Occupational Ergonomics
University of Nottingham
Nottingham, U.K.

The Editors

Nico J. Delleman, Ph.D., is a Senior Researcher at TNO Human Factors in Soesterberg, the Netherlands, and Professor of Simulation-Based Design Ergonomics at the Laboratory of Applied Anthropology, René Descartes University, in Paris, France. Dr. Delleman was a key member of the International Ergonomics Association Technical Committee that published the first international consensus document on exposure assessment of upper limb repetitive movements and is currently in charge of international standardization of evaluation of working postures and movements. He has a broad research experience in the field of physical ergonomics, e.g., studies on office work, various industrial operations, maintenance, building and construction, cars, aircraft, ships, personal protective equipment, ticket vending machines, and supermarket checkouts.

Christine M. Haslegrave, Ph.D., is a Senior Lecturer at the Institute for Occupational Ergonomics at the University of Nottingham, U.K., and an Editor of the journal *Ergonomics*. Her research has included investigation of the biomechanical demands of manual materials handling tasks, musculoskeletal injury risk assessment, work (re)design in relation to health and safety problems in industry, and vehicle ergonomics in design and manufacture. Dr. Haslegrave was Head of the Ergonomics Section at the Motor Industry Research Association, Nuneaton, for several years, where she was involved in vehicle safety and ergonomics legislative testing. In 1995, she received the Otto Edholm Award of the Ergonomics Society for significant contributions to applied research in ergonomics.

Don B. Chaffin, Ph.D., is the R.G. Snyder Distinguished University Professor and the G. Lawton and Louise G. Johnson Professor of Industrial and Operations Engineering and Biomedical Engineering at the University of Michigan, Ann Arbor. He is past Director of the Center for Ergonomics and past Chair of the Department of Industrial and Operations Engineering. His research has resulted in five books, more than 200 papers, and a set of widely used software to assist engineers involved in designing workplaces and vehicles to accommodate various groups of people, and to ensure that people do not suffer overexertion injuries during the performance of manual tasks of all kinds. Dr. Chaffin's work has resulted in his election to Fellow status in five different international professional and scientific organizations, including the American Association for the Advancement of Science. In 1994 he was elected to membership in the U.S. National Academy of Engineering.

Contributors

Arne Aarås, M.D., Ph.D., F.Erg.S. Professor of Occupational Medicine and Optometry, Department of Optometry and Visual Science, Buskerud University College, Kongsberg, Norway

Tom Bendix, M.D., Dr. Med. Sci. Professor of Clinical Biomechanics, Back Research Center, Funen Hospital/University of Southern Denmark, Ringe, Denmark

Bruce Bradtmiller, Ph.D. President, Anthrotech, Inc., Yellow Springs, Ohio, U.S.A.

Bob (R.S.) Bridger, Ph.D. Head, Human Factors, Institute of Naval Medicine, Alverstoke, United Kingdom and External Examiner of U.K. Health and Safety Inspectors' Course, Heriot-Watt University, Edinburgh, U.K.

Margaret I. Bullock, AM, Ph.D., FTSE, FIEA, FESA Professor, School of Health and Rehabilitation Sciences, The University of Queensland, Queensland, Australia

Keith Case, Ph.D., FBCS, F.Erg.S., C.Eng. Professor of Computer Aided Engineering, Mechanical and Manufacturing Engineering, Loughborough University, Loughborough, U.K.

Don B. Chaffin, Ph.D. G. Lawton and Louise G. Johnson Professor, Center for Ergonomics, University of Michigan, Ann Arbor, Michigan, U.S.A.

Rakié Cham, Ph.D. Assistant Professor, Department of Bioengineering, School of Engineering, University of Pittsburgh, Pittsburgh, Pennsylvania, U.S.A.

Marita Christmansson, Ph.D. Associate Professor, Department of Product and Production Development, National Institute of Working Life West, Industry and Human Resources, Chalmers University of Technology, Gothenborg, Sweden

Daniela Colombini, Ph.D., Eur. Erg. Professor of Occupational Medicine, Coordinator, Ergonomics Section, Center for Occupational Medicine (CEMOC), Milan, Italy

Hein A.M. Daanen, Ph.D.   Professor and Head, Department of Performance & Comfort, TNO Human Factors, Soesterberg, the Netherlands

Nico J. Delleman, Ph.D.   Senior Researcher, Department of Performance & Comfort, TNO Human Factors, Soesterburg, the Netherlands and Professor of Simulation-Based Design Ergonomics, René Descartes University, Paris, France

Giovanni De Vito, M.D., M.Sc.   Researcher in Occupational Medicine, Department of Clinical Medicine and Prevention, University of Milan–Bicocca, Monza, Italy

Colin G. Drury, Ph.D., CPE   UB Distinguished Professor of Industrial Engineering, University at Buffalo, State University of New York, Buffalo, New York, U.S.A.

Martin Freer, B.Sc., DPS   Sector Director-Road Transportation, CCD Design & Ergonomics Ltd., Cambridge, U.K.

Claire C. Gordon, Ph.D.   Senior Research Scientist, U.S. Army Natick Soldier Center, Natick, Massachusetts, U.S.A.

Antonio Grieco, M.D., Ph.D.   Late Professor of Occupational Preventive Medicine and Past Director, "Clinica del Lavoro Luigi Devoto," and Chair, Scientific Research Board, Department of Occupational Health, University of Milan, Milan, Italy

M. Susan Hallbeck, Ph.D., PE, CPE   Associate Professor, Industrial and Management Systems Engineering Department, University of Nebraska, Lincoln, Nebraska, U.S.A.

Christine M. Haslegrave, Ph.D., F.Erg.S., C.Eng., MIEE, European Engineer   Senior Lecturer, Institute for Occupational Ergonomics, University of Nottingham, Nottingham, U.K.

Herbert Heuer, Dr.rer.nat.   Professor of Work Psychology and Experimental Psychology, Institute for Occupational Physiology at the University of Dortmund, Dortmund, Germany

Wolfgang Jaschinski, Dr.-Ing.   Head, Research Group Individual Visual Performance, Institute for Occupational Physiology at the University of Dortmund, Dortmund, Germany

Hans W. Jürgens, M.D., Ph.D.   Professor and Head, Physiological Anthropology Research Group, University of Kiel, Kiel, Germany

Roland Kadefors, Ph.D. Professor of Applied Ergonomics, National Institute for Working Life, Gothenborg, Sweden

Makiko Kouchi, Ph.D. Senior Researcher, Digital Human Research Center, National Institute of Advanced Industrial Science and Technology, Tokyo, Japan

Young-Suk Lee, Ph.D. Professor of Thermal Physiology and Clothing Ergonomics, Department of Clothing and Science, College of Human Ecology, Chonnam National University, Gwangju, Korea

Russell Marshall, M.Eng., Ph.D., AMIEE Lecturer, Design Ergonomics Group, Department of Design and Technology, Loughborough University, Loughborough, U.K.

Svend Erik Mathiassen, Ph.D. Professor of Production Ergonomics and Associate Professor of Occupational Medicine, Department of Occupational and Environmental Medicine, Lund University Hospital, Lund, Sweden

Giovanni Molteni, M.D., M.Sc. Associate Professor of Occupational Medicine, Department of Clinical Medicine and Prevention, University of Milan–Bicocca, Monza, Italy

Mark Morrissey, B.Sc., DPS General Manager, Safework, Inc., Montreal, Quebec, Canada

Brian E. Moyer, M.Sc. Research Associate, Department of Bioengineering, University of Pittsburgh, Pittsburgh, Pennsylvania, U.S.A.

Maury A. Nussbaum, Ph.D. Associate Professor, Grado Department of Industrial and Systems Engineering, Virginia Polytechnic Institute and State University, Blacksburg, Virginia, U.S.A.

Enrico Occhipinti, Ph.D, Eur.Erg. Professor of Occupational Medicine, Director, Center for Occupational Medicine (CEMOC), Milan, Italy

Victor L. Paquet, Sc.D., AEP Assistant Professor, Department of Industrial Engineering, University at Buffalo, State University of New York, Buffalo, New York, U.S.A.

J. Mark Porter, B.Sc., Ph.D., F.Erg.S., Eur.Erg. Professor of Design Ergonomics and Head of Department of Design and Technology, Loughborough University, Loughborough, U.K. and Managing Director, SAMMIE CAD Ltd., Loughborough, U.K.

Ulrich Raschke, Ph.D.   Program Manager, Human Simulation Products, UGS PLM Solutions, Ann Arbor, Michigan, U.S.A.

Mark S. Redfern, Ph.D.   Professor of Bioengineering, Department of Bioengineering, University of Pittsburgh, Pittsburgh, Pennsylvania, U.S.A.

Steve Rice, B.Sc.   Associate Technical Fellow, Phantom Works, The Boeing Company, Huntington Beach, California, U.S.A.

Kathleen M. Robinette, Ph.D.   Principal Research Anthropologist, Air Force Research Laboratory, Wright-Patterson Air Force Base, Ohio, U.S.A.

Andreas Seidl, Dr.-Ing.   CEO, Human Solutions GmbH, Kaiserslautern, Germany

Frans C.T. Van der Helm, Ph.D.   Professor of Biomechatronics, Man-Machine Systems and Control Group, Department of Mechanical Engineering, Delft University of Technology, Delft, the Netherlands

Jaap H. Van Dieën, Ph.D.   Professor of Biomechanics, Institute for Fundamental and Clinical Human Movement Sciences, Faculty of Human Movement Sciences, Vrije Universiteit Amsterdam, Amsterdam, the Netherlands

DirkJan (H.E.J.) Veeger, Ph.D.   Assistant Professor, Faculty of Human Movement Sciences, Institute for Fundamental and Clinical Movement Sciences, Vrije Universiteit Amsterdam, Amsterdam, the Netherlands and Associate Professor, Man-Machine Systems and Control Group, Department of Mechanical Engineering, Delft University of Technology, Delft, the Netherlands

Xuguang Wang, Ph.D.   Researcher in Biomechanics, Biomechanics and Impact Mechanics Laboratory (LBMC), French National Institute for Transport and Safety Research (INRETS), Bron, France

Richard Wells, M.Eng, Ph.D.   Professor of Kinesiology, Faculty of Applied Health Sciences, University of Waterloo, Waterloo, Ontario, Canada

Gregory F. Zehner, Ph.D.   Senior Anthropologist, Air Force Research Laboratory, Human Effectiveness Directorate, Wright-Patterson Air Force Base, Ohio, U.S.A.

Contents

1

Introduction

Nico J. Delleman, Christine M. Haslegrave, and Don B. Chaffin

CONTENTS

People adopt postures (mostly without any conscious decision) to deal with the workplaces and surrounding environments that they find. The goal is to perform an action or a task and the posture is chosen to achieve that goal. This adaptability is made possible by the complex anatomy of the musculo-skeletal system, which has so many degrees of freedom that the human body has a great repertoire of postures and postural adjustments. The redundancy provided by the multiple degrees of freedom also means that several alter-native postures may be possible for performance of a given task. Not all, however, are healthy postures, particularly when they are adopted for extended periods of time, when they cause localized loading on joints, mus-cles, ligaments, or other tissues, or when force must be exerted. The chapters of this book discuss and illustrate the implications of working postures and their effects on different parts of the body.

The posture adopted for a task is most directly determined by the dimen-sions and arrangement of the workplace and equipment used (particularly in relation to work height, reach distance, field of view, and space to move freely). The dimensions and arrangement constrain the range of postures that are possible while performing the task and, in poorly designed work-places, are often found to make a healthy posture difficult or impossible.

0-415-27908-9/04/$0.00+$1.50
© 2004 by CRC Press LLC

FIGURE 1.1
Sewing stiff fabric at a machine with knee and foot controls operated by the left leg.

This is clearly illustrated by the postures shown in Figure 1.1 and Figure 1.2. Many other environmental factors, both physical and psychosocial, can also have an influence on posture and these influences are explained in detail in the relevant chapters.

General principles for workplace and work design are outlined, but it is also important to remember that the effects of any of the influencing factors may vary with the individual worker. This is most obvious when considering the anthropometric effects of body size. If, for example, a high workbench is installed, it may well suit taller workers and allow them to adopt healthy working postures, but it will probably mean that shorter workers will have to work with arms raised and, consequently, experience greater fatigue and have difficulty in performing precision tasks. This will disproportionately disadvantage women and older workers (who are on average shorter than young male workers within any national population). The converse situation is of course equally undesirable; a low workbench may suit the smaller members of the work group but taller workers will have to stoop to perform their tasks and, as a result, are likely to experience back muscle fatigue and possibly risk developing back problems. The posture that an individual can

FIGURE 1.2
Press operation.

adopt to perform a particular task is principally determined by the interaction between the layout of their workplace and their own anthropometry. Similar examples could be given for the varying effects of other physical, psychological, and psychosocial influences according to the individual worker's own characteristics, experience, or background.

1.1 Musculoskeletal Injuries

Extreme or awkward postures are recognized as one of the major risk factors for musculoskeletal injury. A critical review of the scientific evidence for causal relationships with physical work factors, carried out for the U.S. National Institute of Occupational Safety and Health (NIOSH) (Bernard 1997), found strong evidence of posture as a risk factor for disorders of the neck and neck/shoulder region and some convincing evidence for shoulder disorders, back injuries, and hand/wrist tendinitis. For example, the odds ratio for low back pain has been found by Punnett et al. (1991) to be in the range of 3 to 6 with exposure to non-neutral trunk postures for 0 to 10% of cycle time. The NIOSH report found insufficient evidence to show a causal relationship with posture for elbow disorders or for carpal tunnel syndrome, but it did not find evidence that there was no effect for these disorders. More research is needed to clarify the effects of posture, based on prospective rather than cross-sectional studies.

Musculoskeletal injuries and disorders (of the back, upper limbs, and lower limbs) are a cause of serious concern for workforces (Hagberg et al. 1995, Bernard 1997, Buckle and Devereux 1999, Op de Beeck and Hermans

2000). They are the most prevalent work-related reason for worker absence in many industries.

Recent studies in the U.S. and European Union provide us with insights into the magnitude of the problem. In the U.S., employers in private industries (which represent 75% of the working population of 135 million workers) report around 7 million cases of work-related musculoskeletal injury each year, with 25% of these leading to lost working days and others to restricted activity (National Research Council 2001). It is also estimated that there are 5 to 6 million cases of work-related back pain each year in the entire U.S. working population, leading to 100 million working days lost.

Also according to the National Research Council (2001), musculoskeletal disorders account for about one third of workers' compensation costs in the U.S., and the total cost of these together with the indirect costs to employers, affected individuals, and society (including lost productivity, uncompensated lost wages, personal losses such as household services, administration of the programs, lost tax revenues, social security replacement benefits, and other related costs) is estimated to be 0.8% of the U.S. gross domestic product. Available estimates for the European Union put the total cost associated with musculoskeletal injuries at between 0.6 and 1% of gross national product (European Agency for Safety and Health at Work 2000). In the European Union (with 150 million workers), 350 million working days are lost due to work-related health problems and approximately half of these involve musculoskeletal injury.

Many of these injuries are caused by static postures, sometimes with forceful exertions or repetitive movements, that have to be maintained for many hours a day — such as when intensively using data entry devices, performing assembly work, food processing, manual materials handling, machinery operations, or vehicle driving, among other activities. The Third European Survey on Working Conditions (Paoli and Merllié 2001) showed the intense nature of work in modern industry and service organizations. This survey showed that 47% of workers work in painful, tiring postures and that 37% carry heavy loads for more than a quarter of their time. The survey also showed that 56% of workers in Europe work at very high speed and that 60% work to tight deadlines. Moreover, despite increasing automation in industry, 15% of tasks have highly repetitive work cycles (with cycle time less than 5 s).

Similar results were found in a study of self-reported working conditions in a representative sample of the labor force carried out for the Health and Safety Executive in the U.K. (Jones et al. 1997). Among the workers, 13% reported working in awkward or tiring positions always or nearly always, 38% for at least a quarter of their time, and 45% sometimes. In relation to exposure to repetitive tasks, 64% of men and 67% of women said that their work involved repeating the same sequence of movements many times — just over 60% for at least a quarter of their working time — and for about 32% their job always or nearly always involved repetition.

1.2 Studies of Posture and Movement

Working postures and movements have long interested scientists, and concern for their influence on occupational health has an equally long history. Leonardo da Vinci (1452–1519) not only drew magnificent sketches of postures for various activities but also undertook related scientific investigations. Observations such as the following appear in his notebooks (MacCurdy 1954):

> The above-mentioned muscles are not firm except at the extremities of their receptacles and at the extremities of their tendons; and this the Master has done in order that the muscles may be free and ready to be able to grow thicker or shorter or finer or longer according to the necessity of the thing which they move.
>
> — **Fogli A2r**

> The centre of the weight of the man who raises one of his feet from the ground rests above the centre of the sole of the foot.
>
> — **Fogli B21v**

In the 18th century, Bernardino Rammazzini wrote of serious musculoskeletal injuries among workers in his own time and linked these to their postures at work.

> Manifold is the harvest of diseases reaped by craftsmen.... As the cause I assign certain violent and irregular motions and unnatural postures ... by which ... the natural structure of the living machine is so impaired that serious diseases gradually develop.
>
> — **Essai sur les maladies des artisans, Chapters 1 and 52, 1777**
> **(translated from the Latin text *De Morbis Artificum***
> **by M. de Fourcroy, as quoted in Tichauer, 1978)**

Toward the end of the 19th century and during the first half of the 20th century, attention turned to the optimization of manufacturing production processes, where motion patterns of industrial tasks were studied to maximize efficiency and production, utilizing techniques developed by Frederick Taylor and by Frank and Lillian Gilbreth. This developed into the modern MTM (motion–time measurement) systems used in many industries today; these are based on video analysis of techniques used in repetitive operations and assembly tasks. Frank Gilbreth, in particular, studied the effects of posture and workplace layout, for example, devising an adjustable workbench for bricklayers to remove the necessity of bending to pick bricks up from floor level (Barnes 1968).

FIGURE 1.3
Lifting techniques suggested for flour mill workers by the Ministry of Labour and National Service (1944).

A more developed understanding of some biomechanical principles had been achieved by the 1940s, as illustrated by a safety pamphlet issued by the Ministry of Labour and National Service (1944) in the U.K. The man at the back of the left-hand photograph in Figure 1.3 illustrates their guidance: "The muscles of the groin should be given some measure of protection from strain by adoption of a stance that is not too wide during lifting, while the muscles of the back are buttressed by utilising the leg and thigh muscles with flexion and extension at the knee joints to raise loads by a short upward thrust." The right-hand photograph shows team handling with a very simple handling aid used to assist flour mill workers in lifting heavy sacks onto their shoulders. General lifting guidelines given in the pamphlet would still be recognized in modern training courses: "Lifting from floor level is achieved by flexion followed by extension of the knee joint; the back muscles are thereby protected." "Here the knees are stiff and the strain of raising the load falls on the back muscles." Not all the advice in the pamphlet would be accepted today (including the unqualified advice to adopt a narrow stance while lifting) because, of course, much greater understanding of the biomechanics of loading on the musculoskeletal system has been achieved.

Some of the earliest applications of cinematographical techniques to record posture came from Eadweard Muybridge, starting in the 1860s (Muybridge 1955). His fascination with the potential of photography led him to capture movements during the performance of tasks such as a woman lifting a young child or a man wielding a pickaxe (Figure 1.4 and Figure 1.5). The photographic sequences show beautifully the fluidity and coordination of the movements. The woman's handholds change several times as she grasps, lifts, and supports the child. The man adopts postures that bring different muscle groups into play and that make complex use of body inertia as he wields the pickaxe. Muybridge's motion sequences were recorded by high-speed

FIGURE 1.4
Detail from Muybridge's "Girl being picked up and held by woman." (From Muybridge, E., 1955. *The Human Figure in Motion*. New York: Dover. With permission.)

FIGURE 1.5
Detail from Muybridge's "Man swinging a pickaxe." (From Muybridge, E., 1955. *The Human Figure in Motion*. New York: Dover. With permission.)

photographs with exposures ranging up to 1/6000th of a second. The Gilbreths later invented micromotion study, which used motion pictures together with a timing device to identify fundamental elements of an operation (Barnes 1968), a far more momentous development than is easily apparent in the current digital age of computerized recording and analysis.

Time study and motion study, as developed by these early practitioners, took account of some ergonomic issues and these were recognized within industry. Attempts were made to develop standards of best practice, for example Frederick Taylor's (1911) "The Principles of Scientific Management," Vernon's (1921) "Industrial Fatigue and Efficiency" as well as his reports for the Industrial Fatigue Research Board, or the pamphlet on "Weight Lifting by Industrial Workers" (Ministry of Labour and National Service 1944), which included the photographs in Figure 1.3. However, later applications within industry tended to give much greater emphasis to productivity gains and to forget the contributions of human factors of fatigue and vulnerability to musculoskeletal injury.

Nevertheless, in scientific research during the second half of the 20th century, systematic biomechanical analysis developed, building on those earlier observations and recordings of posture. This is described in Chaffin et al. (1999). Wilfred Dempster's early work demonstrated postural skills such as the use of body inertia to enhance the strength provided by muscular effort, also showing that the posture of the musculoskeletal framework can be represented as a series of closed and open chains formed by the body segments (Dempster 1955). He showed how critical stability is in maintaining a working posture, forming the hypothesis that the distribution of body mass in relation to its support area forms an anchorage for the body in using its muscles. Extensive work continued at the University of Michigan to develop biomechanical models for analyzing loading on the musculoskeletal system and later for the prediction of work postures (Chaffin et al. 1999). The results of these studies and those of many other research groups can be found in the present book.

At each stage, perhaps, advances in technology have triggered new and significant developments in scientific knowledge of posture and its effects. Today, we have sophisticated capabilities for data capture in posture recording (using photographic, optical, or magnetic systems), equally fantastic animation capabilities to represent human body movements, and powerful computer analysis and simulation to represent and evaluate biomechanical measures of the effects of posture and task parameters on the musculoskeletal system. Researchers are using these techniques to develop our understanding of human physical work and its consequences, but there are obviously still limitations on our knowledge. Not least of these is our limited understanding of the physiological, biochemical, neurological, psychological, and psychosocial influences that affect the onset and severity of, as well as the recovery from, musculoskeletal injury and disease processes, but equally we are still exploring the variability in human behavior and influences of anthropometry and aging on task movements and techniques. The chapters of this book describe these developments and present the current state of the art.

1.3   International Approaches for Evaluation and Engineering

At the international level, the problem of musculoskeletal injuries has received major attention of standards organizations, including the International Organization for Standardization (ISO), a worldwide federation of national standards bodies; the European Committee for Standardization (CEN); and the American National Standards Institute (ANSI).* The most

---

* All standards mentioned in this chapter can be obtained through national standards bodies (ANSI standards through ANSI in Washington, D.C., U.S.A., EN standards also through CEN in Brussels, Belgium, and ISO standards also through ISO in Geneva, Switzerland). Some of the standards mentioned may still be under preparation.

directly relevant standards are ISO 11226 (ISO 2000), a tool for evaluation of existing work situations (refer to Section 1.3.1), and prEN 1005-4 (CEN 2003a), a tool for evaluation during a design/engineering process (refer to Section 1.3.2), although various other health and safety or product safety and design standards are also very relevant (some of which are industry specific). Other standards relevant to repetitive work characterized by the presence of other risk factors for musculoskeletal injury (such as high frequency of movements, force exertion, or long duration) are ISO/CD 11228-3 (ISO 2003), EN 1005-3 (CEN 2002), and prEN 1005-5 (CEN 2003b).

The U.S. National Institute for Occupational Safety and Health (NIOSH) promulgated guidelines as a "Guide to Manual Lifting" in 1981 (which was updated in 1991), based on biomechanical, psychophysical, and epidemiological data and models (Waters et al. 1994). NIOSH has also published a critical review of epidemiological evidence on work-related musculoskeletal disorders (Bernard 1997). Another important, slightly earlier, review of these issues was made by Hagberg et al. (1995). The European Agency for Health and Safety at Work too has published reports on work-related upper limb disorders (Buckle and Devereux 1999) and work-related low back disorders (Op de Beeck and Hermans 2000). All these also address the approaches that can be adopted for assessment, surveillance, and prevention.

The standards and reviews mentioned above represent the consensus of international experts as regards risk assessment of working postures and movements. A program and processes for managing work-related musculoskeletal injuries can be found in the Draft ANSI Z365 standard (ANSI 1997). In many countries, specific legislation on health and safety or working conditions may also apply, but the book does not attempt to give an overview of legislation or standards applicable to particular countries or industries.

The adoption of international standards for evaluating work postures represented an important milestone in prevention and control of work-related musculoskeletal disorders but, as the following chapters in the book show, research is progressing to gain a better understanding of the effects of postures and movements and new, more sophisticated techniques are being developed for assessing risk.

1.3.1 ISO 11226 — Evaluation of Existing Work Situations

ISO 11226 (ISO 2000) establishes ergonomic recommendations for different work tasks, which can be used by those who are familiar with the basic concepts of ergonomics in general and of working postures in particular. The standard contains an approach to determine the acceptability of working postures that are mainly static. The recommendations in the standard address postures with no, or only minimal, external force exertion, while taking into account body angles and time aspects. They are mainly based on experimental studies regarding the musculoskeletal load, discomfort/pain, and endurance/fatigue related to static postures.

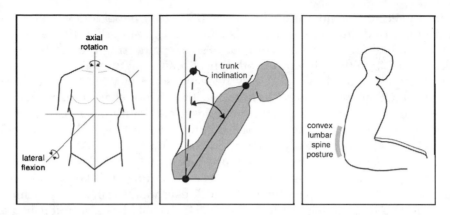

FIGURE 1.6
ISO 11226 variables for evaluation of trunk posture (ISO 2000).

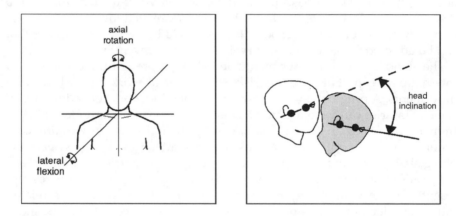

FIGURE 1.7
ISO 11226 variables for evaluation of head posture (ISO 2000).

The main part of the standard consists of specific recommendations for evaluation of posture. The evaluation procedure considers various body segments and joints independently in one or two steps. Figure 1.6 through Figure 1.8 show some examples of body segments evaluated. The first step considers only the body angles (and, for these, recommendations are mainly based on risks for overloading passive body structures such as ligaments, cartilage, and intervertebral discs). An evaluation in the first step may lead to one of the three results of "posture acceptable," "go to step 2 of the evaluation," or "posture not recommended." An evaluation result of "go to step 2" means that the duration of the working posture will also need to be considered (and recommendations in step 2 are based on endurance data).

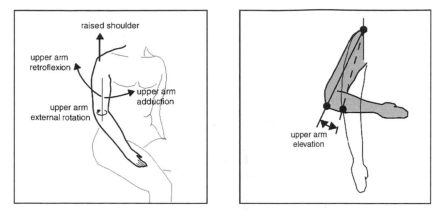

FIGURE 1.8
ISO 11226 variables for evaluation of shoulder and upper arm posture (ISO 2000).

1.3.2 prEN 1005-4 — Evaluation during a Design/Engineering Process

The European standard prEN 1005-4 (CEN 2003b) is intended to present guidance to the designer of machinery in assessing and controlling health risks due only to machine-related postures and movements. As with ISO 11226, the prEN 1005-4 standard specifies recommendations for postures and movements with no or only minimal external force exertion. The standard supports the health and safety requirements of the EEC Machinery Directive (Council of the European Communities 1989), and its subsequent amendments, which applies to all machines traded within the European Union. The technique and principles could, however, be applied elsewhere and for other products as well.

The prEN 1005-4 standard adopts a stepwise risk assessment approach for assessing postures and movements as part of the machinery design process, and provides guidance during the various design stages (as shown in Figure 1.9). The approach is based on the U-shaped model presented in Figure 1.10, which suggests that health risks increase when the task approaches either end of the curve, that is, if there is a static posture (with little or no movement) or if movement frequencies are high.

1.4 Risk Assessment

From the earlier discussion of individual and environmental factors influencing posture, it should be apparent that risk assessments should be made for the full range of individuals within the work group of concern. It is not

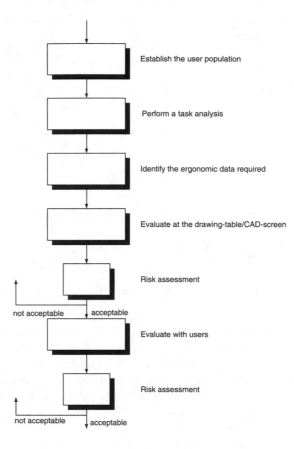

FIGURE 1.9
prEN 1005-4 flowchart illustrating the risk assessment approach (CEN 2003a). In the standard itself the empty boxes contain a reference to a particular section of the standard.

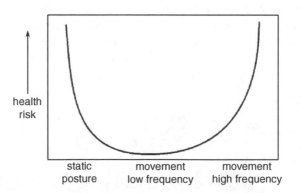

FIGURE 1.10
prEN 1005-4 diagrammatic model of the health risks associated with postures and movements (CEN 2003a).

sufficient to make an assessment from the observation of a single person performing the task without considering how the effects (and assessment) might be altered for members of the work group who are smaller or taller, who are less skilled or have less training, who are wearing different clothing, who are using different tools, who are working in more extreme environmental conditions, or who are working to a deadline or at a high speed determined by a machine or production line. Assessments should consider all aspects of the workers' jobs, with their individual exposures over the working day and working week, and not simply assess a particular task in isolation.

Consideration of representative individuals' exposure during their various work tasks will then allow an assessment of exposure within the workforce to risk factors for musculoskeletal injuries — and considerations of the range of such exposure within the workforce will guide the priorities for implementing improvements. None of the techniques presented can predict a particular individual's likelihood of developing a musculoskeletal injury. There are too many variables involved — in exposure to work conditions and allocation of tasks within a work team as well as in individual susceptibility to musculoskeletal stress.

It should also be emphasized that a healthy workplace is one in which posture can be changed, either while still performing the task or through varying the task(s) performed. Continuous holding of any posture or frequent repetition of the same posture is not healthy.

1.5 Overview of the Book

Today, it is accepted internationally that work-related musculoskeletal injuries constitute an industrial problem, as well as that poor working postures and movements are a key cause of the injuries. In addition to musculoskeletal injury, pain and fatigue may lead to reduced productivity, and deteriorated posture and movement control. The latter can increase the risk of errors and may result in reduced quality and hazardous situations. Obviously, all this is associated with serious costs to both industry and society.

Occupational health professionals, ergonomists, production engineers, and product designers are in the position to avoid the need for painful and tiring postures and movements by creating appropriate working conditions, which in turn will better motivate workers and improve morale, quality, and efficiency for companies. To be well prepared for this task, it is necessary to know how to evaluate postures and movements, and to understand how these are determined by the work tasks and work environment, as well as the design guidelines and tools that are available to achieve less stressful working conditions for different types of task. The chapters following contain the latest scientific knowledge describing the anthropometry, biomechanics, physiology, psychophysics, and human perceptual-motor control related to posture and movement in work activities, together with assessment tools that can be applied to postures of all the major body segments. The latest developments of digital technology for capturing the shape and posture of the body in three dimensions and for simulating work through digital human models are presented with illustrations of their applications in computer-aided design and engineering for a variety of industries. Furthermore, these allow human factors knowledge to be applied at the very earliest stages in the design process. Chapter authors also address seating concepts, hand tool and pedal design, foot–floor interfaces, and aspects of work organization (task duration, breaks, handling frequency) as they affect human performance and interventions to reduce musculoskeletal injury. The book shows the reader which work conditions and individual worker characteristics are important in given situations and which should be considered as part of a risk assessment.

Much of this book is devoted to practical techniques that can be used in the assessment of workplaces and of postures adopted by workers when performing their tasks. With current knowledge, complete guidelines cannot be formulated. The very complexity of human posture (resulting from the interacting effects of workplace and work conditions and the multiple degrees of freedom of human musculoskeletal anatomy) means that research studies to date have not been able to cover all work situations. There are still many gaps in our knowledge and the chapters in the book identify both these and related important research topics, which we hope will stimulate future studies by our readers. The assessment techniques described will allow industrial practitioners to review and improve the design of their workplaces, tasks, and work organization.

References

ANSI, 1997. *Draft ANSI Z365 Control of Cumulative Trauma Disorders*, Washington, D.C.: American National Standards Institute.

Barnes, R.M., 1968. *Motion and Time Study*, New York: John Wiley.

Bernard, B., 1997. *Musculoskeletal Disorders and Workplace Factors: A Critical Review of Epidemiological Evidence for Work-Related Musculoskeletal Disorders of the Neck, Upper Extremity, and Low Back*, DHHS (NIOSH) Publication 97-141, Cincinnati: National Institute for Occupational Safety and Health.

Buckle, P. and Devereux, J., 1999. Work-Related Neck and Upper Limb Musculoskeletal Disorders. Report for European Agency for Safety and Health at Work, Luxembourg: Office for Official Publications of the European Communities.

CEN, 2002. *EN 1005-3 Safety of Machinery — Human Physical Performance*. Part 3: *Recommended Force Limits for Machinery Operation*, Brussels: European Committee for Standardization.

CEN, 2003a. *prEN 1005-4 Safety of Machinery — Human Physical Performance*. Part 4: *Evaluation of Working Postures and Movements in Relation to Machinery*, Brussels: European Committee for Standardization.

CEN, 2003b. *prEN 1005-5 Safety of Machinery — Human Physical Performance*. Part 5: *Risk Assessment for Repetitive Handling at High Frequency*, Brussels: European Committee for Standardization.

Chaffin, D.B., Andersson, G.B.J., and Martin, B., 1999. *Occupational Biomechanics*, 3rd ed., New York: John Wiley.

Council of the European Communities, 1989. Council Directive 89/392/EEC of 14 June 1989 on the approximation of the laws of the Member States relating to machinery, *Official Journal of the European Communities*, No. L183, 29 June 1989.

Dempster, W.T., 1955. Space requirements of the seated operator, WADC Tech. Rep. 55-159, Wright-Patterson Air Force Base, OH: Aerospace Medical Research Laboratories.

European Agency for Safety and Health at Work, 2000. Inventory of socio-economic information about work-related musculoskeletal disorders in the Member States of the European Union, Factsheet Issue 9, Bilbao: European Agency for Safety and Health at Work.

Hagberg, M., Silverstein, B., Wells, R., Smith, M.J., Hendrick, H.W., Carayon, P., and Perusse, M., 1995, *Work Related Musculoskeletal Disorders (WMSDs): A Reference Book for Prevention*, London: Taylor & Francis.

ISO, 2000. *ISO 11226 Ergonomics — Evaluation of Static Working Postures*, Geneva: International Organization for Standardization.

ISO, 2003. *ISO/CD 11228-3 Ergonomics — Manual Handling — Part 3: Handling of Low Loads at High Frequency*, Geneva: International Organization for Standardization.

Jones, I.R., Hodgson, J.T., and Osman, J., 1997. Self-Reported Working Conditions in 1995 — Results from a Household Survey, Norwich, U.K.: HMSO.

MacCurdy, E., 1954. *The Notebooks of Leonardo da Vinci*, Vol. I, London: Reprint Society.

Ministry of Labour and National Service, 1944. Weight lifting by industrial workers (rev. ed., November 1943), Safety Pamphlet 16, London: HMSO.

Muybridge, E., 1955. *The Human Figure in Motion*, New York: Dover.

National Research Council, 2001. Musculoskeletal Disorders and the Workplace: Low Back and Upper Extremities, Washington, D.C.: National Academy Press.

Op de Beeck, R. and Hermans, V., 2000. *Research on Work-Related Low Back Disorders*, Luxembourg: Office for Official Publications of the European Communities.

Paoli, P. and Merllié, D., 2001. *Third European Survey on Working Conditions 2000*, Luxembourg: Office for Official Publications of the European Communities.

Punnett, L., Fine, L.J., Keyserling, W.M., Herrin, G.D., and Chaffin, D.B., 1991. Back disorders and nonneutral trunk postures of automobile workers. *Scandinavian Journal of Work, Environment & Health*, 17, 12, 337–346.

Taylor, F.W., 1911. *The Principles of Scientific Management*, New York: Harper & Bros.

Tichauer, E.R., 1978. *The Biomechanical Basis of Ergonomics*, New York: Wiley-Interscience.

Vernon, H.M., 1921. *Industrial Fatigue and Efficiency*, London: George Routledge & Sons.

Waters, T.R., Putz-Anderson, V., and Garg, A., 1994. *Applications Manual for the Revised NIOSH Lifting Equation*, DHHS (NIOSH) Publication 94-110, Cincinnati, OH: National Institute for Occupational Safety and Health.

2

Anthropometry

CONTENTS

Introduction

Anthropometry is the study of the physical, dimensional measurement of humans. Although it has a long history as a subdiscipline of physical anthropology, its use as an applied science is more recent and dates to the mid-20th century (e.g., Dempster 1955). Currently, a distinction is made between applied anthropometry (sometimes called engineering anthropometry) and the more academic anthropometry that is used in research in conjunction with other areas of human biology (Meredith 1953, Jensen and Nassas 1988, NCHS 1996).

In this chapter, a further distinction is made within applied anthropometry, based on the techniques used to gather the raw data. In the first section, the focus is on anthropometric data collected using traditional techniques and using traditional instruments, such as the anthropometer, calipers, and measuring tape. Traditional anthropometric instruments provide measurements of the body or its parts in one dimension, such as the length of the forearm. They can also provide measurements of a two-dimensional body feature along a plane, such as in head circumference. The second section is devoted to the use of various three-dimensional systems, which can be used to generate surface representations of the body shape. They can also provide measures of surface area or volume, which are generally difficult to obtain with traditional instruments.

2.1   Traditional Anthropometry

Bruce Bradtmiller, Claire C. Gordon, Makiko Kouchi, Hans W. Jürgens, and Young-Suk Lee

Introduction

In applied anthropometry, the dimensions are used for various design tasks. The type of dimension that should be used depends very much on the specific design task. For example, when designing a standard doorway or an emergency door in an aircraft, one-dimensional data are appropriate. Complex designs, such as aircraft cockpits, often require the use of many one-dimensional measurements simultaneously. When designing a soft and flexible cap for the head, the two-dimensional head circumference is appropriate. However, when designing a hardshell helmet for the same head, a three-dimensional representation that takes into account curvatures and shapes would be more appropriate.

With respect to the posture of working positions, usually the simultaneous use of one-dimensional measurements is appropriate, because they can be directly related to the workspace in question. For example, to establish the level of a work surface for a simple standing assembly position, the key dimension would be elbow (or waist) height. If the maximum depth of the work surface is critical, then additional dimensions would be forward arm reach and possibly abdominal depth (because the abdomen contacting the work surface can effectively shorten forward arm reach). However, design for more complex working positions, such as assembly operations in a small enclosed space (as when building components of jet aircraft, for example), or repair operations involving leaning into the work space (in automobile engine repair, for example) usually requires the use of many dimensions and may best be accomplished with digital human models.

Posture and anthropometry are completely linked. The posture of the subject directly influences anthropometric values when data are collected. The reverse is also true — the anthropometric dimensions of an individual directly affect his or her posture. For example, a male who is 200 cm tall and a female who is 150 cm tall have a very different posture when sitting in an identical chair. The tall man's feet will touch the floor and he may sit well back into the chair. The short female's feet may not touch the floor, and she will likely sit forward in the chair to increase the comfort at the knees. Ideally, workspaces would be sufficiently adjustable so that all workers could adopt a posture that minimizes likelihood of injury and fatigue, but to date that goal has not been achieved.

Despite the obvious relationship, however, anthropometry does not always define a precise seated posture in a given space. Specifically, Reed (1998) showed that anthropometry is a good predictor of preferred seat fore-aft location for a driving task, but was a poor predictor of the preferred angle of the torso (with less than 20% of variance explained). Further, there is a considerable variability in posture, even given similar anthropometry. This variability is due to fatigue, the specific workload, and the specific work environment.

2.1.1 Posture in Data Collection

The importance of posture in anthropometric data collection has been recognized since the mid-19th century. European anthropologists, frustrated over an inability to duplicate each other's skull measurements, realized that measurements to the top of the head could be anywhere on that curved surface. They resolved that problem by standardizing the position of the skull so that the lower margin of the orbit and the top of the external auditory meatus (bony space for the ear) were aligned horizontally, and named the orientation the Frankfurt plane, after the city in which the conference was held (Garson 1885, Martin 1914, Hrdlicka 1920). Although this plane ties head orientation to the relative positions of the eyes and ears and is arbitrary, it nevertheless standardized the position so that measurements could be repeated by other observers. The Frankfurt plane is used today as the orientation for the measurement of stature, sitting height, and other measurements taken to or on the head.

The early recognition of the importance of body position, or posture, is reflected in the techniques and procedures for nearly all the dimensions measured by professional anthropologists. Good dimensional definitions will always describe the posture of the subject (e.g., Clauser et al. 1988). For example, the subject's posture in sitting height is shown in Figure 2.1 and described as follows (Clauser et al. 1988):

> The subject sits on a cushionless flat surface with the long axes of the thighs parallel to each other and to the floor. The feet are on an adjustable footrest, and the knees are flexed 90 degrees. The trunk is erect without stiffness; the head is also erect and the subject looks straight ahead in the Frankfurt Plane. The shoulders are relaxed and the upper arms are hanging loosely at the sides with elbows flexed 90 degrees and hands straight.

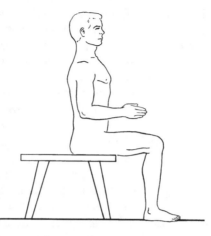

FIGURE 2.1
Posture for measuring sitting height. (From Clauser et al. 1988.)

This is not to suggest that the posture in the example is necessarily the best posture in which to measure all seated dimensions, but it does serve to illustrate a well-defined posture.

The user of anthropometric data should verify the posture of the subject during data collection before the data are used. Further, when using data from a number of different sources — as when international standards are developed, for example — it is important to understand any differences in subject posture before combining or comparing data. For example, the standing posture used in some three-dimensional body scanner surveys stipulates that the feet be 20 cm apart. Stature is usually measured in traditional surveys with the feet together. Although this results in a relatively small difference in the stature measured (of the order of ½ cm), it can still be important where safety is an issue. To the extent possible, the summary data included at the end of this section have been verified to ensure that measurements have been made with the subjects in identical postures. However, subtle, undocumented differences may account for some of the apparent differences in statistical values.

Seated postures are even more difficult to standardize, particularly with regard to the degree of erectness of the spine. A more erect posture will result in higher values for sitting height, eye height sitting, acromial (shoulder) height, elbow height, and so on. A more relaxed position will result in lower values for those dimensions. Traditionally, anthropologists have used the erect posture for data collection because it is easier to standardize, and the observer error and overall variance are therefore reduced. However, the result is that many of the currently available data sets are measured in a posture that few people, if any, actually use in a seated workspace. The alternative approach — measuring in a more relaxed, typical working posture — results in very high observer error and may inflate the estimate of the variability of the dimension within the population.

In a survey of U.S. Air Force women, Clauser et al. (1972) measured both the traditional erect sitting height and a relaxed sitting height. The difference between the two was 1.32 cm at the mean, 1.5 cm at the 5th percentile and 1.2 cm at the 95th percentile. However, a study of U.S. civilians (HES 1962) showed a difference of 4.1 cm at the mean and similar values at the 5th and 95th percentiles. Clearly, the specific verbal instruction to the subject for the relaxed sitting has a significant influence on the resulting posture, and subsequently on the resulting data.

Because most of the available data are taken in the erect posture, the user of seated anthropometric data has a dilemma about whether, and by how much, seated dimensions should be adjusted for use in various seated workspace designs. The final determination should be made using factors such as the amount of adjustability of the workplace, the comfort for the user, and the amount of time the posture is required, as well as whether danger is involved if the seated height is either too high or too low.

2.1.2   Influence of Anthropometry on Posture

The application of anthropometric data to work space design can be done with direct use of one- or two-dimensional measures, or it can be done through the use of digital human models. The same issues apply whether using computer models or applying anthropometric data directly to product design without the use of models. The best models replicate a variety of postures. These postures are achieved through positioning body segments. Body segments have been measured in a number of anthropometric surveys (e.g., Gordon et al. 1989) but they are not exactly equivalent to the links between joint centers. An example of such a segment would be radiale–stylion length (Figure 2.2), which is a measure of the length of the forearm, and is often used to represent the distance between the joints at the elbow and the wrist. Radiale is at the proximal tip of the radius near the elbow and stylion is at the distal tip of the radius near the wrist. In both cases, the bony point is exterior to the actual joint center and this displacement affects the true link length to a certain extent. In addition, radiale is distal to the actual joint center, whereas stylion is approximately level with the actual joint center (although laterally displaced). Thus, radiale–stylion length is the best approximation of the forearm link length, even though we know it is inaccurate. There are similar inaccuracies in other link vs. segment comparisons.

McConville and co-workers (1980) addressed this problem by collecting data on the comparison between segment lengths and actual link lengths, using a cadaveric population (see also Kaleps et al. 1984, and Young et al. 1983). They provided regression equations allowing the estimation of link lengths from the measured segment lengths. These seminal studies, although

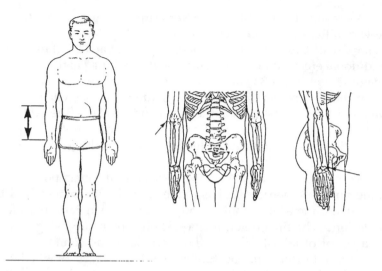

FIGURE 2.2
Radiale–stylion length, showing the dimension on the surface as well as the underlying skeletal points. (From Clauser et al. 1988.)

dated, have formed the basis for much of the internal geometry of the digital human models available commercially today.

The link lengths between joint centers would be the most direct application of anthropometric data to models and would produce the most accurate representation of posture. To the extent that segment lengths are inadequate representations of link lengths, postural representations of the models may be inaccurate. In the complementary case, when recording anthropometric data from humans in a particular posture, clearly measuring segment lengths is the only option, because actual link lengths are internal and cannot be observed. Users may wish to consider correcting those lengths according to published regression equations.Another approach to modeling posture is to infer the location of joint centers from recordings of the movements. This promising approach increases the accuracy of the models when used in conjunction with accurate segmental anthropometry (Schwartz 1999, Seitz and Bubb 1999). Again, users of these data sets should verify whether the data sets contain adequate racial, age, and gender variability. This is true because men and women move differently (due to relatively different musculature at the shoulder, different hip breadth, and so on), and older people move differently from younger people (due to gradual restriction in joint range of movement with age), and it may also be true that people of different racial groups move differently (due to different limb lengths).

With either approach — estimating link lengths using regression equations or estimating joint centers from movement studies — error is present. There is potential error at each segment and it is cumulative. This is a concern when estimating complex postures in workplace settings. The more segments that are involved in establishing the posture, the more error is likely included. Oudenhuijzen (personal communication, 2002) conducted research showing that some digital models misestimated leg reach by as much as 10 cm. Although some of the errors may have been due to incomplete specification of the surrounding environment (i.e., the compressibility of cushions), another source of the apparent error in the anthropometry was incorrect postural placement of the model. Users of digital models to represent working postures should be aware that even when the initial anthropometry informing the models is accurate to within a few millimeters, the resulting dimensions or reach capabilities may have less accuracy when compared to actual humans.

Recording anthropometric data from humans in the posture of interest is clearly the most accurate approach to putting anthropometric data into digital models. The greater the deviance from the data collection posture, the greater the source for error. If new postures arise, either because of new work settings, or new work products, then the model should be validated for the particular work postures in question. This can be done by collecting data on a few people in the new posture and comparing the results to the data on the posture predicted by the model. Existing motion analysis data sets (Park et al. 1999) may be used for this if the postures they contain are appropriate to the new work posture. Validation on larger data sets, including those with racial, gender, and age variability would be the ideal.

2.1.3 Available Data

When designing workspaces to promote good posture, anthropometric data are a key resource. The work space design should include considerations of the population that will work there, what variety of postures will be involved in the work tasks, and whether there are safety as well as comfort concerns if the desired posture is not achieved. In most cases, to accommodate the desired population and still achieve the desired posture, it will be necessary to incorporate adjustability in one or more parameters of the workplace. To gauge the amount of adjustability needed, designers and ergonomists make use of available anthropometric data.

In designing adjustability it is common to address a design range, usually (but incorrectly) designated as the 5th percentile female through the 95th percentile male. The intention is to make sure a wide range of individuals can make use of the workspace. What is incorrect about the designation is that it implies there is a female for whom all her pertinent dimensions are at the 5th percentile value, and that there is similarly a male, all of whose dimensions are at the 95th percentile value. In fact, a person who has a 95th percentile torso height is most likely to have a leg length of considerably less than the 95th percentile. This phenomenon has been amply documented in prior research (e.g., Robinette and McConville 1981).

There are a wide variety of anthropometric data available (e.g., Ohlson et al. 1956, Snow et al. 1975, Buckle 1985, Jürgens et al. 1990, Ashizawa et al.

1994, Molenbroek 1994, Daanen et al. 1997c, Okunribido 2000). The designer must choose which is the most appropriate data set or data sets for his or her design problem. That choice will hinge on the type of user population (military, civilian, domestic, multinational, working age, elderly, youth) and the types of dimensions needed (standing, seated, functional, fixed position). As an example of the kinds of data that are available, a selection of dimensions from four countries has been included. The specific dimensions selected are those that are often used in developing digital models, but of course they can be used directly as well. Some of these are useful in posture prediction, and others are useful in designing workspaces around a healthy posture. These sample data are shown in Table 2.1 through Table 2.4 and illustrated in Figure 2.3. It should be noted that the data in the tables were collected at different times, by different researchers, following somewhat

TABLE 2.1

Anthropometric Summary Statistics from Germany (values in mm)

Dimension	Males			Females		
	P5	P50	P95	P5	P50	P95
Acromial (shoulder) height	1362	1464	1568	1270	1356	1433
Acromial (shoulder) height, sitting	575	630	674	596	636	680
Acromion–radiale length	333	363	397	292	313	346
Bideltoid breadth	441	482	526	395	434	489
Buttock–knee length	566	611	654	543	589	638
Cervicale–top of head						
Digit 3 (middle finger) length	76	85	94	70	77	86
Foot length						
Forearm–forearm breadth						
Hand length	175	190	208	163	177	193
Hip breadth, sitting	336	369	411	345	388	454
Knee height, sitting	502	545	588	461	501	545
Lateral femoral epicondyle height						
Lateral malleolus (ankle) height	83	91	99	77	85	97
Overhead reach (both arms)	1841	1955	2032	1989	2091	2218
Radiale–stylion length						
Sitting height	829	893	935	773	818	862
Stature	1663	1764	1872	1563	1654	1741
Thumb tip (forward) reach						
Trochanterion height	840	907	988	775	842	896
Weight (kg)	63	79	101	51	63	86

Note: P5, P50, and P95 refer, respectively, to the 5th, 50th, and 95th percentiles, commonly used in design. The 5th percentile is the value below which 5% of the observations lie. The 95th percentile is the value below which 95% of the observations lie. The 50th percentile has a corresponding meaning, and is close to the mean (average) value in a normally distributed data set. See the text for cautions about combining percentile values.

Source: Prof. Dr. Hans W. Jürgens, University of Kiel, Germany.

TABLE 2.2

Anthropometric Summary Statistics from Japan (values in mm)

Dimension	Males					Females				
	Mean	Standard Deviation	P5	P50	P95	Mean	Standard Deviation	P5	P50	P95
Acromial (shoulder) height	1380.1	56.7	1293	1382	1484	1277.7	48.6	1203	1274	1371
Acromial (shoulder) height, sitting	593.5	27.6	545	596	638	554.2	22.4	519	554	593
Acromion–radiale length	313.6	17.5	287	313	344	287.2	15.9	262	285	316
Bideltoid breadth	456.2	21.8	423	456	495	407.8	20.1	377	406	443
Buttock–knee length	547.0	24.9	533	574	616	541.6	21.3	511	541	578
Cervicale–top of head	257.7	13	237	258	278	241.1	11	222	242	258
Digit 3 (middle finger) length	79.6	4	73	80	86	72.8	4	67	72	79
Foot length	254.4	11	236	254	273	232.7	10	216	233	246
Forearm–forearm breadth	414.7	30	370	415	466	367.0	26	326	365	411
Hand length	179.6	11	162	180	195	164.9	10	148	165	181
Hip breadth, sitting	345.7	19	317	346	376	357.9	19	330	355	391
Knee height, sitting	519.7	23	482	519	561	475.5	19	446	474	510
Lateral femoral epicondyle height	466.8	21	433	467	501	428.6	19	399	428	461
Lateral malleolus (ankle) height	72.0	5	64	72	82	63.4	5	56	63	71
Overhead reach (both arms)	2149.8	86	2011	2156	2305	1966.7	76	1847	1966	2098
Radiale–stylion length	242.9	13	222	242	265	220.8	11	203	220	239
Sitting height										
Stature	1714.0	63	1612	1713	1820	1591.3	53	1511	1588	1695
Thumb tip (forward) reach	771.0	35.5	715	772	834	682.2	31.4	630	684	731
Trochanterion height	875.7	40.4	802	876	940	814.5	34.8	757	816	875
Weight (kg)	63.3	8.3	50	64	77	52.6	6.2	45	52	63

Note: P5, P50, and P95 refer, respectively, to the 5th, 50th, and 95th percentiles, commonly used in design. The 5th percentile is the value below which 5% of the observations lie. The 95th percentile is the value below which 95% of the observations lie. The 50th percentile has a corresponding meaning, and is close to the mean (average) value in a normally distributed data set. See the text for cautions about combining percentile values.

Source: Dr. Makiko Kouchi, National Institute of Advanced Industrial Science & Technology, Tokyo, Japan. Data collected by National Institute of Biology and Health.

TABLE 2.3

Anthropometric Summary Statistics from Republic of Korea (values in mm)

Dimension	Males					Females				
	Mean	Standard Deviation	P5	P50	P95	Mean	Standard Deviation	P5	P50	P95
Acromial (shoulder) height	1374	47	1292	1373	1452	1269	43	1199	1268	1338
Acromial (shoulder) height, sitting	604	27	561	605	648	568	23	528	568	606
Acromion–radiale length	333	16	305	334	359	307	13	285	308	328
Bideltoid breadth	455	20	420	454	488	415	19	384	414	445
Buttock–knee length	550	24	511	550	588	524	22	489	525	562
Cervicale–top of head	233	11	214	233	251	219	10	203	219	236
Digit 3 (middle finger) length	78	5	70	78	86	73	4	66	73	80
Foot length	247	10	231	247	264	228	10	212	228	244
Forearm–forearm breadth	476	45	401	475	559	434	33	379	431	491
Hand length	181	8	169	181	194	170	8	158	170	181
Hip breadth, sitting	339	19	308	340	371	344	19	312	343	375
Knee height, sitting	497	21	461	496	530	457	19	426	458	488
Lateral femoral epicondyle height										
Lateral malleolus (ankle) height	67	5	59	67	76	60	5	52	60	68
Overhead reach (both arms)	2136	72	2020	2140	2250	1975	67	1860	1975	2090
Radiale–stylion length	254	15	228	254	279	235	15	210	235	260
Sitting height	923	29	876	922	976	868	27	823	867	914
Stature	1684	52	1599	1683	1770	1563	49	1480	1565	1644
Thumb tip (forward) reach	818	36	758	816	881	757	35	698	758	816
Trochanterion height	817	37	759	816	880	753	34	698	753	809
Weight (kg)	66.4	7.9	54	66	80	56.8	6.6	46	55	68

Note: P5, P50, and P95 refer, respectively, to the 5th, 50th, and 95th percentiles, commonly used in design. The 5th percentile is the value below which 5% of the observations lie. The 95th percentile is the value below which 95% of the observations lie. The 50th percentile has a corresponding meaning, and is close to the mean (average) value in a normally distributed data set. See the text for cautions about combining percentile values.

Source: Dr. Young-Suk Lee, Chonnam National University. Data collected by Korea Research Institute of Standards and Science, 1997.

TABLE 2.4

Anthropometric Summary Statistics from United States (values in mm)

Dimension	Males					Females				
	Mean	Standard Deviation	P5	P50	P95	Mean	Standard Deviation	P5	P50	P95
Acromial (shoulder) height	1440.1	64.3	1336	1441	1542	1329.6	60.8	1232	1326	1431
Acromial (shoulder) height, sitting	602.7	29.9	554	603	651	561.4	28.2	514	561	607
Acromion–radiale length	339.7	17.8	311	340	369	309.6	17.0	281	310	340
Bideltoid breadth	491.9	26.0	450	491	535	434.6	23.5	396	434	475
Buttock–knee length	612.7	30.3	563	612	664	583.3	30.6	534	584	634
Cervicale–top of head	235.6	12.2	216	236	256	222.2	12.2	202	222	242
Digit 3 (middle finger) length	83.2	5.4	75	83	93	76.1	4.9	68	76	85
Foot length	268.1	13.0	247	268	290	241.9	12.3	223	241	264
Forearm–forearm breadth	548.6	43.2	479	550	622	475.7	37.6	417	474	538
Hand length	192.9	9.5	178	193	209	178.1	9.5	164	177	195
Hip breadth, sitting	369.5	24.9	332	368	412	393.2	31.7	346	391	450
Knee height, sitting	555.3	28.2	510	555	603	509.9	26.8	468	510	558
Lateral femoral epicondyle height	498.6	26.4	457	498	543	455.0	24.8	417	454	498
Lateral malleolus (ankle) height	67.5	5.5	59	67	77	61.5	5.1	53	61	70
Overhead reach (both arms)	2221.2	100.7	2056	2221	2385	2044.1	94.9	1894	2038	2208
Radiale–stylion length	267.3	15.3	243	267	292	239.0	14.6	217	238	264
Sitting height	916.5	36.2	856	917	974	858.5	34.0	803	858	915
Stature	1751.4	70.0	1638	1752	1867	1625.6	66.6	1521	1623	1746
Thumb tip (forward) reach	798.0	39.7	735	798	863	728.5	36.8	672	727	798
Trochanterion height	919.3	48.0	843	919	1000	849.3	46.1	775	846	932
Weight (kg)	79.1	11.1	62	78	99	63.1	9.4	50	62	80

Note: P5, P50, and P95 refer, respectively, to the 5th, 50th, and 95th percentiles, commonly used in design. The 5th percentile is the value below which 5% of the observations lie. The 95th percentile is the value below which 95% of the observations lie. The 50th percentile has a corresponding meaning, and is close to the mean (average) value in a normally distributed data set. See the text for cautions about combining percentile values.

Source: Dr. Claire C. Gordon, U.S. Army. U.S. Army data, weighted for age and race according to the 2000 U.S. Census, ages 18-45 years. Note that, in the U.S., civilians are often heavier than their military counterparts of a similar height.

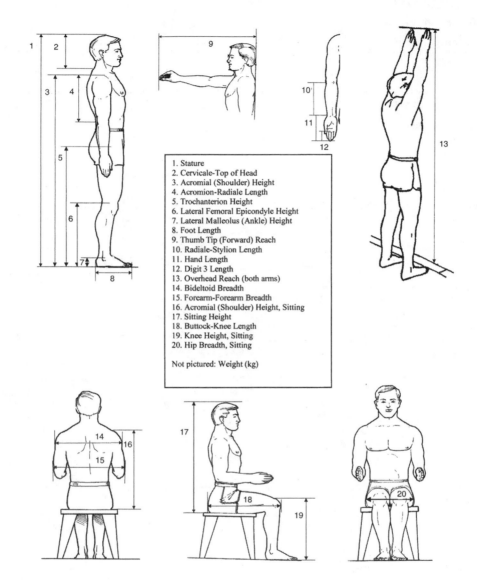

FIGURE 2.3
Visual index to dimensions listed in Tables 2.1 through 2.4. (From Gordon et al. 1989.)

different measurement protocols. To the extent that the data sets are compatible, they provide a way to characterize the postures of individuals from different regions in order to identify the best way to make internationally suitable products and work spaces. To aid in compatibility of future data sets, an international standard ISO 7250:1997 (Basic Human Body Measurements of Technological Design) defining specific measuring techniques has now been agreed upon.

Summary

The influences of posture on anthropometric data collection have been identified. Clearly, placing the human in varying poses will result in varying anthropometry. That some measuring postures have historically not been clearly defined makes comparing data between studies difficult. This is a particular problem in the 21st century as most design is for a global marketplace. It is important that standardized methods, such as those in ISO 7250 (Basic Human Body Measurements of Technological Design) (ISO 1997) and ISO DIS F15535 (General Requirements for Establishing an Anthropometric Database) (ISO 2003) be used for future data collection. These standardized techniques will go a long way toward improving the compatibility of anthropometric data from around the globe. However, even with standardized data collection techniques and well-defined measurement postures, users of the data should be cautious when using anthropometric data in designing workplaces for postures other than those in which the individual subjects were measured.

One way to move from measurement postures to work postures is through the use of digital models. The models are articulated, and can represent an almost infinite variety of work poses. Naturally there are a series of transformations from the original anthropometric data, which largely consist of one- or two-dimensional measurements, into a three-dimensional model. Some of the transformations are extremely accurate, but others are less so. As a result, some of the external anthropometry for postures that a model might achieve is not a completely accurate representation of the external anthropometry of a human in that same posture. Thus, before a final design is selected, analyses done with these models should be tested in a mockup or a laboratory setting with a number of living humans.

Clearly, the key to successful accommodation of a wide variety of humans in various working postures is the inclusion of adjustability in the workplace design. Anthropometric dimensions, either directly or through the use of digital models, can provide valuable guidance about the range of variability necessary to accommodate the population in question.

2.2   Three-Dimensional Anthropometry

Kathleen M. Robinette, Hein A.M. Daanen, and Gregory F. Zehner

Introduction

Advances in optics and computing science have made it possible to digitally capture the size, shape, and position of the different parts of the body in three dimensions in a few seconds. This can be thought of as a posture snapshot in addition to capturing size and shape. After all, what is posture really but the relative location in space of the parts of the body? New methods for tracking the change in posture in three dimensions are also being developed. The study of these three-dimensional snapshots and the tracking of changes in three dimensions are what we refer to as three-dimensional anthropometry. This section describes the most recent three-dimensional anthropometry methods including: (1) why and when to use three-dimensional anthropometry, (2) a general description of three-dimensional anthropometry measurement techniques, (3) data processing methods, and (4) issues with three-dimensional data acquisition and use.

2.2.1 Why and When to Use Three-Dimensional Anthropometry

Traditional anthropometry, as described in the previous section, uses inexpensive tools that are readily transportable, there are extensive databases available, and many current models and other tools are made to incorporate that type of information. However, with three-dimensional anthropometry the tools can be expensive, difficult to move, and may require purchasing or acquiring new methods or software to use the information. So, why use three-dimensional anthropometry? There are many reasons, depending on the application, and, as Jones and Rioux (1997) explain, there are many applications for which three-dimensional anthropometry is beneficial. For posture and motion, which are dependent on location and shape information, three-dimensional anthropometry is essential.

One of the greatest benefits of three-dimensional anthropometry for posture and motion studies is the ability to determine and track the relative location of homologous points, called landmarks. Many traditional measurements by their definition give the impression that landmark locations or shape of the measurement is being identified when in fact this information is not contained in the measurement. For example, measurements to the "top of the head" give the impression that the location of the top of the head is included in the measurement information. In Figure 2.4a the same subject is measured three times for the measurement called "tragion to top of head," and the information contained in the three-dimensional measurement (at the top) is compared to that of the traditional measurement (at the bottom). Each time the subject is measured, the point at the top is in a different place and the size of the measurement is different. The first time the distance measured is 138 mm, the second 142.1 mm, and the third 146.2 mm. What is not always obvious from the measurement description is that after the subject is gone, with the traditional measurement one is left with just a single number. The original point locations cannot be seen or determined. With three-dimensional measurement the location of the points with respect to each other is inherent to the measurement. This same information can be obtained with traditional methods, but three measurements are needed to obtain each point and this is not generally done. For a measurement such as this, which has two points, six measurements would be required. When constructing three-dimensional models from traditional measurements that incompletely define the point locations, such as this one, many assumptions and guesses must be made.

Traditional measurements are also often regarded as measurements in a plane, and as a result many researchers refer to the measurements as two-dimensional when in fact they do not contain information about the plane and are not two-dimensional. For example, measurements such as chest circumference and waist circumference are sometimes taken "horizontally," and thus the description gives the impression that the location or shape in the plane is also obtained. Figure 2.4b illustrates (1) the three-dimensional version of these two measurements both separately and as they are located and shaped on the person, and (2) the traditional measurement information and a hypothetical reconstruction of the measurements from the traditional measurement. As can be seen when reconstructing the three-dimensional figure from the traditional measurements such as these a great deal of conjecture is required and it must be expected that the resulting figure will inaccurately characterize the shape as well as the posture. Again, traditional measurement methods can be used to obtain the information in the plane, with many additional measurements, but the additional measurements are usually not taken.

Three-dimensional scanning or surface measurement is also capable of capturing some other types of measurements that traditional tools cannot, such as volume of a body segment, surface area, center of volume or gravity, and moments of inertia. These types of measurements are needed for biomechanical assessments and modeling. Stereophotogrammetric studies done in the 1960s to the 1980s were used to calculate segmental moments that were then used in many ergonomic and biomechanical models (McConville et al. 1980, Mollard et al. 1982, Young et al. 1983). Several investigators have found that calculation of surface area and surface area differences can be performed fairly accurately from three-dimensional scans (Tikuisis et al. 2001, Tan et al. 2001).

In addition to the fact that three-dimensional information can be derived from single three-dimensional scans, the comparison of two or more scans offers insight in movement artifacts, apparel fit, and design ranges or specifications. The ability to superimpose scans and calculate the differences can be used within subjects, between subjects, and between subjects and equipment or apparel.

The difference between two consecutive scans of an individual yields information on body position differences and movement artifacts (Daanen et al. 1997a). In Figure 2.5, two scans are aligned by points on the chest and back. The radial difference below the waist level exceeds 1 cm, indicating different hip flexion angles between the two scans. Reducing the time between consecutive scans (obtained through digital surface photogrammetry) enables motion capture and shape capture simultaneously (e.g., Dekker 2000). Subtraction of a scan with maximal inspiration and maximal expiration shows the difference in chest position and can be used as an estimator

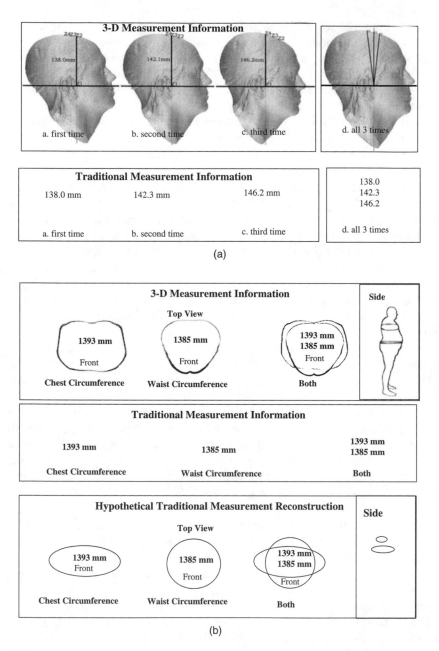

FIGURE 2.4

Illustration of the problem with re-creation of a person in three dimensions with traditional measurements. (a) Tragion (Z1) to top of head measured three times with the subject looking straight ahead. (b) Comparison of three-dimensional vs. traditional information for modeling.

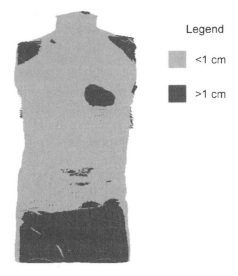

Legend

☐ <1 cm

■ >1 cm

FIGURE 2.5
Mathematical comparison of two consecutive three-dimensional scans of the same person.

of the functional lung capacity (Kovatz et al. 1988). The scans in Figure 2.5 were taken in a different phase of the breathing cycle, indicated by the radial differences at shoulder level and the chest exceeding 1 cm. This shows that three-dimensional scanners are sensitive, even more than one may need for a specific application. Therefore, data reduction is often necessary. Comparisons of scans with a longer time span in between may give information about such changes in human shape as pregnancy (Perkins 1999), aging, disease, or the locations where body fat increases or decreases during changes in body weight; see Robinette et al. (1997) for a review. Figure 2.6 illustrates one method that Perkins (1999) used to compare the subjects at different points during their pregnancy. Midsagittal and transverse cross sections of scans of the subject taken at five intervals during the pregnancy are shown.

The differences between the scans of two subjects show the body locations where similarities and differences are most prominent. Figure 2.7 illustrates that the locations of two traditional measurements (bust height and waist height) on two subjects in the same posture can be very different. Even though they have the same torso height, the one is not just a larger copy of the other, and shape differences affect the relative locations of the traditional measurements. This means the traditional measurement magnitude is confounded with shape. In other words, two people with the same waist measurement can actually be very different in size because their shapes are also different. These things can only be truly tracked with three-dimensional measurement. Figure 2.8 shows the radial difference map from the front and

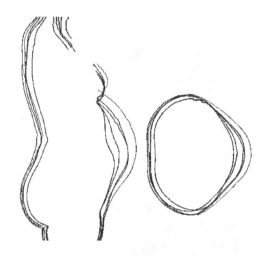

FIGURE 2.6
Midsagittal and transverse cross sections of a woman scanned at five different points during pregnancy.

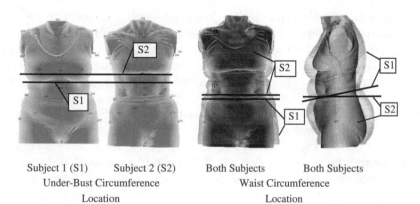

Subject 1 (S1) Subject 2 (S2)	Both Subjects Both Subjects
Under-Bust Circumference	Waist Circumference
Location	Location

FIGURE 2.7
Comparison of locations of two traditional measurements on two women with the same torso height.

back of two subjects of similar torso height but different shapes. The two subjects were centered at the neck and thighs to align them. The user can define which part of the body is kept invariant by using this part as the alignment basis. If, for example, the bust points and belly were used for alignment, the radial difference map would be completely different. The points for alignment depend on the application.

Besides the subject alone, the relative location of the seat, the workspace, and the apparel or equipment worn can also be measured, and changes in these locations can be tracked. This kind of information is used for design, evaluation, or specification of the workstation. For the design of a standing

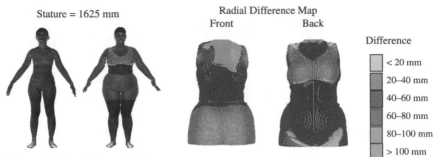

Stature = 1625 mm

Radial Difference Map
Front   Back

Difference

☐ < 20 mm
☐ 20–40 mm
☐ 40–60 mm
☐ 60–80 mm
☐ 80–100 mm
☐ > 100 mm

Mass = 51.9 kg   Mass = 91.4 kg

FIGURE 2.8 (Color figure follows p. 42.)
Radial differences between the torsos of two women with the same torso height but different masses.

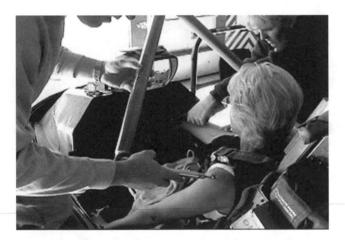

FIGURE 2.9
Digitizing three-dimensional landmarks on a subject within a cockpit.

workstation, for example, the subjects should stand against it as closely as possible. Thus, the subjects can be lined up at the belly (or whatever point is closest to the workstation in three dimensions) and we can then calculate and visualize the spread of the eye position in three-dimensional space or the required space for accommodation.

For the evaluation of seated workstations, the workstation geometry can be recorded in three dimensions, and the three-dimensional locations of the subject's landmarks when seated in the workstation can also be recorded. In the following example, a Faro® Arm was used to take anatomical landmark location data on several test subjects in two different ejection seats within a cockpit mockup (Figure 2.9). Anatomical landmark location data were generated from the traces of the landmarks as the seat was moved throughout its range of adjustment. Using live subjects not only allows the variability

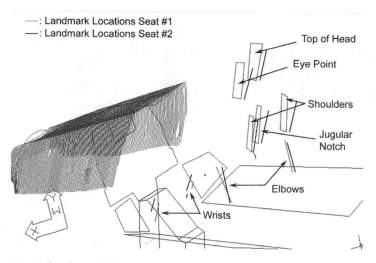

FIGURE 2.10 (Color figure follows p. 42.)
Comparison of one subject's landmark movements through seat adjustments in two different seats within the same cockpit.

between subjects to be summarized quantitatively, but can also illustrate the displacements for the subjects due to seat-specific adjustment. The traces for one subject in two different seats are illustrated in Figure 2.10. From the four-sided traces in this figure, it is obvious that seat number 1 has adjustability fore and aft that is independent of the up and down adjustment, while in seat number 2 adjustment fore and aft is linked to the up and down adjustment. Also, seat number 1 always places the subject farther forward in the cockpit than seat number 2 and can position the subject higher as well. Reach and clearance data were also recorded for the subjects when the restraint harness inertial reels were locked and when unlocked. This was done to determine what landmark locations within the structure provide good accommodation in terms of reach and clearance. By using actual subjects in the measuring process, instead of simply calculating geometric differences between the seats, the potential effects of different restraint systems, postural changes due to seat back and pan angles, seat cushion compression, and human body tissue compression can be taken into account.

Subject mobility can also be demonstrated by lining up the subject's location in the seat and tracking the arm and foot motions (Duncan et al. 2000). An example of this is shown in Figure 2.11. This figure shows a three-dimensional scan of the seated subject, along with the subject's hand and foot motion paths obtained from three-dimensional landmark tracking. As can be seen, the relative size and shape of the body, its location in the seat, and the subject's movements are all measured and are visualized simultaneously. The presence of the three-dimensional body gives the motion paths more visual meaning and makes the data easier to interpret.

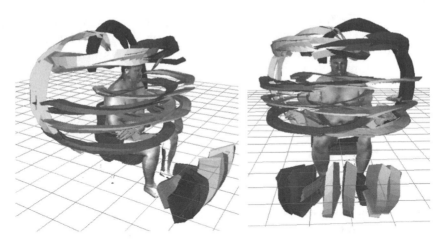

FIGURE 2.11 (Color figure follows p. 42.)
Motion paths for hands and feet of a seated subject when reaching to points within the work space. (Courtesy of Deere & Company.)

2.2.2 Measurement Techniques

The methods used to collect three-dimensional anthropometry vary depending on the application. For capturing posture and posture change or movement, it is often desirable to combine more than one type of three-dimensional anthropometric device. For example, surface scanning with landmarks might be useful for establishing the details of the body surface and size and landmark tracking for establishing changes or movements. Both sets of data might then be collected on each subject. This section begins with a discussion of the different types of measuring devices, followed by a discussion of how the devices are used in data collection.

2.2.2.1 Three-Dimensional Devices

There are many different devices for three-dimensional anthropometric measurement. Some capture predefined landmarks, some capture body surface contours as well as surface landmarks, and some capture subsurface anatomy. Rioux and Bruckart (1997) provide a good review of the different types with more detail than we can provide here. We modify their categories slightly and categorize them as follows: (1) contact devices, (2) target trackers, (3) stereophotogrammetry, (4) optical scanners, and (5) volume scanners.

Contact devices can be described as pointers that are moved until they touch the surface or point of interest and that point is recorded. One commonly used device is the Faro Arm. Some investigators such as Annis and Gordon (1988) have also made custom devices, which include some method for securing the subject in position while the points are digitized. One advantage of these devices is that points can be digitized in difficult-to-see areas such as inside an aircraft cockpit. One disadvantage is that each point takes a minimum of several seconds to digitize and subjects can move between points without some device to secure the subject in place, introducing errors into the measurement.

Target trackers are devices that record and follow a small number of points over time. The points are physically marked and tracked by sensors placed around the body. These can use stereo camera sensors, magnetic tracking sensors, optical tracking sensors, or sonar sensors. Perhaps obviously, the ability to track changes or motions is the biggest advantage of these devices. One disadvantage is that usually the set of points that can be independently tracked with any reliability is relatively small. A typical set might include 16 to 30 points. Also, the resolution of the points is limited by the size of the markers and the need to have them spaced far enough apart so as to be able to view them from at least two sensors at all times while maintaining separation of one marker from another.

Stereophotogrammetry is the use of photographs taken simultaneously from two or more points of view. The photographs are then analyzed to extract points. Some early photogrammetric systems used film or glass plate images that were then viewed in a special stereo viewer or stereoscope for recording. Herron (1972) describes one such system. More recently, techniques have been developed to extract three-dimensional information from the images using computers, and digital photography and video have also been added. This has included simple photographic images of the outer surface of the body, as well as X-ray image pairs for information about the internal structures and surfaces. One advantage to using photogrammetry is that the images are collected almost instantaneously so the subject has little time to move. This makes it possible to take sequences or video to combine three-dimensional scanning with motion tracking, such as described by Probe (1990). A disadvantage is that the process to convert the image to three-dimensional data can be slow and unreliable. Also, if more than two cameras are used, as in the case where more than a 180° view is desired, it can be difficult to combine the multiple views with accuracy.

Optical scanners are devices that capture dense points on the surface of the body directly, using some sort of optical recording device such as a CCD camera. Two of the first whole-body scanners developed were those by Jones et al. (1989) and Rioux et al. (1987). The first effective scanner designed for measuring human subjects in surveys was a scanner built by Cyberware, Inc. for the U.S. Air Force (Brunsman et al. 1997, Daanen et al. 1997b). This scanner was used in the survey of the civilian populations of NATO countries, called the CAESAR survey (Blackwell et al. 2002, Robinette et al. 2002). This scanner was also the first designed to be able to record the locations of premarked landmarks, as well as to be able to measure both standing subjects and subjects seated in a seat with a back angle of up to 30° (such as an ejection seat). These are two features that are very important for posture studies. Optical scanners collect thousands or hundreds of thousands of points reasonably quickly, usually within a few seconds. There can be some movement issues, but they are generally manageable. One disadvantage is that they can only scan what they can see. Therefore, points such as the axilla (armpit), which is hidden under the arm when the arms are hanging by the sides, cannot be directly recorded without putting the body in a different posture.

Volume scanners are devices that are focused on measuring organs and structures of the internal body rather than the external body surface, although some do both. These include X-ray computed tomography (CT), magnetic resonance imaging (MRI), and ultrasound, among others. The National Library of Medicine's Visible Human Project® employed both CT and MRI technology along with photographs of the actual anatomic cross sections of two individuals, one male and one female (Patrias 2000). Many new methods for extracting information using these technologies have resulted from that effort. One advantage of volume scanners is the ability to actually locate anatomic landmarks on the bone in three-dimensional space, rather than having to rely on markers placed on the skin. These methods are impractical for use on large numbers of healthy subjects, however. Also, the supine posture is the only position that can be scanned at this point.

2.2.2.2 Data Collection Methods

There are several important aspects to three-dimensional data collection, in addition to the devices themselves, that are sometimes overlooked. For example, the positioning of the subjects prior to data collection must be well defined and standardized from subject to subject or it can be impossible to compare across subjects. Also, for most applications the ability to identify landmarks in three dimensions is very important; yet some types of scanning systems do not have a landmark identification mechanism built in and additional procedures must be followed both prior to and after data collection to be able to find the landmarks with accuracy and consistency. Burnsides et al. (2001) describe the process used to identify landmarks for the CAESAR project, which used optical scanners with color cameras in addition to the range cameras that determine the distance to the object. This process involved premarking the subjects with 72 white stickers, 12 of which were special three-dimensional stickers, and using specially created software called INTEGRATE (Burnsides et al. 2000) with quality control checking to identify the landmarks after the scan.

The type of apparel worn can be important, depending on the system. If it is important to measure a person in three dimensions in the normal apparel worn during a workday, then certain types of scanners might be ruled out, or the apparel color will have to be controlled. Dark clothing will not reflect light and will therefore not provide good scan data with optical scanning devices, nor will dark hair. Apparel, such as spectacles that have shiny metallic surfaces, can also be problematic for both optical scanners and motion trackers. Metal can cause interference and inaccuracy problems for magnetic trackers as well. If capturing the body surface is important, then apparel that follows the body contours but minimally constricts the body may be necessary.

Perhaps one of the most important things often overlooked is the ability to calibrate the system and keep it in an acceptable calibration range. Some three-dimensional data collection systems do not come with calibration devices or methods. If this is the case, then some method must be created to ensure that the system is consistently and accurately measuring over time, particularly after any change is made to the system. Daanen et al. (1997) discuss some of the first calibration and accuracy testing of the Cyberware WB4 scanner used for the CAESAR project.

Finally, it must be noted that in some cases the needs of a project cannot be met by a single three-dimensional technology. Some require the use of several different types of three-dimensional data collection devices, used on the same subjects. For example, a Faro Arm might be used to locate the subject within a work space and to digitize the relative locations of key controls within the work space. The same subject might then be separately scanned in the same seat to obtain a full three-dimensional body image, and the subject's landmarks might be tracked during typical motions with a motion tracking system. This can give a complete picture of the posture, shape, and motion within the workstation such as that shown in Figure 2.11. The bottom line is that it is important to consider both the goals and the data analysis methods in addition to the three-dimensional anthropometric devices when planning data collection.

2.2.3   Data Processing Methods

Once three-dimensional data are collected, it is often the case that the real work just begins. Generally, to use the information some data processing and analysis is required. The sorts of work that must be done include (1) three-dimensional landmark identification, (2) data fusion (merging of either scan views or different data types), (3) subject visualization, (4) translation to computer-aided design (CAD) or model formats, (5) statistical shape or movement characterization, (6) segmentation, (7) data reduction, (8) scan comparisons within and between subjects, (9) solid modeling for rapid prototyping, and (10) visualization of differences within populations.

There is a common misperception that scanners automatically do all or many of these things, but in reality these tools are in the early stages of development. Some three-dimensional scanners do not even include software to merge the views from different cameras. To accomplish the needed analysis some software tools are needed. One tool used in many of the studies mentioned thus far is called INTEGRATE (Burnsides et al. 2000). It can be used to manually pick landmarks from a scan, to calculate distances and volumes, to translate to various CAD and modeling formats, to reduce data, to segment, and to compare scans and three-dimensional points, among many others. However, it is not a commercial package, and as a result, it is not very user friendly and runs on a limited number of computer platforms.

There are a few commercial packages available, such as Polyworks® by Innovmetric, and Surfacer® by Imageware, which have some of the analysis capabilities of INTEGRATE, such as data reduction, comparison of two objects, and data editing. These have much better human interfaces and are continually upgraded to run on the latest operating systems. However, they were not developed for human data analysis, but were developed for engineering and quality assurance assessment of manufactured objects. So issues like comparison of populations and saving and using three-dimensional landmark data are not addressed in these packages. To exploit the benefits of the latest three-dimensional technologies, new analysis software tools are needed.

With regard to landmark identification, many efforts have been made to fully automate landmark recognition without premarking subjects (Pollak 1989, Nurre 1997). Burnsides et al. (2001) found that automated landmark recognition, even with premarked points and with 500 subjects in the training set, was only about 70% accurate. This meant that all subjects had to be checked to ensure the landmarks were correctly identified. This was a time-consuming process and subject to human error in the checking process given the large number of scans to check. Therefore, an interactive process was developed where the human operator points out the landmarks in a flat image, and the software takes it from that point and indicates any potential errors to the operator. Heuristic checks were also built in. This sped up the process of landmarking while improving accuracy and reliability to more than 98.8%. The remaining 1.2% of inaccuracy was due to such problems as subject deformities (scoliosis), artificial limbs (where the system cannot find a bone that is not there), and missing markers rather than to the software.

Most workplace design and evaluation applications do not require a static three-dimensional scan, but need a movable three-dimensional manikin that interfaces with the CAD environment. Generally, digital human models are used for this purpose (refer to Chapter 14). A digital human model consists of an underlying skeleton, often with kinematic properties, and a "skin" on top. These models are constructed on the basis of 10 to 120 one-dimensional body dimensions that have to be entered, and the skeleton and skin are estimated from these. Figure 2.12 shows the differences between a digital human model and a scan that was superimposed. The digital human model has been manipulated so that the posture resembles the scan posture. As can be seen, the shoulders of the digital human model (green) are wider than those of the subject's scan (blue). Now that three-dimensional data are becoming available, some companies are beginning to convert three-dimensional scans into their human models. The underlying skeleton of the human model can be generated from the three-dimensional coordinates from the landmark files, while the outer layer may be constructed from the surface of the scans. Promising progress has been made.

Also, it is now possible to segment, link, and reposition standing posture three-dimensional scans to known seated postures, creating a simple kinematic model that can be used for illustrative purposes, as well as to study

FIGURE 2.12 (Color figure follows p. 42.)
Visualization of the difference between an electronic human model and the three-dimensional contours of the subject being modeled.

posture change. Figure 2.13 illustrates one example. The subject shown in this figure was scanned twice, once in a standard standing pose and once seated in an ejection seat. In the standing pose, 73 labeled landmarks were also obtained. The landmarks were used to segment the standing pose into 19 segments and to link the segments at centers of rotation of the joints. This was done automatically using the INTEGRATE software tool. Each segment is shown in a different color. The segmentation and link system used was taken from McConville et al. (1980). With this linking if one pulls on the hand the arm will follow, if one pulls on the foot the leg will follow, etc. This linked figure was then seated in the same position as the scan of the subject in the seat. As can be seen, the correspondence between the re-posed standing figure and the seated figure is very good. The change in the soft tissue and the change in the spinal curvature can be measured by measuring the differences between the positions of the segments in the standing pose and those of the seated pose. It seems that the change in the torso or spine is relatively small, although the pelvic segment seems to be slightly rotated clockwise if viewed from the left side. The small amount of spine change indicated is in accordance with seating studies such as that in Reed et al. (2000). They found that seat height, angle, and steering wheel location affected the limb postures significantly but had little effect on the torso or

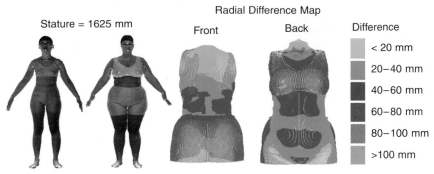

Radial Difference Map

Stature = 1625 mm

Front   Back   Difference

< 20 mm

20–40 mm

40–60 mm

60–80 mm

80–100 mm

>100 mm

Mass = 51.9 kg   Mass = 91.4 kg

COLOR FIGURE 2.8
Radial differences between the torsos of two women with the same torso height but different masses.

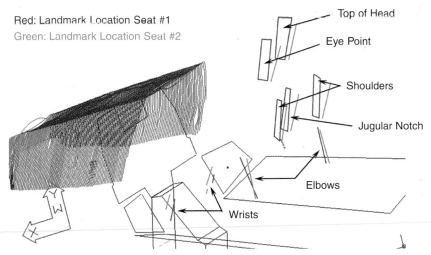

Red: Landmark Location Seat #1
Green: Landmark Location Seat #2

Top of Head

Eye Point

Shoulders

Jugular Notch

Elbows

Wrists

COLOR FIGURE 2.10
Comparison of one subject's landmark movements through seat adjustments in two different seats within the same cockpit.

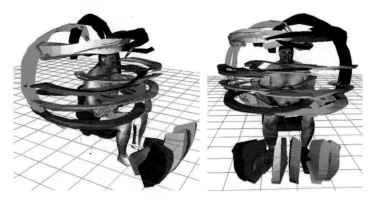

COLOR FIGURE 2.11
Motion paths for hands and feet of a seated subject when reaching to points within the work space. (Courtesy of Deere & Company.)

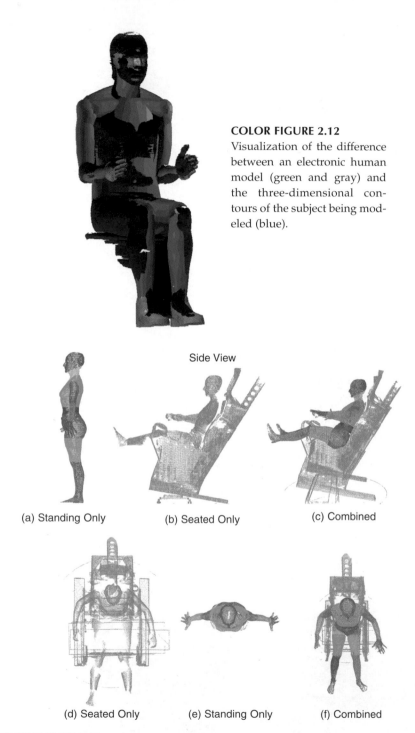

COLOR FIGURE 2.12
Visualization of the difference between an electronic human model (green and gray) and the three-dimensional contours of the subject being modeled (blue).

Side View

(a) Standing Only   (b) Seated Only   (c) Combined

(d) Seated Only   (e) Standing Only   (f) Combined

COLOR FIGURE 2.13
Comparison of the segmented and joint-center linked standing scan of a subject with the same subject repositioned in a seated scan.

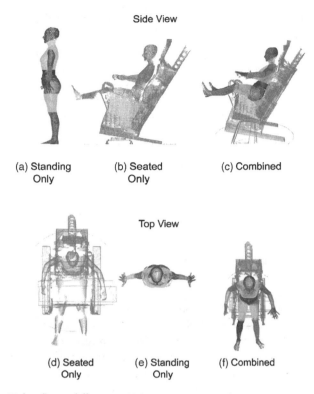

Side View

(a) Standing
Only

(b) Seated
Only

(c) Combined

Top View

(d) Seated
Only

(e) Standing
Only

(f) Combined

FIGURE 2.13 (Color figure follows p. 42.)
Comparison of the segmented and joint-center linked standing scan of a subject with the same
subject repositioned in a seated scan.

spine posture. Investigators are developing automated animations of figures
such as these, and using them to develop realistic tissue deformation as well.

2.2.4 Three-Dimensional vs. One-Dimensional Scanning

When some people think of three-dimensional scanning, they visualize a
scanner digitizing a body and "spitting out" hundreds of traditional (one-
dimensional) body measurements. Some companies provide automated
traditional measurement extraction tools with their scanners and, in fact,
some scanners do not provide three-dimensional data at all but only the
estimated one-dimensional traditional measurements that are extracted
using embedded software. In these instances, scanning is viewed as a quick
noncontact way to obtain these measurements. While in some instances this
may be true, it may not be the best way to obtain those measurements.
Several studies have compared traditional measurements taken with traditional tools (tape measures, calipers, and the like), with traditional measurements extracted from scans (Daanen and Brunsman 1998, Bradtmiller and

Gross 1999, and Perkins et al. 2000). These studies indicate that caution should be exercised when scan extraction software is used to emulate tape measures. Bradtmiller and Gross (1999) indicated that for some measurements, such as waist circumference and sleeve length, the difference between the measurement taken the traditional way with a tape measure and that extracted from a scan on the same subject was often greater than the difference between sizes of garments. One sleeve length measurement difference they reported was 240 mm. This is more than nine times larger than the difference between sleeve length sizes (25 mm), which means the measuring difference spans nine sizes. Perkins et al. (2000) found similar problems with circumferences and other measurements taken with a tape measure, although they also found that point-to-point distances between premarked landmarks were comparable in magnitude and repeatability. It is clear that tape measures and scanners measure in very different ways. Tape measures span some parts of the surface and compress others, and are limited in their placement by the material properties of the tape (stiffness, etc.). Scanners follow the nooks and crannies, do not compress at all, have no material properties, and have to be "told" where to go along the surface in some manner. Therefore, the three-dimensional points from the scanner can be very precise, as Daanen et al. (1997b) demonstrate, but the resulting emulation of a tape measure can be fraught with human error.

If all that are needed are one-dimensional measurements, three-dimensional scanning may be a waste of time and money. If the measurements obtained are to be compared with historical or other data taken with a tape measure, then a three-dimensional scanner will probably not do as good a job as a tape. However, for posture and motion studies one-dimensional measurements will probably not fulfill the requirements. If the three-dimensional locations of landmarks, or the shape of a body surface, or body segments, volumes, moments of inertia, etc. are needed, then three-dimensional anthropometry will be the obvious method of choice. Furthermore, sometimes even for traditional applications three-dimensional data may be better than one-dimensional. Meunier (2000) demonstrated that for predicting apparel sizes, three-dimensional landmarks performed better than the original one-dimensional measurements used to create the garments.

Summary

Anthropometry is the study of human body measurement and three-dimensional anthropometry is the study of all types of three-dimensional measurement of the human body, from static scans to tracking three-dimensional points over time. It is true that three-dimensional data collection is expensive and does not have the centuries-long history of traditional one-dimensional tools and methods. However, a complete understanding of human posture and motion is dependent on the three-dimensional location of key anatomical surfaces and points. Therefore, three-dimensional anthropometry is critical for posture study and the digital modeling of humans. Three-dimensional anthropometry has capabilities that traditional methods do not have, and its application removes much of the guesswork in characterizing human size, shape, posture, and motion.

Three-dimensional anthropometry is a few decades old, and there is a need for new analytical and visualization software. These tools would improve the process of three-dimensional data collection as well as its application to a variety of problems involving size, shape, kinematics, human modeling, and interactive databases. A few tools are mentioned here, but more of these tools are being developed, as the new types of three-dimensional data become more widely available.

References

Annis, J.F. and Gordon, C.C., 1988. Development and Validation of an Automated Headboard Device for Measurement of Three-dimensional Coordinates of the Head and Face. Final report, Natick-TR 88/048, Natick, MA: U.S. Army Natick Research, Development and Engineering Center.

Ashizawa, K., Okada, A., Kouchi, M., Horina, S., and Kikuchi, Y., 1994. Anthropometrical data of middle-aged Japanese women for industrial design applications. *Journal of Human Ecology*, 23, 73–80.

Blackwell, S., Robinette, K., Daanen, H., Boehmer, M., Fleming, S., Kelly, S., Brill, T., Hoeferlin, D., and Burnsides, D., 2002. Civilian American and European Surface Anthropometry Resource (CAESAR), Final report, Vol. II: Descriptions, AFRL-HE-WP-TR-2002-0173, Wright-Patterson Air Force Base, OH: U.S. Air Force Research Laboratory, Human Effectiveness Directorate, Crew System Interface Division.

Bradtmiller, B. and Gross, M., 1999. 3D Whole Body Scans: Measurement Extraction Software Validation, SAE Technical Papers Series 1991-01-1892, Warrendale, PA: Society of Automotive Engineers.

Brunsman, M.A., Daanen, H., and Robinette, K.M., 1997. Optimal postures and positioning for human body scanning, in *Proceedings of International Conference on Recent Advances in 3-D Digital Imaging and Modeling*, Los Alamitos: IEEE Computer Society Press, 266–273.

Buckle, P.W., 1985. Self-reported anthropometry. *Ergonomics*, 28, 1575–1577.

Burnsides, D.B., Files, P., and Whitestone, J.J., 2000. INTEGRATE 1.28: A Software Tool for Visualizing, Analyzing and Manipulating Three-Dimensional Data (U), Technical Rep. AFRL-HE-WP-TR-2000-0100, Wright-Patterson Air Force Base, OH: Aerospace Medical Research Laboratory.

Burnsides, D.B., Boehmer, M., and Robinette, K.M., 2001. 3-D landmark detection and identification in the CAESAR Project, in *Proceedings of the Third International Conference on 3-D Digital Imaging and Modeling Conference*, Los Alamitos: IEEE Computer Society Press, 393–398.

Clauser, C.E., Tucker, P.E., McConville, J.T., Churchill, E., Laubach, L.L., and Reardon, J.A., 1972. Anthropometry of Air Force Women. AMRL-TR-70-5, AD 743 113 Wright-Patterson Air Force Base, OH: Aerospace Medical Research Laboratory.

Clauser, C., Tebbetts, I., Bradtmiller, B., McConville, J., and Gordon, C., 1988. Measurer's Handbook: U.S. Army Anthropometric Survey, 1987–1988, TR-88-043, AD A202 721, Natick, MA: U.S. Army Natick Research, Development and Engineering Center.

Daanen, H. and Brunsman, M.A., 1998. Difference between manual anthropometric measurements and anthropometric measures derived manually from whole body scans: a pilot study, TNO-report TM-98-A004, Soesterberg, the Netherlands: TNO Human Factors Research Institute.

Daanen, H.A.M., Brunsman, M.A., and Robinette, K.M., 1997a. Reducing movement artifacts in whole body scanning, paper presented at International Conference on Recent Advances in 3-D Digital Imaging and Modeling, Ottawa, May 12–15, 262–265.

Daanen, H., Brunsman, M.A., and Taylor, S.E., 1997b. Absolute Accuracy of the Cyberware WB4 Whole Body Scanner, AL/CF-TR-1997-0046, ADA 327818. Wright Patterson Air Force Base, OH: Armstrong Laboratory, Air Force Materiel Command.

Daanen, H.A.M., Oudenhuijzen, A.J.K., and Werkhoven, P.J., 1997c. Anthropometry of High School Graduates, TNO Rep. TM-97-A007, Soesterberg, the Netherlands: TNO Human Factors Research Institute.

Dekker, L., 2000. Automated model generation from instantaneous whole-body capture, paper presented at the 3D Human Imaging and Application Meeting, Glasgow, November 22, 2000. Available at http://faraday.dcs.gla.ac.uk/bmvasc/ events.htm.

Dempster, W.T., 1955. Space Requirements of the Seated Operator, WADC Tech. Rep. 55-159, Wright-Patterson Air Force Base, OH: Wright Air Development Center.

Duncan, J., Keleher, B., Newendorp, B., Ryken, M., Chipperfield, K., Raschke, U., and Brunsman, M., 2000. Designing for populations of people using tools describing more of their dimensions, in *Proceedings of the 14th Triennial Congress of the International Ergonomics Association and 44th Annual Meeting of the Human Factors and Ergonomics Society*, Vol. 6, Santa Monica, CA: Human Factors and Ergonomics Society, 739.

Garson, J.G., 1885. The Frankfort Craniometric Agreement, with critical remarks thereon. *Journal of the Anthropological Institute of Great Britain and Ireland*, 14, 64–83.

Gordon, C.C., Bradtmiller, B., Clauser, C.E., Churchill, T., McConville, J.T., Tebbetts, I., and Walker, R.A., 1989. 1987–1988 Anthropometric Survey of U.S. Army Personnel: Methods and Summary Statistics, TR-89-044, Natick, MA: U.S. Army Natick Research, Development and Engineering Center.

Herron, R.E., 1972. Biostereometric measurement of body forms. *Yearbook of Physical Anthropology*, 16, 80–121.

HES (Health Examination Survey), 1962. User Data Set 3: Physical Measurements. Health Examination Survey, Washington, D.C.: U.S. Department of Health, Education and Welfare.

Hrdlicka, A., 1920. *Anthropometry*, Philadelphia: Wistar Institute of Anatomy and Biology.

ISO, 1997. *ISO 7250: 1997 Basic Human Body Measurements of Technological Design*, Geneva: International Standards Organization.

ISO, 2003. *ISO/FDIS 15535: General Requirements for Establishing an Anthropometric Database*, Geneva: International Standards Organization.

Jensen, R.K. and Nassas, G., 1988. Growth of segment principal moments of inertia between 4 and 20 years. *Medicine and Science in Sports and Exercise*, 20, 594–604.

Jones, P.R.M. and Rioux, M., 1997. Three-dimensional surface anthropometry: applications to the human body. *Optics and Lasers in Engineering*, 28, 89–117.

Jones, P.R.M., West, G.M., Harris, D.H., and Read, J.B., 1989. The Loughborough Anthropometric Shadow Scanner LASS. *Endeavour*, 13(4), 162–168.

Jürgens, H.W., Aune, I.A., and Pieper, U., 1990, *International Data on Anthropometry*, Occupational Safety and Health Series, No. 65, Geneva: World Labour Office.

Kaleps, I., Clauser, C.E., Young, J.W., Chandler, J.F., Zehner, G.F., and McConville, J.T., 1984. Investigation into the mass distribution properties of the human body and its segments. *Ergonomics*, 27, 1225–1237.

Kovatz, F., Boszormenyi-Nagy, G., Nagy, G.G., and Ordig, L., 1988. Morphometry of the upright trunk during breathing. *SPIE Proceedings, Biostereometrics*, 1030, 255–262.

Martin, R., 1914. *Lehrbuch der Anthropologie*, Jena: Verlag von Gustav Fischer.

McConville, J.T., Churchill, T.D., Kaleps, I., Clauser, C.E., and Cuzzi, J., 1980. Anthropometric Relationships of Body and Body Segment Moments of Inertia, AFAMRL-TR-80-119, AD-A097-238, Wright-Patterson Air Force Base, OH: Aerospace Medical Research Laboratory.

Meredith, H.V., 1953. Growth in head width during the first twelve years of life. *Pediatrics*, 12, 411–429.

Meunier, P., 2000. Use of body shape information in clothing size selection, in *Proceedings of the 14th Triennial Congress of the International Ergonomics Association and 44th Annual Meeting of the Human Factors and Ergonomics Society*, Vol. 6, Santa Monica, CA: Human Factors and Ergonomics Society, 715–718.

Molenbroek, J.F.M., 1994. *Made to Measure: Human Body Dimensions for Designing and Evaluating Consumer Durables*, Physical Ergonomics Series 3, Delft: Delft University Press.

Mollard, R., Sauvignon, M., and Pineau, J., 1982. Biostereometric study of a sample of 50 young adults by photogrammetry. *SPIE Proceedings, Biostereometrics*, 361, 234–240.

NCHS (National Center for Health Statistics), 1996. Third National Health and Nutrition Examination Survey, 1988–1994, NHANES III Laboratory Data File (CD-ROM). U.S. Department of Health and Human Services (Public Use Data File Documentation 76200), Hyattsville, MD: Centers for Disease Control and Prevention.

Nurre, J., 1997. Locating landmarks on human body scan data, in *Proceedings of International Conference on Recent Advances in 3-D Digital Imaging and Modeling*, Los Alamitos, CA: IEEE Computer Society Press, 289–295.

Ohlson, M.A., Biester, A., Brewer, W.D., Hawthorne, B.E., and Hutchinson, M.B., 1956. Anthropometry and nutritional status of adult women. *Human Biology*, 28, 189–202.

Okunribido, O., 2000. A survey of hand anthropometry of female rural farm workers in Ibadan, Western Nigeria. *Ergonomics*, 43, 282–292.

Park, W., Woolley, C., Foulke, J., Chaffin, D.B., Raschke, U., and Zhang, X., 1999. Integration of Electromagnetic and Optical Motion Tracking Devices for Capturing Human Motion Data, SAE Technical Paper 1999-01-1911, Warrendale, PA: Society of Automotive Engineers.

Patrias, K., 2000. Visible Human Project, Current Bibliographies in Medicine 2000-5, Bethesda, MD: U.S. Department of Health and Human Services, Public Health Service, National Institutes of Health. Available at http://www.nlm.nih.gov/pubs/resources.html.

Perkins, T., 1999. Tracking Size and Shape Changes During Pregnancy, SAE Technical Papers Series 1991-01-1889, Warrendale, PA: Society of Automotive Engineers.

Perkins, T., Burnsides, D.B., Robinette, K.M., and Naishadham, D., 2000. Comparative Consistency of Univariate Measures from Traditional and 3-D Scan Anthropometry, SAE Technical Papers Series 2000-01-2145, Warrendale, PA: Society of Automotive Engineers.

Pollak, R., 1989. An Investigation into Techniques for Landmark Identification on 3D Images of Human Subjects, AAMRL-SR-90-500, Wright-Patterson AFB, OH: Armstrong Aerospace Medical Research Laboratory, Human Systems Division, Air Force Systems Command.

Probe, J.D., 1990. Quantitative assessment of human motion using video motion analysis, in *Proceedings of the Third Annual Workshop on Space Operations Automation and Robotics (SOAR 1989)*, Houston, TX: National Aeronautics and Space Administration, Lyndon Johnson Space Center, 155–157.

Reed, M.P., 1998. Statistical and biomechanical prediction of automobile driving posture, UMTRI-90957, Ann Arbor: University of Michigan Transportation Research Institute/Department of Industrial and Operations Engineering.

Reed, M.P., Manary, M.A., Flannagan, C.A.C., and Schneider, L.W., 2000. Effects of vehicle interior geometry and anthropometric variables on automobile driving posture. *Human Factors*, 42(4), 541–552.

Rioux, M. and Bruckart, J., 1997. Data collection, in *3-D Surface Anthropometry: Review of Technologies*, Robinette, K.M., Vannier, M.W., Rioux, M., and Jones, P.R.M., Eds., AGARD Advisory Rep. 329, Neuilly-sur-Seine, France: Advisory Group for Aerospace Research and Development, chap. 3.

Rioux, M., Bechthold, G., Taylor, D., and Duggan, M., 1987. Design of a large depth of view three-dimensional camera for robot vision. *Optical Engineering*, 26(12), 1245–1250.

Robinette, K.M. and McConville, J.T., 1981. An Alternative to Percentile Models, SAE Technical Paper Series 810217, Detroit: Society of Automotive Engineers.

Robinette, K.M., Vannier, M.W., Rioux, M., and Jones, P.R.M., 1997. 3-D Surface Anthropometry: Review of Technologies, AGARD Advisory Rep. 329, Neuilly-sur-Seine, France: Advisory Group for Aerospace Research and Development.

Robinette, K., Blackwell, S., Daanen, H., Fleming, S., Boehmer, M., Brill, T., Hoeferlin, D., and Burnsides, D., 2002. Civilian American and European Surface Anthropometry Resource (CAESAR), Final Report, Vol. I: Summary, AFRL-HE-WP-TR-2002-0169, Wright-Patterson Air Force Base, OH: U.S. Air Force Research Laboratory, Human Effectiveness Directorate, Crew System Interface Division.

Schwartz, M., 1999. Determination of kinematically consistent joint centers from segment orientations, in *Proceedings of the First Joint BMES/EMBS Conference* (IEEE Engineering in Medicine and Biology 21st Annual Conference), Atlanta.

Seitz, T. and Bubb, H., 1999. Measuring of Human Anthropometry, Posture and Motion, SAE Technical Paper 1999-01-1913, The Hague: *Proceedings of Digital Human Modeling for Design and Engineering*, Warrendale, PA: SAE International.

Snow, C.C., Reynolds, H.M., and Allgood, M.A., 1975. Anthropometry of Airline Stewardesses, FAA-AM-75-2, Oklahoma City: Federal Aviation Administration.

Tan, T.K., Brandsma, M.G., and Daanen, H.A.M., 2001. Body surface area determination using 3D whole body scans, in *Proceedings of Numerisation 3D Scanning Conference*, Paris, 4–5 April.

Tikuisis, P., Meunier, P., and Jubenville, C.E., 2001. Human body surface area: measurement and prediction using three dimensional body scans. *European Journal of Applied Physiology*, 85(3–4), 264–271.

Young, J.W., Chandler, R.F., Snow, C.C., Robinette, K.M., Zehner, G.F., and Lofberg, M.S., 1983. Anthropometric and Mass Distribution Characteristics of Adult Females, FAA-AM-83-16, Oklahoma City: Office of Aviation Medicine, Federal Aviation Administration.

3

Motor Behavior

Nico J. Delleman

CONTENTS

Introduction

Working posture is determined by the characteristics of the worker, the workstation, and the operation (Table 3.1). Concerning the sense systems as determinants of working posture, the main attention in this chapter will be on the eyes and vision. Regarding the capabilities of motor systems one may, for example, think of joint ranges of motion, strength, and endurance. It is recognized that many other determinants may be involved in posture selection in general, e.g., the level of skill of the worker and psychosocial factors such as stress, boredom, tradition, and culture.

    Determinants of working posture may be active in relation to the postural space or act as a postural strategy. Both concepts are defined below and exemplified following the determinants listed in Table 3.1.

TABLE 3.1
Determinants of Working Posture

Worker

Dimensions, spatial position and orientation, and mass (body segment, whole body)
Capabilities of motor systems
Capabilities of sense systems

Workstation (means of transport, machines, furniture, tools/objects)[a]

Dimensions, spatial position and orientation, and mass

Operation

Vision
   Gaze direction and viewing distance (spatial position of target with respect to head/eyes)
   Interface with visual target (angle between gaze direction and target surface, visual
    interference)
Control (hand, foot)
   Reach direction and reach distance (spatial position of target with respect to trunk)
   Interface with target, i.e., actuator/tool/object (type and orientation of grip/contact)
   Direction and magnitude of external force exertion
Stability (body segment, whole body)
   Interface with workstation (type of body support)

---

[a] In a wider sense, a workstation includes fixtures, fittings, floor, walls, and ceiling.

Postural Space

Postural space is defined as all working postures that can be adopted voluntarily and momentarily, given a set of physical limitations. Basically, each space is determined by the ranges of motion of body joints and eyes, which ranges may be reduced by personal protective equipment, such as glasses and clothing. Furthermore, the capability of the eyes in terms of visual acuity may affect the space by limiting the range of viewing distances. Dimensions, spatial position, and orientation of a workstation and a worker may affect the postural space, simply by mutual physical interference. Worker dimensions also include personal protective equipment, such as shoes, clothing, helmet, etc. The operation to be performed may pose demands on vision, hand/foot control, and/or stability. That is, all three types of demand can be characterized by interfaces with the workstation, affecting the postural space. For example, vision requires a minimum angle between the line of sight (gaze direction) and the surface of a visual target, as well as the absence of interference of the line of sight. Control may require a certain type and orientation of grip/contact. Stability in terms of a balanced posture of the whole body always requires a base of one or more points/areas of support at the workstation, while stability in terms a fixed posture of a body segment may require additional support.

Postural Strategy

Within the postural space, a worker must meet vision and/or control demands concerning the position of the target with respect to the body, as described in Table 3.1 in terms of spherical coordinates by a direction and a distance. A hand position, for example, may be realized by many combinations of orientations of the forearm, the upper arm, the shoulder girdle, and the trunk. Most likely, a worker will prefer a selection of these, guided by an underlying principle (see Bernstein 1967; for analogous considerations concerning external force exertion, see Haslegrave 1994). A postural strategy is defined as a systematic relationship disclosed between the determinant in question and the working posture. Evershed (cited in Kilpatrick 1972), Korein (1985), and Case et al. (1990) hypothesize that, when reaching to a target, a body segment will be moved only if a target cannot be reached by movement of the segments located more distally (closer to the hand). Hsiao and Keyserling (1991) hypothesize that a proximal segment (i.e., one closer to the buttocks) would show a greater tendency to stay close to a neutral posture (within its neutral range) than a distal segment, whenever movement of segments was necessary to view or reach a target. A neutral range is defined as the part of the maximum range of motion that presents minimal discomfort to the joints and adjacent body segments. Several other researchers have incorporated hypothetical postural strategies into their kinematic models. For example, the model by Jung et al. (1992, 1995) incorporates an optimization algorithm that calculates, for each range of motion separately, the deviation of the segment from the center of the range of motion, divides it by the maximum deviation from the center, takes the square, and multiplies by a penalty. The resulting scores for eight ranges of motion are summed. The model predicts that in reality the upper body posture with the lowest total score will be chosen by subjects. No anatomical/physiological explanation was found for the calculation procedure described. The authors state only that "the sizes of the penalties were selected using simulations based on anthropometric characteristics of the segments involved, in order to prevent discontinuous (non-smooth) joint motions." The BOEMAN model (Springer 1969, Healy 1971, Katz 1972) and the model by Kilpatrick (1970, 1972) resemble the model by Jung and colleagues (Jung et al. 1992, 1995). Rosenbaum et al. (1995) postulated that reaching behavior is guided by knowledge gained by a subject about postures adopted earlier at final hand positions ("stored postures"), that is, knowledge of spatial accuracy costs (the extent to which stored postures miss the current target) and travel costs (how much effort it will be to move to the stored postures from the starting posture). According to Soechting et al. (1995), final arm postures can be predicted on the basis of a strategy that aims to minimize the work done to transport the hand from the starting position to the final position. It is concluded that the hypothetical postural strategies described above refer to the upper body (the positions of the feet and buttocks being fixed during standing and sitting, respectively). Little is stated about what makes people decide to move the whole body (changing from a sitting posture to a standing posture, or taking a step).

In Sections 3.1 and 3.2 studies on postural strategies used during gazing and reaching are described.

3.1   Gazing

Gaze is directed toward a target by reorienting the eye in the head/orbit or by reorienting various body segments (e.g., head, chest, pelvis). Studies of the relationship between gaze direction and working posture focus on the vertical gaze direction (upward/downward), on the horizontal gaze direction (to left/right), or on a combination of both, i.e., an oblique gaze direction. In the sections below, oblique, vertical, and horizontal gaze direction are discussed in succession. Eye movements and viewing distance are dealt with in Chapter 4.

3.1.1   Oblique Gaze Direction

Nakayama (1983) described quite clearly two fundamental laws of human eye rotation. Donders' law states that each gaze direction is associated with one and only one orientation of the eye in the orbit. While obeying Donders' law, Listing's law for the eyes in the head (Figure 3.1) is much more specific, stating that the eye uses only two of its three degrees of rotational freedom while redirecting gaze. The eye rotates around axes that all lie in one plane, which is fixed to the head. This so-called Listing's plane (Figure 3.1) is somewhat tilted forward with respect to the frontal plane of the upright head (Hore et al. 1992, Radau et al. 1994; for a review of conditions affecting Listing's plane orientation, refer to Crawford and Vilis 1995), as well as perpendicular to a primary gaze direction. Gaze changes occur without rotation around the latter axis (the primary gaze direction) when viewing distant targets.

Straumann et al. (1991) and Tweed and Vilis (1992) have shown an analogue of Listing's law for the head. During fixed trunk gaze changes (i.e., the trunk upright and stationary; only head and eyes free to move), the direction in which the head faces, that is, the direction in which the nose points, changes by rotations around axes that all also lie in one plane. This plane is perpendicular to the primary facing direction, which is approximately the normal head posture for viewing a target at eye level straight ahead. No rotation around the axis of the primary facing direction was demonstrated. The results of both studies were obtained for target positions up to 70° eccentric from a central target. However, for somewhat more

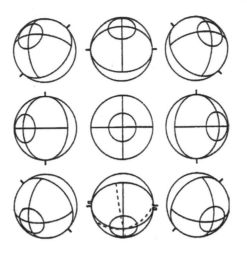

FIGURE 3.1

Listing's law for the eye in the head. The nine eye orientations drawn in solid lines accord with Listing's law, because they are attainable by rotating from a primary orientation (center) around axes lying in Listing's plane (the plane of the paper). The orientation drawn in dashed lines at bottom center does not fit Listing's law, because the rotation to this orientation from the primary orientation occurs around an axis tilted out of Listing's plane. (Reprinted from *Vision Research*, 30, Tweed, D. and Vilis, T., Geometric relations of eye position and velocity vectors during saccades, 111–127, Copyright (1990), with permission from Elsevier Science.)

eccentric targets (70° to 90°), Glenn and Vilis (1992) disclosed for each gaze direction a unique amount of rotation of the head around the axis of the primary facing direction. Changes of head orientation are constrained to a twisted surface of rotation axes instead of to a plane (apparently, while studying small sections of a curved surface, the surface easily appears flat). This means that the analogue of Listing's law for the head is violated, while the analogue of Donders' law for the head is upheld. It seems that the head behaves more like a so-called Fick gimbal, i.e., a rotational system in which a horizontal axis is nested within a fixed vertical axis (Figure 3.2).

Straumann et al. (1991) and Glenn and Vilis (1992) found that during oblique gaze shifts (without chest/trunk movement), the eye makes predominantly vertical movements (within the head/orbit), whereas the head makes predominantly horizontal movements, presumably to minimize work done against gravity. Glenn and Vilis (1992) positioned visual targets at each corner and at the center of a square (with sides vertical or horizontal). For the head the average ratio of vertical and horizontal components (v/h) was 0.54 for movements to targets at 70° eccentricity and 0.50 for targets at 90° eccentricity, whereas for the eye (within the orbit/head) the average v/h was 2.51 for movements to targets at 70° eccentricity and 1.42 for targets at 90° eccentricity.

Hsiao and Keyserling (1991) studied postural behavior of seated subjects, during continuous/static viewing for 2 min at each of nine target positions (either 0°, 30°, or 60° to the right, and either 60° down, 0°, or 60° up). It was

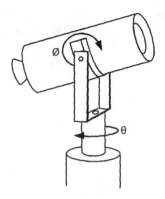

FIGURE 3.2
Fick gimbal system, exemplified by a telescope. Rotations around vertical and horizontal axes
are denoted by θ and Ø, respectively. (From Nakayama, K., in *Vergence Eye Movements: Basic
and Clinical Aspects*, Schor, C.M. and Ciuffreda, K.J., Eds., Boston: Butterworths, 1983, 543–564.
With permission.)

hypothesized that a proximal body segment (i.e., one closer to the buttocks–seat interface) would show a greater tendency to stay close to a neutral
posture than a distal segment, whenever movement of segments was necessary to view a target. The posture of body segments (including the eyes)
was measured and classified as within or outside neutral ranges. Indeed,
looking at the overall result for all target positions tested, the trunk was
found to have a greater neutral tendency than the head/neck. However,
contrary to the hypothesis, the eyes (line of sight with respect to the head)
had the greatest neutral tendency of all. The least neutral tendency of all
was found for the pelvis; on average about 80% of the movement to direct
the gaze onto a target to the right was created by rotation of the pelvis around
the vertical axis (it is not clear whether subjects rotated the pelvis with respect
to the seat or used a swiveling seat).

Radau et al. (1994) studied eye, head, and chest orientation of standing
subjects (with only foot position fixed), while fixating a visual target at each
of 12 positions (either 45°, 90°, or 135° to the left or right, and either 45°
down or 45° up). It was found that the ratio of vertical to horizontal components for the eye (within the orbit/head) was generally greater than one.
The chest, by contrast, moved almost entirely in the horizontal direction,
whereas the head performed an intermediate role. The horizontal component
of gaze is created on average 50 to 60% by the chest and 40 to 50% by the
head with respect to the chest, leaving the eye with little to contribute and
near its center position within the head/orbit (Vilis 1996). The results of
Radau and colleagues concerning vertical gaze direction support the main
hypothesis of Hsiao and Keyserling (1991), stating that a proximal segment
would show a greater tendency to stay close to a neutral posture (upright
trunk, upright head/neck) than a distal segment, whenever movement of
segments is necessary to view a target. Bearing in mind the experimental

results by Radau and colleagues, it is very likely that Hsiao and Keyserling were not able to demonstrate that the eyes have the least neutral tendency, because of the extremely large neutral range defined, i.e., 60°. The results of Radau and colleagues concerning horizontal gaze direction disclosed that a distal segment shows a greater tendency to stay close to a neutral posture than a proximal segment (i.e., eyes vs. head, as well as head vs. chest), which is totally opposite to the main hypothesis by Hsiao and Keyserling (1991).

3.1.2 Vertical Gaze Direction

The vertical gaze direction is usually described as gaze inclination, i.e., the angle between the gaze direction and the horizontal plane. Under fixed trunk conditions, Brues (1946) measured head inclination through a range of gaze inclinations between straight down and straight up, with intervals of 22.5°. At gaze inclinations –90°, –67.5°, –45°, and –22.5° downward, and 22.5°, 45°, 67.5°, and 90° upward, the contribution of head inclination to gaze inclina-tion is approximately 60, 52, 49, 42, 47, 56, 61, and 66%, respectively (the remainder comes from a reorientation of the eyes with respect to the head/ orbits). From a typical result of a gaze-pursuit task (with instructions to keep the back against the chair), as presented by Straumann et al. (1991), it was estimated that the contribution of head inclination to gaze inclination is about 55 to 60% for targets positioned between 25° below and 25° above the horizontal. This result was confirmed for gaze inclinations between horizon-tal and 50° below horizontal by data on a reading task with an upright trunk (Conrady et al. 1987). For the gaze range studied, the data of Conrady and colleagues showed a linear relationship between gaze inclination and head inclination. All three studies show about the same relationship between gaze inclination and head inclination, given a stationary trunk.

Delleman and Hin (2000) described postural behavior in static gazing upward/downward, without postural constraints concerning the trunk/ chest, and studied the role of the range of motion of the eyes, the head, the chest, and the pelvis. Eight subjects were asked to fixate various targets at a distance of 130 cm from the initial mid-eye position, i.e., the position halfway between the two eyes with an upright trunk, an upright head, and looking straight ahead along the horizontal. The task was executed while standing (feet fixed). There were 12 vertical target angles, ranging from straight upward (denoted +90°) to 75° downward (denoted –75°), using steps of 15° on a circle with the center at the initial mid-eye position (Figure 3.3). For the pelvis and chest segments, no major change of posture was measured. The relationships between vertical gaze direction and head inclination for all subjects were described best by a linear regression equation (Table 3.2). The average contribution of the head to direct the gaze onto target was 60% (which means 40% by the eyes), which matches relatively well with the figures found in the majority of the studies done under trunk/chest-fixed conditions.

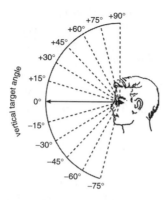

FIGURE 3.3
The vertical target angles tested by Delleman and Hin (2000).

TABLE 3.2

Relationship between the Head Inclination and
the Vertical Gaze Direction for Individual Subjects

Subject Number	A (slope)	B (intercept)	R
1	0.59	1.0	0.996
2	0.59	0.9	0.987
3	0.56	−0.9	0.995
4	0.61	2.8	0.997
5	0.61	6.7	0.987
6	0.61	−1.4	0.997
7	0.81	0.3	0.992
8	0.57	2.9	0.991

Note: Head inclination = A × vertical gaze direction + B
($n = 12$ gaze directions). R = Pearson correlation coefficient. The results presented by Delleman and Hin
(2000) have been corrected slightly.

Table 3.3 contains the contribution of the head to direct the gaze onto
target, i.e., head inclination/vertical gaze direction, which is in fact the factor
A (slope) from Table 3.2. Furthermore, Table 3.3 contains the contribution of
the head range of motion to gaze range of motion (defined as the sum of
the eye range of motion and the head range of motion). The resemblance of
the results in the two columns of Table 3.3 (except for subject number 7)
suggests that the head and the eyes use the particular ratio mentioned above
(60:40) for approaching the end of both ranges of motion involved at about
the same relative rate, thereby sharing the musculoskeletal load equally.

3.1.3   Horizontal Gaze Direction

Most studies on postural strategies in gazing sideward have been carried
out under fixed trunk/chest conditions. In a review, Delleman (1999) states

TABLE 3.3

Contribution of the Head to Direct the Gaze onto Target (head inclination/vertical gaze direction), and the Contribution of Head Range of Motion (ROM) to Gaze ROM (sum of eye ROM and head ROM) for Individual Subjects

Subject Number	Head Inclination/ Vertical Gaze Direction	Head ROM/ Gaze ROM
1	0.59	0.56
2	0.59	0.63
3	0.56	0.51
4	0.61	0.65
5	0.61	0.55
6	0.61	0.60
7	0.81	0.64
8	0.57	0.55
Average	0.62	0.59

Note: The results presented by Delleman and Hin (2000) have been corrected slightly.

that the relationship between target eccentricity in the horizontal direction and head posture is linear for subjects on a group level. It also turned out that the head contributes a relatively large share of the horizontal gaze direction (with group averages in various studies roughly between 60 and 100%). In particular, the studies by Delreux et al. (1991), Fuller (1992), and Rossetti et al. (1994) disclosed quite a considerable intersubject variability. According to Delreux et al. and Fuller, this supports the existence of so-called head movers and eye movers (non-head movers), as introduced by Afanador and Aitsebaomo (1982) and confirmed by Roll et al. (1986). Furthermore, various possible reasons for the variability observed were mentioned, such as the behavioral/experimental situation, target eccentricity, and initial alignment of head and gaze (i.e., whether the eyes were in the straight-ahead position in the orbit). Given less restricted circumstances, Hsiao and Keyserling (1991) and Radau et al. (1994) showed that the trunk/chest and pelvis, respectively, contribute considerably to direct the gaze onto an eccentric target (refer to Section 3.1.1).

Delleman et al. (2001) described postural behavior in static gazing sideward, and studied the role of the range of motion of the eyes, the head, the chest, and the pelvis. They asked 11 subjects to fixate various targets at a distance of 160 cm from the initial mid-eye position, i.e., the position halfway between the two eyes with an upright trunk, an upright head, and looking straight ahead along the horizontal. The effects of target angle, main posture (standing, sitting), and hand position on the body posture were tested. The target angles ranged from straight forward (denoted 0°) to straight backward (denoted 180°), using steps of 15° on a circle with the center at the initial mid-eye position (Figure 3.4). Four hand positions were distinguished, i.e., both hands free (arms hanging down along the trunk),

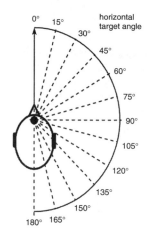

FIGURE 3.4
The horizontal target angles tested by Delleman et al. (2001).

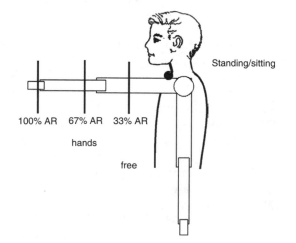

FIGURE 3.5
The two main postures (standing, sitting) and four hand positions tested by Delleman et al. (2001). These were with both hands free (arms hanging down along the trunk), and both hands fixed at 33, 67, and 100% arm reach (AR), which was defined as the horizontal distance between the top of the sternum and the hand grip axes while stretching the arms out horizontally and straight forward in a comfortable way.

and both hands fixed at 33, 67, and 100% arm reach (AR), which was defined as the horizontal distance between the top of the sternum and the hand grip axes while stretching the arms out horizontally and straight forward in a comfortable way (Figure 3.5). Skin-mounted sensors were positioned on the forehead, at the sternum (chest), and on the sacrum (pelvis). A fourth sensor was placed temporarily on the visual target. The yaw output of the skin-mounted sensors in space (rotation around the vertical) was used to calculate the heading of each segment in space, i.e., yaw while gazing at the destination

TABLE 3.4

Relationship between the Heading of Head-in-Space
and the Horizontal Gaze Direction for the Experimental
Conditions Tested

Experimental Conditions	A (slope)	B (intercept)	R
Standing, hands free	0.84	−2.6	0.98
Standing, hands 33% AR	0.75	0.3	0.99
Standing, hands 67% AR	0.80	−1.3	0.99
Standing, hands 100% AR	0.78	−0.2	0.99
Sitting, hands free	0.75	0.3	0.98
Sitting, hands 33% AR	0.74	1.3	0.98
Sitting, hands 67% AR	0.75	1.1	0.98
Sitting, hands 100% AR	0.71	2.5	0.98

Note: Heading of head-in-space = A × horizontal gaze
direction + B (n = 11 subjects and 13 gaze directions per
subject). R = Pearson correlation coefficient. AR (arm
reach) = horizontal distance between the top of the sternum and the hand grip axes while stretching the arms out
horizontally and straight forward in a comfortable way.

target minus yaw while gazing at the initial target. The horizontal gaze
direction was calculated from the position of the sensor at the destination
target and the mid-eye position. The latter was calculated from the position
of the forehead sensor and the position of mid-eye with respect to the
forehead sensor.

For all experimental conditions, a significant linear relationship was found
between heading of head-in-space and horizontal gaze direction (Table 3.4).
Figure 3.6 shows heading of head/heading of gaze (column *A* from Table 3.4)
for the experimental conditions. Figure 3.7 shows maximal heading of head/
maximal heading of gaze for the experimental conditions. For these two
variables, the same statistically significant differences among experimental
conditions were found. That is, with hands free, hands at 67% AR, and hands
at 100% AR, the score for sitting was lower than the score for standing. For
standing, the scores for hands at 33% AR and hands at 100% AR were lower
than the score with hands free, and the score for hands at 33% AR was lower
than the score for hands at 67% AR.

The results above show that for each of the eight experimental conditions
the head (supported by underlying segments) contributes at a particular rate
to direct the gaze onto target (Table 3.4, Figure 3.6). These particular rates
are reflected by the ratios of maximal heading of head (sum of ranges of
motion of the head and underlying segments) and maximal heading of gaze
(sum of ranges of motion of the eye, the head, and underlying segments)
(Figure 3.7). Similarly to the conclusion of the study on postural behavior in
static gazing upward and downward (Delleman and Hin, 2000), this suggests
some sort of musculoskeletal load sharing, taking account of the restricted
ranges of motion of the pelvis (in sitting) and the chest (due to fixed hand
positions). However, two remarks have to be made. First, for almost all

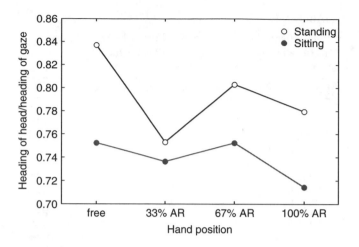

FIGURE 3.6
Heading of head/heading of gaze for the experimental conditions tested (average group scores, where $n = 11$). AR (arm reach) = horizontal distance between the top of the sternum and the hand grip axes while stretching the arms out horizontally and straight forward in a comfortable way.

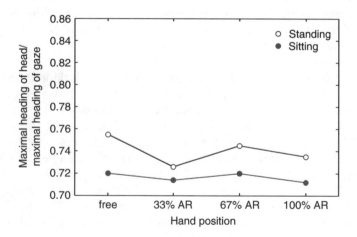

FIGURE 3.7
Maximal heading of head/maximal heading of gaze for the experimental conditions tested (average group scores, where $n = 11$). AR (arm reach) = horizontal distance between the top of the sternum and the hand grip axes while stretching the arms out horizontally and straight forward in a comfortable way.

experimental conditions the scores in Figure 3.6 are higher than the scores in Figure 3.7. This may be attributed to the way that the maximal heading of the eyes in the orbits/head was measured. That is, the duration of maximal heading of the eyes in the procedure described above was much shorter than the duration of gazing at the target angles in the main experiment. It is likely that the subjects tended to stay away from the unfavorable maximal heading of the eyes in the orbits/head during sustained gazing. As a consequence, sustained gazing during measurement of maximal heading of the eyes in the orbits/head would then reduce the denominator part of "maximal heading of head/maximal heading of gaze" in Figure 3.7 and bring the level of the scores closer to the ones in Figure 3.6. Second, for relatively small target angles the heading of head-in-space is created solely by neck twisting, while at increasing target angles the chest and pelvis take over. Delleman et al. (2001) stated that a first look at the data on relatively small target angles showed that for the "hands-free" condition there was no significant difference on "heading of head/heading of gaze" between standing and sitting. When calculating the ratio of the range of motion of the head (maximal heading of head-vs.-chest) and the sum of the ranges of motion of the head and the eyes (maximal heading of head-vs.-chest plus maximal heading of eye-in-head), also no difference was found between sitting and standing. This result suggests that musculoskeletal load sharing is also present at the subsystem of head and eyes.

3.2   Reaching

For manual operations the hand is directed toward a target by reorienting one or more body segments (e.g., forearm, upper arm, shoulder girdle, chest, pelvis). Here the position of a target is described with respect to the upper body in terms of spherical coordinates by a reach direction and a reach distance. In the following discussion, attention is on studies describing how the working posture is determined by reach direction, reach distance, and hand trajectories, successively. Hand trajectories usually include simultaneous changes of reach direction and reach distance.

3.2.1   Reach Direction

Straumann et al. (1991; also Hepp et al. 1992) showed an analogue of Listing's law for the arm; i.e., the reach direction of the arm changes by rotations around axes that all lie in one plane. This result was obtained for a small range of target positions, i.e., up to 25° from a central target. However, for more eccentric targets, Hore et al. (1992) and Miller et al. (1992) disclosed that changes of arm orientation are constrained to a curved surface of rotation axes instead of to a plane, as found for the head (see Section 3.1.1). Here also the arm seems to behave like a Fick gimbal. The studies by Straumann et al. (1991), Hore et al. (1992), and Miller et al. (1992) were performed with a fully extended or slightly flexed elbow, while the trunk was kept fixed. According to Hepp et al. (1992), at these reach distances, the orientation of the arm's constraining plane changed little with reach distance, while upper extremity kinematics were similar to "visual grasping" with eye and head (see Section 3.1.1). Hepp and colleagues also stated that great rotations around the longitudinal axis of the upper arm are made while reaching with the upper arm close to the body. These rotations indicate that changes of arm orientation are not always constrained to a surface of rotation axes. Donders' law, as applied to the arm, states that every spatial position of the hand corresponds to a unique posture of the arm. Hepp et al. (1992), Soechting et al. (1995), Gielen et al. (1997), and Desmurget et al. (1998) demonstrated that the arm posture at a final hand position depends on the starting position of the hand, and that, consequently, Donders' law is violated. Fischer et al. (1997) quantified the travel costs in the model by Rosenbaum (1995, refer to the introduction on postural strategy) on the basis of the weight of the body segments moved by a rotation at the hip, by a rotation at the shoulder, as well as by a rotation at the elbow (all in the sagittal plane). Some of the predictions from the model were supported by experiments on subjects moving the hand from one location to another, whether via a third location or not. That is, the least rotation was found at the joint moving the greatest weight, i.e., the hip. Furthermore, postures at a preceding hand location affected postures at a succeeding location. Subjects seemed to minimize the differences between hip joint positions at successive locations reached by the hand. It was also found that the velocity of the movement affected the postures at the third location mentioned above, which was located away from the straight line between the initial and final locations of the hand.

Hsiao and Keyserling (1991) studied postural behavior of seated subjects during continuous/static reaching for 2 min at each of nine target positions (either 60° to the right, 0°, or 60° to the left and either 60° down, 0°, or 60° up). These positions were tested at three reach distances, i.e., approximately 40, 70, and 100 cm away from vertebra T2. It was hypothesized that a proximal segment (i.e., one closer to the buttocks–seat interface) would show a greater tendency to stay close to a neutral posture than a distal segment, whenever movement of segments was necessary to reach a target. Indeed, looking at the overall result for all combinations of reach directions and distances tested, the trunk showed a greater neutral tendency than the upper extremity segments. However, contrary to the hypothesis, the least neutral tendency of all was found for the pelvis. More specifically, the vertical position of the target clearly showed most effect on shoulder flexion, while the target's horizontal position mostly affected the rotation of the pelvis around the vertical axis (although it is not clear whether subjects rotated the pelvis with respect to the seat or used a swiveling seat). The main effects of reach distance are discussed in the next section. According to the experimental data presented by Hsiao and Keyserling (see above) subjects are more likely to show rotation of the pelvis around the vertical axis than to twist their trunk during continuous/static reaching. The model on reach posture of the upper body proposed by Jung et al. (1992, 1995) does not include the possibility of rotation of the pelvis around the vertical axis. However, the model

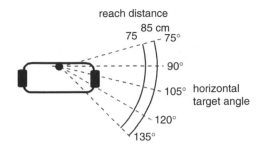

FIGURE 3.8
The five reach directions and two reach distances tested by Delleman et al. (2003).

makes trunk twisting (and trunk lateral bending) the least likely option to be used by subjects to reach a target. In that way, the model supports the postural behavior observed by Hsiao and Keyserling.

Delleman et al. (2003) described postural behavior in static reaching sideward, and studied the role of the ranges of motion of the arm, the chest, and the pelvis. Eight subjects were asked to touch various spatial targets with the tip of the right index finger. Ten targets were positioned on concentric circles in the horizontal plane with the center at the top of the sternum (Figure 3.8). The reach directions were 75°, 90°, 105°, 120°, and 135° to the right (where 0° = straight forward, and 180° = straight backward). The reach distances were 75 and 85 cm. For all subjects, full arm reach sideward (i.e., from the top of the sternum to the right index finger) ranged from 88 to 99 cm. Skin-mounted sensors were positioned at the top of the sternum (chest), on the sacrum (pelvis), and on the tip of the right index finger. The yaw output of the sternum and sacrum sensors in space (rotation around the vertical) was used to calculate the headings of the chest and pelvis, i.e., yaw while touching the target minus yaw while adopting the reference posture. Heading of chest-vs.-pelvis is defined as heading of chest-in-space minus heading of pelvis-in-space. The horizontal arm reach direction (heading of the arm) was calculated from the horizontal plane positions of the sensor on the tip of the right index finger and the sensor at the top of the sternum, while touching the target. Heading of arm-vs.-chest is defined as heading of arm-in-space minus heading of chest-in-space. For both reach distances, significant linear relationships were found between heading of arm-vs.-chest and horizontal arm reach direction (Table 3.5), between heading of chest-vs.-pelvis and horizontal arm reach direction (Table 3.6), and between heading of pelvis-in-space and horizontal arm reach direction (Table 3.7). The arm, chest, and pelvis each contribute at a particular rate to direct the index finger onto the target. These particular rates are reasonably similar to the average contributions of the arm (58%), chest (12%), and pelvis (30%) to the total of the ranges of motion of the arm, chest, and pelvis (Table 3.8). Similarly to the conclusion of the studies on postural behavior in static gazing (Delleman and Hin 2000, Delleman et al. 2001), this suggests that the arm, chest, and pelvis share the musculoskeletal load equally. On the basis of a

TABLE 3.5

Relationship between the Heading of Arm-vs.-Chest and the Horizontal Arm Reach Direction for the Reach Distances Tested

Reach Distance	A (slope)	B (intercept)	R
75 cm	0.50	26.7	0.78
85 cm	0.55	18.9	0.86

Note: Heading of arm-vs.-chest = $A \times$ horizontal reach direction + B ($n = 8$ subjects and 5 reach directions per subject). R = Pearson correlation coefficient.

TABLE 3.6

Relationship between the Heading of Chest-vs.-Pelvis and the Horizontal Arm Reach Direction for the Reach Distances Tested

Reach Distance	A (slope)	B (intercept)	R	O
75 cm	0.15	−9.3	0.77	62°
85 cm	0.13	−6.5	0.69	50°

Note: Heading of chest-vs.-pelvis = $A \times$ horizontal arm reach direction + B ($n = 8$ subjects and 5 reach directions per subject). R = Pearson correlation coefficient. O = horizontal arm reach direction at which heading of chest-vs.-pelvis starts, estimated from the regression equation.

TABLE 3.7

Relationship between the Heading of Pelvis-in-Space and the Horizontal Arm Reach Direction for the Reach Distances Tested

Reach Distance	A (slope)	B (intercept)	R	O
75 cm	0.35	−17.4	0.61	50°
85 cm	0.32	−12.4	0.63	39°

Note: Heading of pelvis-in-space = $A \times$ horizontal arm reach direction + B ($n = 8$ subjects and 5 reach directions per subject). R = Pearson correlation coefficient. O = horizontal arm reach direction at which heading of pelvis-in-space starts, estimated from the regression equation.

pilot study, it was expected that the trunk would be involved when reaching sideward more than 75°. However, Table 3.6 and Table 3.7 show that the trunk becomes involved earlier (refer to column *O*); at the 75-cm reach distance the pelvis started to move at an estimated horizontal arm reach

TABLE 3.8

Average Contributions of the Arm, Chest, and Pelvis to
Direct the Index Finger onto Target (motor behavior at reach
distances 75 and 85 cm), and the Average Contributions of
the Arm, Chest, and Pelvis to the Total of the Ranges of
Motion of the Arm, Chest, and Pelvis (range of motion)

		Contribution		
	Reach Distance	Arm	Chest	Pelvis
Motor behavior	75 cm	0.50	0.15	0.35
Motor behavior	85 cm	0.55	0.13	0.32
Range of motion		0.58	0.12	0.30

direction of 50°, while the chest (with respect to the pelvis) started to move
at 62°. At the 85-cm reach distance the values were 39° and 50°, respectively.

3.2.2   Reach Distance

Several researchers have studied the effects of reach distance on upper body
posture. For seated subjects, Hsiao and Keyserling (1991, refer to the previous
section for details of the experimental arrangement) showed that the reach
distance affected trunk flexion, shoulder flexion, elbow flexion, and pelvis
shifting (although it is not clear whether subjects shifted the pelvis with
respect to the seat or shifted the seat). In particular, subjects shifted their
pelvis when a target location was too far away (roughly more than two thirds
of the length of the upper extremity). Snyder et al. (1971) published experimental data on the relationship between the spatial end position of the
elbow and the spatial positions of various skin surface markers on the upper
body (with positions of the feet and buttocks fixed during standing and
sitting, respectively). The spatial orientation of the trunk and the upper arm
was studied, while moving the elbow straight forward virtually along a
series of end positions, starting from an upright trunk posture, with the
upper arm hanging down. It was found that initially most of the forward
"movement" of the elbow was created by upper arm elevation forward,
while the trunk inclined forward at a slow rate. While "moving" the elbow
further forward, the upper arm reached a maximum and the trunk gradually
took over (see Kaminski et al. 1995, Mark et al. 1997, Zhang and Chaffin
1997, Fischer et al. 1997, Vaughan et al. 1998). It appeared that during standing the contribution of the trunk was a little smaller than when sitting.

3.2.3   Hand Trajectories

Soechting and Flanders (1991a,b) reviewed the literature on the trajectory
(path) of the hand during point-to-point movements. They concluded that
some trajectories are straight while others are (slightly) curved, but that there

is little intra- and intersubject variability (e.g., Morasso 1981, Soechting and Lacquaniti 1981, Lacquaniti and Soechting 1982, Atkeson and Hollerbach 1985, Lacquaniti et al. 1986). Uno et al. (1989) have shown that a model based on the minimum change of joint torque can account for the various trajectories observed. For horizontal movements, Dornay et al. (1996) showed similar results when using a minimum muscle-tension change model. The equilibrium point hypothesis (e.g., Fel'dman 1966, Bizzi et al. 1984), as well as the minimum jerk model (e.g., Hogan 1984, Flash and Hogan 1985), predicts straight trajectories only. In their reviews, Soechting and Flanders (1991a,b) also concluded that hand trajectories are invariant of movement velocity (Soechting and Lacquaniti 1981, Lacquaniti et al. 1982, Atkeson and Hollerbach 1985), as well as invariant of a weight of up to 2.5 kg in the hand (Lacquaniti et al. 1982, Atkeson and Hollerbach 1985). However, Pollick and Ishimura (1996) showed that for point-to-point movements at maximum velocity the curvature of the hand trajectory is inversely related to movement velocity. Furthermore, the two studies on the effect of weight in the hand concern movements in a vertical plane. For movements in a horizontal plane, Uno et al. (1989) showed that a variable spring force on the hand (between 3.3 and 10.4 N) drastically changed the hand trajectory.

Summary

A hand position and a gaze direction may be realized by many combinations of orientations of body segments and eyes. The literature reviewed reveals various hypotheses on the strategies that people use. For example, it is hypothesized that (1) a body segment will be moved only if a target cannot be reached by the segments located more distally, (2) a proximal segment will show a greater tendency to stay close to its neutral joint position than a distal segment, to minimize discomfort, and (3) segments will be moved in such a way that the amount of work done is minimized. Recent studies show that the contributions of body segments and eyes to direct the hand/finger or gaze onto a target reflect the relative sizes of their ranges of motion, suggesting that musculoskeletal load is shared equally.

References

Afanador, A.J. and Aitsebaomo, A.P., 1982. The range of eye movements through progressive multifocals. *Optometric Monographs*, 73, 82–88.
Atkeson, C.G. and Hollerbach, J.M., 1985. Kinematic features of unrestrained vertical arm movements. *Journal of Neuroscience*, 5, 2318–2330.

Bernstein, N., 1967. *The Co-ordination and Regulation of Movements.* Oxford: Pergamon Press.

Bizzi, E., Accornero, N., Chapple, W., and Hogan, N., 1984. Posture control and trajectory formation during arm movement. *Journal of Neuroscience, 4,* 2738–2744.

Brues, A., 1946. Movement of the head and eye in sighting, in *Human Body Size in Military Aircraft and Personal Equipment,* Randall, E., Damon, A., Benton, R., and Patt, D., Eds., AF-TR-5501, Wright-Patterson Air Force Base, OH: Aero Medical Laboratory, Wright Air Development Center, 211–219.

Case, K., Porter, J.M., and Bonney, M.C., 1990. SAMMIE: a man and workplace modelling system, in *Computer-Aided Ergonomics,* Karwowski, W., Genaidy, A.M., and Asfour, S.S., Eds., London: Taylor & Francis, 31–56.

Conrady, P., Krueger, H., and Zülch, J., 1987. *Untersuchung der Belastung bei Lupen- und Mikroskopierarbeiten,* Fb 516, Dortmund, Germany: Bundesanstalt für Arbeitsschutz.

Crawford, J.D. and Vilis, T., 1995. How do motor systems deal with the problems of controlling three-dimensional rotations? *Journal of Motor Behavior, 27,* 89–99.

Delleman, N.J., 1999. Working Postures — Prediction and Evaluation, Ph.D. thesis, Soesterberg, the Netherlands: TNO Human Factors.

Delleman, N.J. and Hin, A.J.S., 2000. Postural behaviour in static gazing upwards and downwards, SAE Technical Papers Series 2000-01-2173, Warrendale, PA: Society of Automotive Engineers.

Delleman, N.J., Huysmans, M.A., and Kuijt-Evers, L.F.M., 2001. Postural behaviour in static gazing sidewards, SAE Technical Papers Series 2001-01-2093, Warrendale, PA: Society of Automotive Engineers.

Delleman, N.J., Hin, A.J.S., and Tan, T.K., 2003. Postural behaviour in static reaching sidewards, SAE Technical Papers Series 2003-01-2230, Warrendale, PA: Society of Automotive Engineers.

Delreux, V., Van den Abeele, S., Lefèvre, P., and Roucoux, A., 1991. Eye-head co-ordination: influence of eye position on the control of head movement amplitude, in *Brain and Space,* Paillard, J., Ed., Oxford: Oxford University Press, 38–48.

Desmurget, M., Gréa, H., and Prablanc, C., 1998. Final posture of the upper limb depends on the initial position of the hand during prehension movements. *Experimental Brain Research, 119,* 511–516.

Dornay, M., Uno, Y., Kawato, M., and Suzuki, R., 1996. Minimum muscle-tension change trajectories predicted by using a 17-muscle model of the monkey's arm. *Journal of Motor Behavior, 28,* 83–100.

Fel'dman, A.G., 1966. Functional tuning of the nervous system with control of movement or maintenance of a steady posture: II. Controllable parameters of the muscles. *Biophysics, 11,* 565–578.

Fischer, M.H., Rosenbaum, D.A., and Vaughan, J., 1997. Speed and sequential effects in reaching. *Journal of Experimental Psychology: Human Perception and Performance, 23,* 404–428.

Flash, T. and Hogan, N., 1985. The coordination of arm movements: an experimentally confirmed mathematical model. *Journal of Neuroscience, 5,* 1688–1703.

Fuller, J.H., 1992. Head movement propensity. *Experimental Brain Research, 92,* 152–164.

Gielen, C.C.A.M., Vrijenhoek, E.J., Flash, T., and Neggers, S.F.W., 1997. Arm position constraints during pointing and reaching in 3D space. *Journal of Neurophysiology, 78,* 660–673.

Glenn, B. and Vilis, T., 1992. Violations of Listing's law after large eye and head gaze shifts. *Journal of Neurophysiology,* 68, 309–318.

Haslegrave, C.M., 1994. What do we mean by a "working posture"? *Ergonomics, 37,* 781–799.

Healy, M.J., 1971. Cockpit Geometry Evaluation: Phase II-A, Final Report, Vol. IV, Mathematical Model, Report D162-10128-2A, JANAIR Report 701215, Seattle, WA: Boeing-JANAIR.

Hepp, K., Haslwanter, T., Straumann, D., Hepp-Reymond, M.-C., and Henn, V., 1992. The control of arm-, gaze-, and head-movements by Listing's law, in *Control of Arm Movement in Space,* Caminiti, R., Johnson, P.B., and Burnod, Y., Eds., Berlin: Springer-Verlag, 307–319.

Hogan, N., 1984. An organizing principle for a class of voluntary movements. *Journal of Neuroscience,* 4, 2745–2754.

Hore, J., Watts, S., and Vilis, T., 1992. Constraints on arm position when pointing in three dimensions: Donders' law and the Fick gimbal strategy. *Journal of Neurophysiology,* 68, 374–383.

Hsiao, H. and Keyserling, W.M., 1991. Evaluating posture behavior during seated tasks. *International Journal of Industrial Ergonomics,* 8, 313–334.

Jung, E.S., Kee, D., and Chung, M.K., 1992. Reach posture prediction of upper limb for ergonomic workspace evaluation, in *Proceedings of the Human Factors Society 36th Annual Meeting,* Vol. 1, Santa Monica CA: Human Factors Society, 702–706.

Jung, E.S., Kee, D., and Chung, M.K., 1995. Upper body reach posture prediction for ergonomic evaluation models. *International Journal of Industrial Ergonomics,* 16, 95–107.

Kaminski, T.R., Bock, C., and Gentile, A.M., 1995. The coordination between trunk and arm motion during pointing movements. *Experimental Brain Research,* 106, 457–466.

Katz, R. 1972. Cockpit Geometry Evaluation: Phase III, Final Report, Vol. III, Computer Program System, Report D162-10127-3, JANAIR Report 720402, Seattle, WA: Boeing-JANAIR.

Kilpatrick, K.E., 1970. A Model for the Design of Manual Work Stations, Doctoral dissertation, Ann Arbor, MI: University of Michigan.

Kilpatrick, K.E., 1972. A biokinematic model for workplace design. *Human Factors,* 14, 237–247.

Korein, J.U., 1985. *A Geometric Investigation of Reach,* Cambridge, MA: MIT Press.

Lacquaniti, F. and Soechting, J.F., 1982. Coordination of arm and wrist motion during a reaching task. *Journal of Neuroscience,* 2, 399–408.

Lacquaniti, F., Soechting, J.F., and Terzuolo, C.A., 1982. Some factors pertinent to the organization and control of arm movements. *Brain Research,* 252, 394–397.

Lacquaniti, F., Soechting, J.F., and Terzuolo, S.A., 1986. Path constraints on point-to-point arm movements in three-dimensional space. *Neuroscience,* 17, 313–324.

Mark, L.S., Nemeth, K., Gardner, D., Dainoff, M.J., Paasche, J., Duffy, M., and Grandt, K., 1997. Postural dynamics and the preferred critical boundary for visually guided reaching. *Journal of Experimental Psychology: Human Perception and Performance,* 23, 1365–1379.

Miller, L.E., Theeuwen, M., and Gielen, C.C.A.M., 1992. The control of arm pointing movements in three dimensions. *Experimental Brain Research,* 90, 415–426.

Morasso, P., 1981. Spatial control of arm movements. *Experimental Brain Research,* 42, 223–227.

Nakayama, K., 1983. Kinematics of normal and strabismic eyes, in *Vergence Eye Movements: Basic and Clinical Aspects,* Schor, C.M. and Ciuffreda, K.J., Eds., Boston: Butterworths, 543–564.

Pollick, F.E. and Ishimura, G., 1996. The three-dimensional curvature of straight-ahead movements. *Journal of Motor Behavior,* 28, 271–279.

Radau, P., Tweed, D., and Vilis, T., 1994. Three-dimensional eye, head, and chest orientations after large gaze shifts and the underlying neural strategies. *Journal of Neurophysiology,* 72, 2840–2852.

Roll, R., Bard, C., and Paillard, J., 1986. Head orienting contributes to the directional accuracy of aiming at distant targets. *Human Movement Science,* 5, 359–371.

Rosenbaum, D.A., Loukopoulos, L.D., Meulenbroek, R.G.J., Vaughan, J., and Engelbrecht, S.E., 1995. Planning reaches by evaluating stored postures. *Psychological Review,* 102, 28–67.

Rossetti, Y., Tadary, B., and Prablanc, C., 1994. Optimal contributions of head and eye positions to spatial accuracy in man tested by visually directed pointing. *Experimental Brain Research,* 97, 487–496.

Snyder, R.G., Chaffin, D.B., and Schutz, R.K., 1971. Link System of the Human Torso, HSRI-71-112, Ann Arbor, MI: Highway Safety Research Institute, University of Michigan; or AMRL-TR-71-88, Wright-Patterson Air Force Base, OH: Aerospace Medical Research Laboratory.

Soechting, J.F. and Flanders, M., 1991a. Arm movements in three-dimensional space: computation, theory, and observation. *Exercise and Sport Sciences Reviews,* 19, 389–418.

Soechting, J.F. and Flanders, M., 1991b. Deducing central algorithms of arm movement control from kinematics, in *Motor Control: Concepts and Issues,* Humphrey, D.R. and Freund, H.-J., Eds., Chichester, U.K.: John Wiley & Sons, 293–306.

Soechting, J.F. and Lacquaniti, F., 1981. Invariant characteristics of a pointing movement in man. *Journal of Neuroscience,* 1, 710–720.

Soechting, J.F., Buneo, C.A., Herrmann, U., and Flanders, M., 1995. Moving effortlessly in three dimensions: does Donders' law apply to arm movement? *Journal of Neuroscience,* 15, 6271–6280.

Springer, W.E., 1969 Cockpit Geometry Evaluation: Phase I, Final Report, Vol. I, Program Description and Summary, Report D162-10125-1, JANAIR Report 690101, Seattle, WA: Boeing-JANAIR.

Straumann, D., Haslwanter, Th., Hepp-Reymond, M.-C., and Hepp, K., 1991. Listing's law for eye, head, and arm movements and their synergistic control. *Experimental Brain Research,* 86, 209–215.

Tweed, D. and Vilis, T., 1992. Listing's law for gaze-directing head movements, in *The Head-Neck Sensory Motor System,* Berthoz, A., Graf, W., and Vidal, P.P., Eds., New York: Oxford University Press, 387–391.

Uno, Y., Kawato, M., and Suzuki, R., 1989. Formation and control of optimal trajectory in human multi-joint arm movement. *Biological Cybernetics,* 61, 89–101.

Vaughan, J., Rosenbaum, D.A., Harp, C.J., Loukopoulos, L.D., and Engelbrecht, S., 1998. Finding final postures. *Journal of Motor Behavior,* 30, 273–284.

Vilis, T., 1996. Personal communication.

Zhang, X. and Chaffin, D.B., 1997. Task effects on three-dimensional dynamic postures during seated reaching movements: an investigative scheme and illustration. *Human Factors,* 39, 659–671.

4

Vision and Eyes

Wolfgang Jaschinski and Herbert Heuer

CONTENTS

Introduction

Back pain can arise if a subject moves heavy objects with a weight exceeding the capability of the individual's musculoskeletal system. Similarly, an unfavorable position of the "objects of regard" relative to the eyes, e.g., a visual display at a computer workstation, can impose strain on the muscular systems of the eyes, causing the subject to experience visual fatigue. The two relevant parameters of screen position are (1) the viewing distance from the eyes to the screen and (2) the angle of vertical inclination of gaze direction, which depends on the height of the screen relative to eye level. We describe in this chapter how these two parameters can be designed ergonomically to avoid visual fatigue at the workplace. The relevant physiological mechanisms are the resting states of the ocular muscles and the accuracy of the adjustment of vergence. We start with a brief description of the basic ocular muscle functions and their operating ranges.

4.1   Basic Ocular Muscle Functions and Their Operating Ranges

Because in each eye only the center of the retina, the fovea, has a high spatial resolution that allows us to see small details, the extraocular muscles have to move our eyes so that the point of regard in space is projected onto the fovea (Ciuffreda and Tannen 1996). The different ocular muscle systems have particular ranges of operation. As the limits of these ranges are reached or exceeded, visual fatigue will arise sooner or later.

Saccadic eye movements are high-velocity movements from word to word during reading or from one object to the other when searching on a visual display. Smooth pursuit eye movements occur when tracking a slowly moving object. Saccadic and pursuit eye movements are conjunctive, i.e., made in the same direction by the two eyes. Conjunctive horizontal eye movements as large as 45° relative to straight ahead are possible, but not comfortable; therefore, horizontal eye movements usually do not exceed 35° to 40° (Weston 1949). In the vertical direction, a declination of the eyes within the orbit by more than 25° is described to result in fatigue (Weston 1953). For any horizontal or vertical change in gaze direction from the primary horizontal straight ahead direction, both the head is moved relative to the neck and the eyes are rotated in the orbit accordingly, as described in Chapter 3. Despite large ranges of possible eye movements, in natural vision saccades are not larger than about 15° (Bahill et al. 1975); larger excursions of the eyes are avoided as they tend to be uncomfortable.

A change in fixation from a distant to a near target requires the two eyes to converge, i.e., to move inward toward the nose; this is referred to as disjunctive eye movement. The angle of convergence of the visual axes increases as the viewing distance is shortened: if one fixates a fingertip that is moved from arm length to the nose, one will notice discomfort at a certain point. In most subjects, the maximal angle of convergence, i.e., the point where double vision occurs, corresponds to a viewing distance of less than 0.1 m (Scheiman and Wick 1994, p. 225). As illustrated in Figure 4.1, for a midsagittal target at a viewing distance d (in meters) and an interpupillary distance p (in meters), each eye has to converge the visual axis by an angle of

$$\phi \text{ [radians]} = \text{arc tan } [(p/2)/d)] \qquad (4.1)$$

This can reasonably be approximated by $(p/2)/d$, since usual vergence angles are rather small in workplace conditions. Thus, for $d = 1$ m we have ϕ [rad] = $p/2$. This angle, referred to as the meter angle (ma), is a convenient unit for the convergence angle as it is simply related to the viewing distance by

$$\phi \text{ [ma]} = 1/d \qquad (4.2)$$

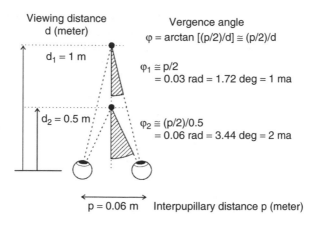

FIGURE 4.1
Examples of vergence angles as a function of viewing distance and interpupillary distance; both are given in meter units to result in a vergence angle in radians. A meter angle (ma) is the vergence angle (of one eye) corresponding to a viewing distance of 1 m (not to scale).

Changes in viewing distances also require a readjustment of the ocular focus to maintain a sharp retinal image and, thus, clear vision. This autofocus mechanism, i.e., accommodation, is driven by the intraocular ciliary muscle that adjusts the refraction of the lens. Relative to optical infinity, the refraction is increased by an accommodative amount (in diopters, D) of $1/d$ for a target at a viewing distance of d (meters). Thus, the amount of convergence and accommodation both increase as a function of the proximity of the target expressed as the inverse of the viewing distance (in meters). The closest target that can be focused sharply, i.e., the near point of accommodation, corresponds to the maximal accommodative response; it decreases about linearly from 10 to 1 D as one ages from 20 to 60 years (Weale 1992). To compensate for this effect, known as presbyopia, sooner or later reading glasses become necessary to provide clear near vision. To avoid accommodative fatigue in the long run, the viewing distance should not be closer than that corresponding to about half the dioptric amount of maximal accommodation; e.g., for a subject with a maximal accommodation of 4 D the comfortable long-term viewing distance should be not closer than 50 cm, corresponding to 2 D.

These descriptions of the oculomotor mechanisms show that vision at a short viewing distance can be the origin of visual fatigue. However, near vision can also induce neck muscular tension (Lie and Watten 1987): Electromyographic (EMG) recordings from six muscles of head, neck, and shoulder showed increased activity at a short viewing distance of 40 cm (compared to 6 m); further, a clinical population with severe and long-lasting neck and shoulder problems and inappropriate optometric corrections experienced decreased EMG activity when the individuals were given more appropriate corrections in their glasses. Optometric corrections were found to reduce the visual discomfort in the intervention field studies of Aarås et al. (1998, 2001). In general, however, there is no evidence that visual work may lead to ocular diseases. Fatigue effects after many hours of visual work usually disappear after relaxation. Sometimes, subjects report that visual symptoms occur after they have been working at the computer screen; such observations can often be explained by (even small) individual anomalies in refraction or binocular vision, which may lead to discomfort only when more difficult visual tasks are performed for a longer period of time. In most cases, these ocular anomalies can be corrected with glasses. However, using short viewing distances for many hours a day may in the long run increase the risk of becoming short-sighted, as shown by recent studies on the etiology of myopia (Rosenfield and Gilmartin 1998, Hepsen et al. 2001). "Virtual reality" displays with stereoscopic presentations may have the problem that conditions of vergence and accommodation differ from those in natural vision (Barfield and Furness 1995).

4.2   Viewing Distance

4.2.1   Concept of Ocular Resting Positions

The ocular muscles move the eyes depending on where the point of regard (e.g., on a visual display) is located relative to the eyes. However, in a condition where no visual stimulus is presented in the visual field, the eyes assume a continuous state that depends on the remaining tonus of the ocular muscles; only when paralyzed do the eyes return to their anatomical position of rest (Millodot 1986). One way of measuring tonic positions is to test the eyes in a completely dark visual field; however, it is not the darkness itself that is important, but the fact that any effective visual stimulus is absent. It is reasonable to assume that the most comfortable screen position is the one that corresponds to the tonic state of ocular muscles, as this state is assumed by the eyes with no effort induced by any external stimulus. As far as the viewing distance is concerned, we must consider the mechanisms of vergence and accommodation; these have tonic states as illustrated in Figure 4.2, also referred to as dark vergence and dark focus, respectively. The averages within groups of subjects are equivalent to a viewing distance of about 1 m (Owens and Leibowitz 1983, Owens 1984). However, it is important that these tonic states vary reliably among subjects, but are not correlated: some subjects have a distant dark vergence and a close dark focus (or the other way round); in other subjects, these two positions may be similar.

With respect to ocular tonic positions, two predictions can be made. First, the amount of visual fatigue should depend on the extent to which the visual target, i.e., the computer screen, is closer relative to the individual tonic state.

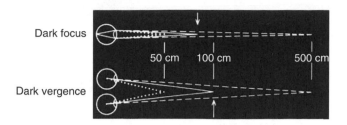

FIGURE 4.2

Illustration of dark focus and dark vergence: for each function, three positions are shown, as examples of individual states of accommodation and vergence in a dark visual field, used to exclude any visual stimulus. The arrows indicate the averages within groups of subjects. (From Jaschinski, W., *Arbeitsmedizin Sozialmedizin Umweltmedizin*, 34, 225–230, 1999a. With permission.)

In fact, visual fatigue at near screens — measured with questionnaires — tends to be larger in subjects with a distant dark vergence (Tyrrell and Leibowitz 1990, Jaschinski-Kruza 1991, Jaschinski 1998). Similarly, Best et al. (1996) found that subjects with a distant dark vergence had worse performance in a near vision task. Dark focus, however, appears to be a less relevant factor for near vision symptoms (Jaschinski-Kruza 1988, Jaschinski 2002), although, in degraded viewing conditions, e.g., dim lighting or low contrast, targets are blurred at positions that deviate from dark focus (Owens 1984).

The second prediction is that subjects may notice which workplace conditions induce visual fatigue and which do not; from this experience, subjects might be able to adjust the screen toward the most comfortable condition, presumably the tonic positions of the ocular muscles. In fact, when subjects have the ability to adjust freely, the average preference for viewing distance is about 75 to 90 cm (Grandjean 1987, Jaschinski et al. 1998a, 1999), which compares well with the tonic positions illustrated in Figure 4.2. Further, individual differences also play a role. Heuer et al. (1989) recorded the actual viewing distance while subjects worked at a visual display: a subgroup of subjects with a more distant dark vergence tended to change to a longer viewing distance in the course of the 1.5-h task, presumably to reduce the discrepancy between the screen position and the distant resting vergence. These studies provided evidence that the tonic state of vergence is a resting state in the sense that the corresponding viewing distance is comfortable for work at visual displays.

4.2.2 Fixation Disparity: The Accuracy of Ocular Vergence

A muscular system can be evaluated by measuring how accurately it is able to operate. Binocular vision is optimal when the fixated target is imaged onto the center of the fovea in each eye, so that the principal visual directions of both eyes intersect at the fixation point. However, even subjects who have normal binocular vision (with good stereoscopic acuity) may have slight

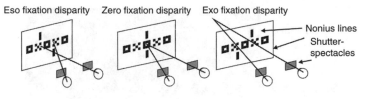

FIGURE 4.3
Illustration of the method for measuring fixation disparity with two nonius lines that were presented dichoptically, i.e., one to each eye, by means of liquid crystal shutter glasses in front of the eyes. In the case of eso (or exo) fixation disparity, the upper nonius line has to be placed to the left (or right) of the lower nonius line, in order to be perceived collinear. Then, each nonius line is lying on the principal visual direction that determines the direction "straight ahead" of each eye. The fixation disparity is the visual angle corresponding to the resulting physical nonius offset. The central fusion stimulus, a string of characters OXOXO, was visible to both eyes.

deviations from this optimal state. These small errors in convergence typically amount to a few minutes of arc and are called exo or eso fixation disparity depending on whether the eyes converge slightly behind or in front of the fixation point, respectively. As shown in Figure 4.3, fixation disparity can be measured by presenting upper and lower nonius lines, which are visible only for the right and left eye, respectively. A central fusion stimulus is visible for both eyes. A fixation disparity is present when a physical nonius offset has to be adjusted on the screen, in order for the subject to perceive subjectively that nonius lines coincide. Measures of fixation disparity have been used for a long time in optometric research and practice: subjects with larger exo fixation disparity (tested conventionally at a reading distance of 40 cm) tend to suffer from visual fatigue during near work or reading (Scheiman and Wick 1994, Evans 1997).

For the computer workstation it is relevant to know to what extent fixation disparity may depend on the position of the screen relative to the eyes. Fixation disparity is strongly affected by the viewing distance. When the latter is plotted in the unit 1/m (corresponding to the required vergence angle, see above) the linear curves in Figure 4.4 show that small or zero fixation disparity is observed at a viewing distance corresponding to the mean dark vergence, i.e., the vergence resting position at about 1 m. The more the actual viewing distance is shortened, the more the fixation disparity increases in the exo direction, which means that the vergence near response (relative to the more distant resting position) does not reach the near stimulus level. These proximity–fixation-disparity curves have been confirmed to be a reliable individual parameter of the vergence system in young adults with normal binocular vision (Jaschinski 1997). The slope of fixation disparity curves as a function of vergence stimulus (i.e., viewing distance in the present case) reflects the gain factor of the fusional vergence mechanism, as shown in vergence control studies (Schor 1983, Hung 1992, Jaschinski 2001): a subject with a low vergence gain has a steep fixation disparity curve, i.e., a more exo fixation disparity at near viewing distance. The measurement of

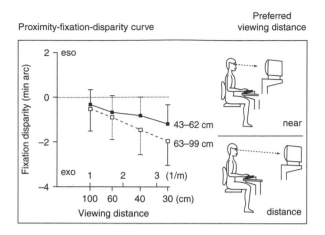

FIGURE 4.4

Proximity–fixation-disparity curves for two groups (each with $n = 20$) with ranges of the preferred viewing distance longer and shorter than the median (63 cm). Error bars indicate standard deviations. See text. (From W. Jaschinski, The proximity-fixation-disparity curve and the preferred viewing distance at a visual display as an indicator of near vision fatigue, *Optometry and Vision Science*, 79, 158–159, © The American Academy of Optometry, 2002. With permission.)

fixation disparity is possible with consumer computer technology as described by the Institut für Arbeitsphysiologie (2002).

The individual slopes have been shown to be related to the viewing distance preferred as comfortable: those subjects with a steep proximity–fixation-disparity curve tended to avoid near vision in two laboratory studies: they moved more quickly away from a near screen (which was initially located at 40 cm) in Jaschinski (1998) and assumed longer preferred viewing distances in Jaschinski (2002). This latter result is illustrated in Figure 4.4: the complete sample of 40 subjects had preferred viewing distances in the range of 43 to 99 cm, and subjects with preferred viewing distances longer or shorter than the median (63 cm) had steeper or flatter proximity–fixation-disparity curves, respectively.

To summarize, a number of studies have provided evidence that near vision fatigue can arise in young adults, even at a 50 cm viewing distance, which is long compared to their accommodative near point. These subjects tend to have a distant resting position of vergence and/or a larger exo fixation disparity at near viewing distance. These optometric parameters are within the range of the interindividual variability of subjects with normal binocular vision, i.e., without clinical anomalies. For these subjects, longer viewing distances may be helpful to reduce vergence stress. In principle, it would be possible to measure optometric parameters like dark vergence and/or fixation disparity curves for individuals in order to recommend the appropriate viewing distance. Although such a diagnostic optometric procedure might be justified in single cases when subjects suffer from otherwise unexplained visual fatigue, it is certainly too expensive and not necessary for each user (see below for an alternative).

4.3   Vertical Gaze Direction

After viewing distance, the second important parameter for an ergonomic position of computer screens is the height of the display (relative to eye level), which results in a certain angle of vertical gaze direction (e.g., Sommerich et al. 2001). The screen height is relevant for the musculoskeletal system of head, neck, and shoulder, as described in Chapters 3 and 5. Additionally, the following visual functions may play a role.

At a fixed vertical head position, the ocular muscles adjust the vertical direction of gaze toward the target. In a visual field without any target, e.g., in darkness, a vertical resting inclination of the eyes can be observed. At an upright head posture, the vertical direction of gaze at rest is declined relative to horizontal by about 10° on the average, with an interindividual range between about +5° upward and −25° downward. This was confirmed by measurements of the gaze inclination where a minimal exertion of ocular muscles is perceived (Levy 1969, Menozzi et al. 1994). The vertical gaze direction has also an effect on dark vergence (Heuer and Owens 1989) and fixation disparity (Jaschinski et al. 1998b): on the average, both these measures shift into the eso direction, i.e., closer, when the gaze direction is lowered. When this effect was measured over a large range of eye inclination from +15° to −45°, it appeared that some subjects had a strong and others a weak inclination effect, which was confirmed in repeated sessions (Jaschinski et al. 1998b). Consequently, lower screens tend to avoid large exo fixation disparity. However, at computer workstations and other workplaces, the possible range of vertical eye inclination is much smaller and the resulting changes in fixation disparity with gaze inclination were found to be smaller and not reliable within individuals (Jainta and Jaschinski 2002). Therefore, fixation disparity may be less relevant for the choice of screen height. Finally, it should be mentioned that lowering the screen reduces the ocular surface area and, thus, the risk of dry eyes (Abe et al. 1995).

These visual functions suggest that a slightly declined gaze direction appears to be optimal at visual display units. Accordingly, field studies have found that computer screens at eye level resulted in more visual fatigue than lower screens (Bergqvist and Knave 1994, Jaschinski et al. 1998a).

4.4   User Preferences

The following studies (already mentioned above) suggest that subjects are able to experience within rather short periods of time whether a particular screen position will induce visual fatigue so that they can choose the more comfortable condition as the preferred one. In two laboratory studies, a tendency to assume longer viewing distances was observed in subjects with a distant dark vergence (Heuer et al. 1989) and in subjects with a steeper fixation disparity curve (Jaschinski 1998). In these two studies the time required to reach a final choice of the preferred viewing distance was 0.5 to 1.5 h. The relationship between the preferred viewing distance and fatigue due to near vision was investigated by Jaschinski (2002): the subjects performed a series of three blocks of a visual task of 30 min each. In Block A, the viewing distance was 100 cm, as a reference without near vision. In Block B, the viewing distance of 50 cm induced a defined near vision load. This order of conditions was chosen, because a previous study (Jaschinski-Kruza 1991) had shown that subjects tend to be more sensitive to near vision fatigue (at 50 cm) if they have earlier experienced the same task without the requirement of near vision (i.e., at 100 cm). This fixed order is appropriate for evaluating interindividual differences in fatigue. In Block C, subjects were free to choose a comfortable viewing distance: they arrived at their individual preferred viewing distance within only 15 min, presumably because of their prior experience at different viewing distances. Positive correlations were found between the following aspects of visual fatigue: those subjects who indicated more near vision fatigue at 50 cm (compared to at 100 cm) rated the screen at 50 cm strongly as too near in Block B and preferred longer viewing distances in Block C. Therefore, a long preferred viewing distance can be used as an indicator of near vision fatigue that the subjects had experienced in the earlier Block B.

The relevance of prior experience was also confirmed in two field studies that led to a trial procedure for determining the preferred screen position (Jaschinski et al. 1998a, 1999, Institut für Arbeitsphysiologie 2002): employees were instructed to use the screen in four rather extreme positions relative to the eyes on four separate days: near and at eye level, near and low, distant and at eye level, distant and low. After having experienced these conditions, some of the subjects changed their screen to a position that they found more comfortable. Figure 4.5 illustrates the distribution of the resulting preferred

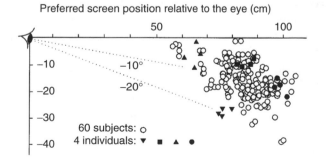

FIGURE 4.5
Preferred screen positions relative to the eyes, assumed after a trial procedure to determine the most comfortable condition in two field studies (Jaschinski et al. 1998a, 1999). See text for details. (From Jaschinski, W., *Arbeitsmedizin Sozialmedizin Umweltmedizin*, 34, 277–281, 1999b. With permission.)

screen positions relative to the eyes in the complete sample of subjects (Jaschinski et al. 1998a, 1999). For the 60 subjects, each open symbol indicates the mean of the individual preferred positions measured on 1 day per week (for up to 4 weeks per subject; the screens had been disarranged after each working day). The closed symbols show the preferred adjustments of four subjects as the mean position in each of the 4 weeks. Thus, it is important to note that, within the large interindividual range (with viewing distance between 55 and 105 cm and vertical gaze direction between horizontal and 20° downward), subjects chose the screen within a much smaller intraindividual range. This means that most subjects preferred the screen at a particular individual position, with respect to viewing distance and screen height.

The adjustment of screen position on the basis of user preferences appears to be useful because most users are able to find comfortable and physiologically reasonable solutions. However, the technical facilities and sufficient space at the workplace have to be provided for a free adjustment. The results shown in Figure 4.5 with rather longer viewing distances were observed in offices where there was enough space, which often may not be available. However, viewing distances around 100 cm may not be necessary in most cases: if one wishes to reduce the convergence load by moving the display backward, the first centimeters are relatively more effective because the vergence angle is proportional to the inverse of the viewing distance in meters (see above). Further, the size of characters on the screen must be increased for comfortable vision at longer viewing distances to subtend 20 to 22 min arc per ISO 9241-3 (ISO 1992). This character size is about four times larger than the minimal character size of about 5 min arc that is necessary just to discriminate characters at the limit of ocular resolution with normal visual acuity. Accordingly, symbols and icons should be sufficiently large that they can just be discriminated at four times the viewing distance at work.

So far, we have considered a static position of the object of regard, the computer screen, relative to the eyes. However, some working conditions may require repetitive changes of gaze direction from the screen to paper documents or to the surroundings. The question arises whether such dynamics in eye position may result in visual fatigue. For this condition, we must consider that in natural vision the eyes are always in motion: we permanently change fixation from one object to the next and in most cases this will not lead to visual fatigue. In particular, small conjunctive eye movements, e.g., saccades during reading can be performed by most subjects for a long time without particular complaint. However, very frequent changes in eye position over large visual angles should be avoided. Repetitive vergence changes especially are more difficult to perform; thus, objects that have to be fixated frequently should not be located at very different viewing distances. On the other hand, a constant stationary eye position is quite unnatural: it is fatiguing to fixate on a point in space for a long period of time, especially in near vision. Thus, the two extreme ways of using one's eyes should be avoided, both excessive eye movements and a very stationary eye position. The compromise may constitute the condition of natural eye movement pattern, which, however, is difficult to specify quantitatively. In an experimental study, a change in viewing distance every 2 s between 50 and 70 cm did not result in more visual fatigue than the same task at a constant viewing distance of 50 cm (Jaschinski-Kruza 1990), at least for subjects younger than 50 years. Presbyopic subjects have more difficulties with visual targets at different viewing distances. In should be mentioned that individuals with problems in near convergence may benefit from vergence training methods that involve repetitive fixations at near and distant objects (Scheiman and Wick 1994).

All visual functions tend to improve when the luminance of the visual field increases; thus, room lighting at the workplace should be sufficiently high, but glare must be eliminated (Weston 1953). Lighting has a correlation to pain intensity of the neck and shoulder (Aarås et al. 1998); further, an intervention field study showed that visual discomfort was significantly reduced by increasing the illuminance level from 300 lux to more than 600 lux and the luminance levels of the ceiling and walls from 30 cd/m² to more than 80 cd/m² with a uniform luminance distribution (Aarås et al. 2001).

Summary

Discomfort and strain at visual work, e.g., with visual display units, can be reduced when the position of the screen relative to the eyes is chosen according to physiological properties of the muscular systems of the eyes. Both the resting positions of ocular muscles and the accuracy of the vergence angle between the visual axes suggest that short viewing distances around 50 cm may induce visual strain, also in young adult subjects with normal vision. A slightly declined gaze direction is favorable. However, individual differences in these physiological functions should be taken into account; thus, the recommendation of the same screen position for all users may not be appropriate. Rather, the majority of subjects may discover their individual preference within the range of viewing distances between about 50 and 100 cm and the range of vertical gaze directions between horizontal and 15° downward, provided they have already had experience of the advantages and disadvantages of different screen positions. These recommendations hold for young adults, who are able to focus on the screen easily at these viewing distances. Starting at the age of 40 to 50 years, reading glasses for clear vision of documents and the visual display become necessary sooner or later. Prescriptions for these glasses depend on the subject's vision and the actual visual task, and should be made after the workplace has been ergonomically arranged.

References

Aarås, A., Horgen, G., Bjorset, H-H., Ro, O., and Thorsen, M., 1998. Musculoskeletal, visual and psychosocial stress in VDU operators before and after multidisciplinary ergonomic interventions. *Applied Ergonomics*, 29, 335–354.

Aarås, A., Horgen, G., Bjorset, H-H., Ro, O., and Walsoe, H., 2001. Musculoskeletal, visual and psychosocial stress in VDU operators before and after multidisciplinary ergonomic interventions. A 6 years prospective study. *Applied Ergonomics*, 32, 559–571.

Abe, S., Sotoyama, M., Taptagaporn, S., Saito, S., Villanueva, M.B.G., and Saito, S., 1995, Relationship between vertical gaze direction and tear volume, in *Work with Display Units 94*, Grieco, A., Molteni, G., Piccoli, B., and Occipinti, E., Eds., Amsterdam: Elsevier Science, 95–99.

Bahill, A.T., Adler, D., and Stark, L., 1975. Most naturally occurring human saccades have magnitudes of 15 degrees or less. *Investigative Ophthalmology and Vision Science*, 14, 468.

Barfield, W. and Furness, T.A., 1995, *Virtual Environments and Advanced Interface Design*, New York: Oxford University Press.

Bergqvist, U.O.V. and Knave, B.G., 1994. Eye discomfort and work with visual display terminals. *Scandinavian Journal of Work, Environment & Health*, 20, 27–33.

Best, P.S., Littleton, A.K., Gramopadhye, A.K., and Tyrrell, R.A., 1996. Relations between individual differences in oculomotor resting states and visual inspection performance. *Ergonomics*, 39, 35–40.

Ciuffreda, K.J. and Tannen, B., 1996. *Eye Movement Basics for the Clinician*, St. Louis: Mosby.

Evans, B.J.W., 1997. *Pickwell's Binocular Vision Anomalies: Investigation and Treatment*, London: Butterworths.

Grandjean, E., 1987. Design of VDT workstations, in *Handbook of Human Factors*, Salvendy, G., Ed., New York: Wiley, 1359–1397.

Hepsen, I.F., Evereklioglu, C., and Bayramlar, H., 2001. The effect of reading and near-work on the development of myopia in emmetropic boys: a prospective, controlled, three-year follow-up study. *Vision Research*, 41, 2511–2520.

Heuer, H. and Owens, D.A., 1989. Vertical gaze direction and the resting postures of the eyes. *Perception*, 18, 363–377.

Heuer, H., Hollendiek, G., Kröger, H., and Römer, T., 1989. Die Ruhelage der Augen und ihr Einfluss auf Beobachtungsabstand und visuelle Ermüdung bei der Bildschirmarbeit. *Zeitschrift für experimentelle und angewandte Psychologie*, 39, 538–566.

Hung, G.K., 1992. Quantitative analysis of associated and disassociated phorias: linear and nonlinear static models. *IEEE Transactions on Biomedical Engineering*, 39, 135–145.

Institut für Arbeitsphysiologie, 2002. Eye-Test PC; Where to Place the Computer Monitor? Dortmund, Germany: Institut für Arbeitsphysiologie. Available at www.ifado.de/vision.

ISO, 1992, ISO 9241-3 *Ergonomic Requirements for Office Work with Visual Display Terminals (VDTs)*. Part 3: *Visual Display Requirements*, Geneva: International Organization for Standardization.

Jainta, S. and Jaschinski, W., 2002. Fixation disparity: binocular vergence accuracy for a visual display at different positions relative to the eyes. *Human Factors*, 44, 443–450.

Jaschinski, W., 1997. Fixation disparity and accommodation as a function of viewing distance and prism load. *Ophthalmic and Physiological Optics*, 17, 324–339.

Jaschinski, W., 1998. Fixation disparity at different viewing distances and the preferred viewing distance in a laboratory near-vision task. *Ophthalmology and Physiological Optics*, 18, 30–39.

Jaschinski, W., 1999a. Die Bedeutung von Sehabstand und Blickneigung für individuelle Sehfunktionen and visuelle Ermüdung am Bildschirmarbeitzplatz. *Arbeitzmedizin Sozialmedizin Umweltmedizin*, 34, 225–230.

Jaschinski, W., 1999b. Zur individuellen ergonomischen Gestaltung am Bildschirmarbeitzplatz: Sehabstand und Blickneigungswinkel. *Arbeitzmedizin Sozialmedizin Umweltmedizin*, 34, 277–281.

Jaschinski, W., 2001. Fixation disparity and accommodation for stimuli closer and more distant than oculomotor tonic positions. *Vision Research*, 41, 923–933.

Jaschinski, W., 2002. The proximity-fixation-disparity curve and the preferred viewing distance at a visual display as an indicator of near vision fatigue. *Optometry and Vision Science*, 79, 158–169.

Jaschinski, W., Heuer, H., and Kylian, H., 1998a. Preferred position of visual displays relative to the eyes: a field study of visual strain and individual differences. *Ergonomics*, 41, 1034–1049.

Jaschinski, W., Koitcheva, V. and Heuer, H., 1998b. Fixation disparity accommodation dark vergence and dark focus during inclined gaze. *Ophthalmic and Physiological Optics*, 18, 351–359.

Jaschinski, W., Heuer, H., and Kylian, H., 1999. A procedure to determine the individually comfortable position of visual displays relative to the eyes. *Ergonomics*, 42, 535–549.

Jaschinski-Kruza, W., 1988. Visual strain during VDU work: the effect of viewing distance and dark focus. *Ergonomics*, 31, 1449–1465.

Jaschinski-Kruza, W., 1990. On the preferred viewing distances to screen and document at VDU workplaces. *Ergonomics*, 33, 1055–1063.

Jaschinski-Kruza, W., 1991. Eyestrain in VDU users: viewing distance and the resting position of ocular muscles. *Human Factors*, 33, 69–83.

Levy, J., 1969. Physiological position of rest and phoria. *American Journal of Ophthalmology*, 68, 706–713.

Lie, I. and Watten, R.G., 1987. Oculomotor factors in the aetiology of occupational cervicobrachial diseases (OCD). *European Journal of Applied Physiology*, 56, 151–156.

Menozzi, M., v. Buol, A., Krueger, H., and Miège, C., 1994. Direction of gaze and comfort: discovering the relation for the ergonomic optimization of visual tasks. *Ophthalmic and Physiological Optics*, 14, 393–399.

Millodot, M., 1986. *Dictionary of Optometry*, London: Butterworths.

Owens, D.A., 1984. The resting state of the eyes. *American Scientist*, 72, 378–387.

Owens, D.A. and Leibowitz, H.W., 1983. Perceptual and motor consequences of tonic vergence, in *Vergence Eye Movements: Basic and Clinical Aspects*, Schor, C.M. and Ciuffreda, K.J., Eds., Boston: Butterworths, 25–74.

Rosenfield, M. and Gilmartin, B., 1998. *Myopia and Nearwork*, Oxford: Butterworth-Heinemann.

Scheiman, M. and Wick, B., 1994. *Clinical Management of Binocular Vision*, Philadelphia: Lippincott.

Schor, C.M., 1983. Fixation disparity and vergence adaptation, in *Vergence Eye Movements. Basic and Clinical Aspects,* Schor, C.M. and Ciuffreda, K.J., Eds., Boston, MA: Butterworths, 465–516.

Sommerich, C.M., Joines, S.M.B., and Psihogios, J.P., 2001. Effects of computer monitor viewing angle and related factors on strain, performance, and preference outcomes. *Human Factors, 43,* 39–55.

Tyrrell, R.A. and Leibowitz, H.W., 1990. The relation of vergence effort to reports of visual fatigue following prolonged near work. *Human Factors, 32,* 341–357.

Weale, R.A., 1992. *The Senescence of Human Vision,* Oxford: Oxford University Press.

Weston, H.C., 1949. *Sight Light and Efficiency,* London: Lewis.

Weston, H.C., 1953. Visual fatigue with special reference to lighting, in *Symposium on Fatigue,* Floyd, W.F and Welford, A.T., Eds., London: Lewis, 117–135.

5

Head and Neck

Nico J. Delleman

CONTENTS

Introduction

On the basis of a systematic literature review, Ariëns (2001) stated that neck pain is a major musculoskeletal health problem in modern society, occurring in many different occupational groups. Furthermore, it was found that neck posture is a risk factor for neck pain. In this chapter it will be shown that the earlier studies reporting the effects of head and neck posture (e.g., neck load, health complaints) focus solely on head inclination (i.e., the amount of deviation from the upright posture or from the vertical; see Figure 5.1 and Section 5.1). This can be seen as a measure of musculoskeletal load, an equivalent of the force delivered by the neck muscles to counteract the gravitational force on the head. It should be noted that the way of measuring head inclination varies among studies, with some using reference points on the skull only while others also include the whole or parts of the cervical spine in their definition. Furthermore, some researchers measure head inclination with respect to the upright head posture, while others measure head inclination with respect to the vertical. In more recent years, attention has also been given to another posture variable that was seen as a possible risk

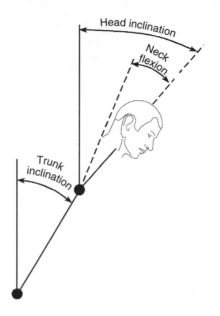

FIGURE 5.1

The concepts of head inclination, trunk inclination, and neck flexion. Neck flexion is defined as head inclination minus trunk inclination (when the outcome of the subtraction is greater than 0°); neck extension is defined as head inclination minus trunk inclination (when the outcome of the subtraction is smaller than 0°). It should be noted that the method of measuring head inclination varies among studies, with some using reference points on the skull only while others also include the whole or parts of the cervical spine in their definition. Furthermore, some researchers measure head inclination with respect to the upright head posture, while others measure inclination with respect to the vertical (the same is true for trunk inclination as regards upright trunk posture and the vertical).

factor for neck pain. That is neck flexion/extension (i.e., the amount of inclination of the head with respect to the amount of inclination of the trunk; Figure 5.1). This is discussed in more detail in Section 5.2. Section 5.3 contains a discussion of the recommended gaze inclination for visual display unit (VDU) operation, as this relies on the information presented up to this point in the chapter. Only a little is known about the effects of neck twisting (discussed in Section 5.4) and lateral flexion, although Straumann et al. (1991) and Tweed and Vilis (1992) showed that subjects normally do not use lateral flexion during gaze changes. It is stressed that when using the results and data mentioned in the sections below, the definitions and measurement procedures used by the authors of particular papers should always be taken into account. To reduce musculoskeletal discomfort, it is very important to evaluate psychosocial factors in addition to posture and movements. Psychological stress is reported to increase the static muscle load of the neck and shoulder girdle muscles (see Section 9.2.5).

Sections 5.1 and 5.2 contain short descriptions of studies on the effects of head inclination and of neck flexion/extension, respectively. In each of these two sections, the studies are presented in chronological order, to demonstrate

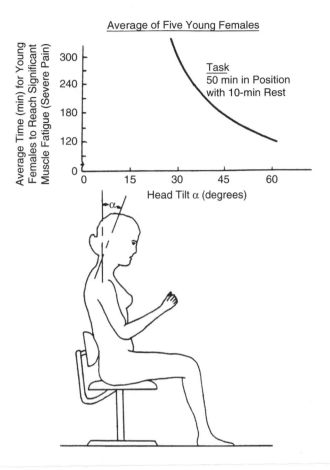

FIGURE 5.2
Neck extensor fatigue vs. head inclination (head tilt angle α). (From Chaffin, D.B., Andersson, G.B.J., and Martin, B.J., *Occupational Biomechanics*, 3rd ed., Copyright © 1999. This material is used by permission of John Wiley & Sons, Inc.)

the early attention given to head inclination, and the subsequent growing attention to neck flexion/extension.

5.1 Head Inclination

Chaffin (1973) conducted a study to ascertain the time it would take to reach class II muscle fatigue (defined below) when subjects held their heads at specific degrees of tilt (inclination) for 50-min periods, with a 10-min rest between these periods. The holding times are shown in Figure 5.2. Class II muscle fatigue is identified as reports of "cramping continuous with deep hot pain intermittent" and substantial electromyographic spectral changes

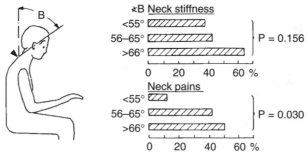

FIGURE 5.3

Incidence of impairments in the neck related to head inclination (angle B) of 57 accounting machine operators. (Reprinted from *Applied Ergonomics,* 11, Hünting, W., Grandjean, E., and Maeda, K., Constrained postures in accounting machine operators, 145–149, Copyright (1980), with permission from Elsevier Science.)

from the upper portion of the trapezius muscle. It was found that a head tilt angle of 15° produced no sensations of discomfort or electromyographic changes after 6 h.

In a field study on accounting machine operators, Hünting et al. (1980) found that the incidence of neck stiffness and pains increased with the degree of head inclination (Figure 5.3). The main task of the operators was to read figures from coupons and to type them into a keyboard. The average daily work on the machine was estimated to be 5 to 6 h and the work was intensive. Keying speed varied between 8000 and 12000 strokes per hour. The effects of head inclination found are supported by Hünting et al. (1981) for VDU workplaces. In that study, a greater head inclination was related to a higher incidence of medical findings in the neck and shoulders, including painfully limited head movement and painful pressure points in the neck muscles as well as at tendon insertions in the shoulder area. The significant effects found for the data-entry subgroup in particular indicate that a long work duration and a high typing speed affect health in a negative way.

Kilbom et al. (1986) studied 96 workers in the electronics manufacturing industry for their medical history and status, particularly in relation to musculoskeletal disorders, and for an evaluation of their working technique. Head inclination, classified as 0 to 20° or >20°, turned out to be significantly related to the severity of symptoms of neck disorders (correlation coefficients $r = 0.41$ for average time per work cycle, $r = 0.27$ for percentage of work cycle time). After a 1- and 2-year follow-up the following predictors for deterioration to severe disorders were found: head inclination as average time per work cycle and as percentage of work cycle time, as well as the number of head inclinations per hour.

Lee et al. (1986) investigated the musculoskeletal load experienced by microscopists for different head inclinations. Figure 5.4 shows that in the cervical region average electromyographic activity (RMS amplitude) for the

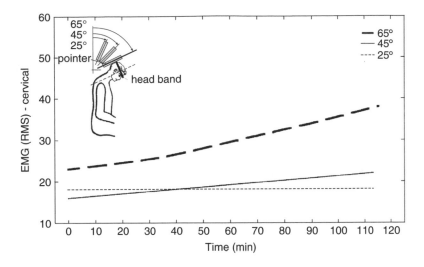

FIGURE 5.4
Electromyographic activity at the cervical region for head inclinations 25°, 45°, and 65° over time (regression lines).

subjects at 65° of head inclination exhibited a steeper slope compared to the lines at the other head inclinations. The regression line for 25° of head inclination showed hardly any increase over time. The electromyographic amplitude increases were only slight for the 45° inclination. The muscle activities for the three head inclinations showed a high correlation with levels of load in the same muscles, as reported by the subjects on a 5-point scale ranging from "comfortable" to "very painful" (correlation coefficients between 0.91 and 0.94).

5.2   Neck Flexion/Extension

Bendix and Hagberg (1984) are about the first known to mention that neck flexion/extension (i.e., the inclination of the head with respect to the inclination of the trunk; see Figure 5.1) may play a role with respect to neck load. This notion was supported by various subsequent studies. From the data on typists presented by Lepoutre et al. (1986) it can be concluded that a greater spinal curvature at the 1st thoracic vertebra (equivalent to more neck flexion) is associated with reduced fatigue and pain at the neck (on the basis of the data presented it may be assumed that head inclination did not change). Kumar (1994) found that an increase of forward head inclination (by lowering the monitor at a desktop) led to a decrease of discomfort for bifocal lens wearers (Figure 5.5). The results of the latter two studies are contradictory to the results presented in Section 5.1, which showed that an increase of head inclination led to an increase of discomfort, fatigue, and pain.

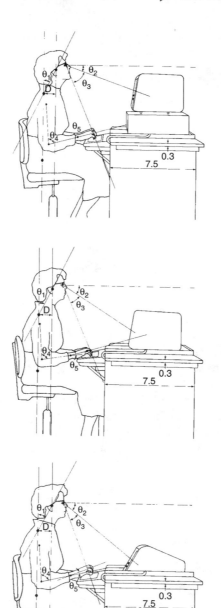

FIGURE 5.5
Bifocal lens wearer working at three monitor heights (raised, level, and sunken with respect the desktop). (From Kumar, S., A computer desk for bifocal lens wearers, with special emphasis on selected telecommunication tasks, *Ergonomics*, 37, 1669–1678, 1994. With permission from Taylor & Francis (http://www.tandf.co.uk).)

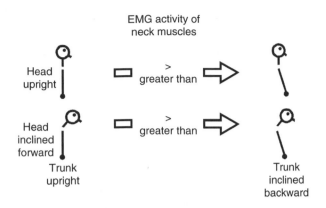

FIGURE 5.6
Electromyographic activity of neck muscles for two head inclinations: at each an increase of neck flexion, created by a backward inclination of the trunk (thoraco-lumbar spine), leads to a decrease of electromyographic activity.

   Schüldt et al. (1986, 1987) showed that, at a constant head inclination, electromyographic activity of various neck muscles was reduced by an increase of neck flexion (created by a backward inclination of the trunk) (Figure 5.6). It is likely that the lower electromyographic activity at a greater neck flexion represents a more favorable neck posture from a biomechanical viewpoint (e.g., concerning muscle moment arms and length–tension characteristics). According to Harms-Ringdahl and Schüldt (1988), the maximum neck extensor moment around the axis through the C7–T1 joint was highest at a slightly flexed neck posture (half the motion range between the individual's neutral and maximum flexed positions) (Figure 5.7). That was significantly different from the moment in an extended posture, but not significantly different from either the neutral posture (subject sitting upright and looking straight forward) or the highly flexed posture (near maximum neck flexion).
   Considering (1) the studies described in this section so far, (2) the fact that head inclination is associated with inclination of the thoracic region of the trunk (Nakaseko et al. 1993) and complaints at the low back (Grandjean et al. 1982, Lee et al. 1986), and (3) the fact that systematic electromyographic effects of trunk inclination were found, not only in the lumbar region but also up into the cervicothoracic region of the head–neck–trunk system (Andersson and Örtengren 1974, Andersson et al. 1974), there is sufficient reason to study neck flexion/extension in addition to head inclination. Focusing on this issue, Delleman (1999) conducted a series of studies on visual-manual operations using a standardized research approach. The experiments on sewing machine operation, touch-typing VDU operation, and three maintenance operations will be described below.
   In the first experiment, five sewing machine operators worked for 45 min at each of four workstation adjustments, i.e., two pedal positions × two table slopes (Figure 5.8). One pedal position was the average of the pedal positions

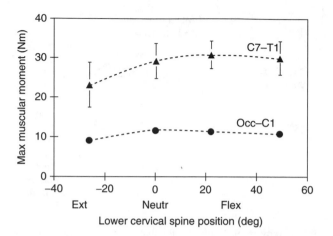

FIGURE 5.7

Maximum neck extensor moments around the axes through the intervertebral disc C7–T1 and the Occ–C1 joint for four neck flexion/extension positions (denoted "lower cervical spine positions"). Vertical bars show the 95% confidence intervals. C7 = 7th cervical vertebra, T1 = 1st thoracic vertebra, Occ = occipital bone, C1 = 1st cervical vertebra (atlas). (From Harms-Ringdahl, K. and Schüldt, K., *Clinical Biomechanics*, 4, 17–24, 1988. With permission.)

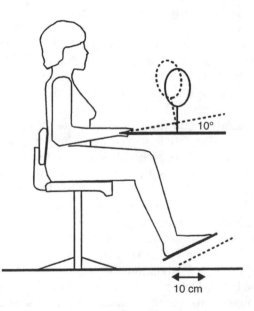

FIGURE 5.8

Sewing machine operation: the four workstation adjustments tested (two table slopes × two pedal positions). One pedal position was the average of the pedal positions at the operators' own industrial workstations (solid line), while the other was 10 cm farther away from the front edge of the table (broken line). One table slope was 0° (flat table, solid line), while the other was 10° inclined toward the operator (broken line).

at the operators' own industrial workstations, while the other was 10 cm farther away from the front edge of the table. The latter condition was introduced to allow the operator to sit closer to the table, and to create a more upright trunk posture. One table slope was 0° (flat table), which is in accordance with the operators' own industrial workstations. The other table slope was 10° (inclined toward the operator), which was introduced to create a more upright head posture. It was found that the results on perceived neck posture were reflected by the neck flexion measured and not by head inclination (Figure 5.9, average results for two table heights). The main difference between the two posture variables was found when working at a workstation with a 10° table slope and a pedal position 10 cm farther away from the front edge of the table as compared to the average pedal position at the operators' own industrial workstations. The resemblance of the results on perceived posture of the neck and neck flexion indicates that operators favor a greater neck flexion. Apparently, during sewing machine operation, neck flexion/extension plays a dominant role with respect to the workers' perceptions of neck posture, as compared to head inclination.

In the second experiment, eight touch-typing VDU operators worked for 25 min at each of eight workstation adjustments, i.e., four screen heights × two chair backrest inclinations (Figure 5.10). The screen heights covered the whole range of existing guidelines on favorable gaze inclination. Screen height was expected to affect head inclination, and thereby neck flexion/extension. One chair backrest inclination was 0° to create the upright trunk posture, which is generally advised, while the other backrest was inclined 15° backward, which is in accordance with the typical trunk posture seen in practice. The resulting trunk inclination was also expected to affect neck flexion/extension. Workers' perceptions revealed an optimum (i.e., most favorable) visual target height at 10 cm below eye height, for an upright backrest as well as for a backward inclined backrest (Figure 5.11a). The perceptions for visual target heights of 40 and 25 cm below eye height were worse than for the optimum target height. It is very likely that the more forward inclined head (Figure 5.11b) at the latter two target heights is counteracted by a higher force of the muscles in the neck and upper back. In addition, at these head inclinations the muscles may be stretched beyond their optimum (i.e., most favorable) length, reducing their active force capacity. Both a higher force level and a reduced force capacity lead to an earlier onset of muscle fatigue. At the optimum visual target height, the chair backrest inclinations showed no significant differences with regard to workers' perceptions. However, especially for visual target heights of 40 cm below and 5 cm above eye height, remarkable differences were found between the two backrest inclinations. For a visual target height of 40 cm below eye height, the perceptions for the backward inclined backrest tended to be worse than for the upright backrest. An explanation was found in the postural data, in that a considerably greater neck flexion was measured for the backward inclined backrest than for the upright backrest (Figure 5.11c). Gravitational

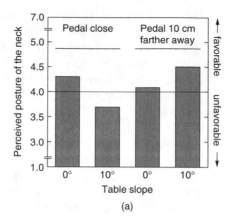

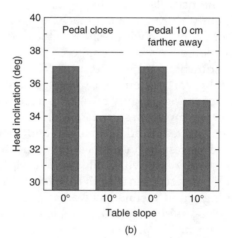

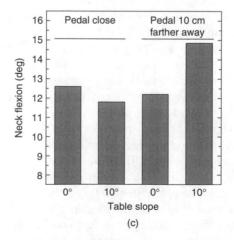

FIGURE 5.9

Sewing machine operation: the four workstation adjustments tested vs. (a) perceived posture of the neck, (b) head inclination, and (c) neck flexion.

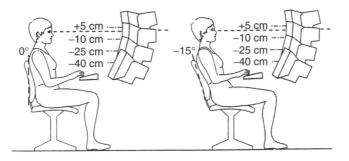

FIGURE 5.10
Touch-typing VDU operation: the eight workstation adjustments tested (four screen heights ×
two backrest inclinations). Screen height is defined as the height of the middle of the screen.

effects are not considered a likely cause, because the head inclination for the
backward inclined backrest was even slightly less than for the upright backrest. The latter result suggests that further forward inclination of the head
is limited with the backward inclined backrest because neck flexion has
reached its maximum (Table 5.1). For a visual target height of 5 cm above
eye height, the workers' perceptions for the upright backrest tended to be
worse than for the backward inclined backrest. Gravitational effects can be
excluded as a causal factor, because the same backward inclination of the
head was found at both backrest inclinations. Apart from the fact that a
backward inclined backrest may be more favorable because it supports body
weight somewhat more, here too neck flexion/extension is the most likely
cause for the differing workers' perceptions. A considerably smaller neck
flexion was found with the upright backrest than with the backward inclined
backrest. This smaller angle may shorten the muscles in the neck and upper
back below their optimum lengths, reducing active force capacity. The same
explanation most probably holds for the workers' perceptions for a visual
target at 5 cm above eye height, which were worse than for the optimum
visual target height at 10 cm below eye height. The above considerations
stress the role of neck flexion/extension with respect to musculoskeletal load
on the neck structures, in addition to the role of head inclination. The study
suggests that a slightly/moderately flexed neck is most favorable; Conrady
et al. (1987) and Van der Grinten and Smitt (1992) provide similar evidence.
About 15° flexion seems to be the most favorable neck position (Figure 5.11).
The same value can be deduced from the study on sewing machine operation
(Figure 5.9). On the basis of an analysis of the data on the individual VDU
operators it is recommended to keep neck flexion between 0° and 25°, which
avoids unfavorable neck positions for the majority of subjects. This recommendation is supported by the analysis of the data on individual maintenance workers (see below).

In the third experiment, seven or eight maintenance workers were
involved in pneumatic wrenching, oxy-gas cutting, and grinding, each at
five different heights. For all three maintenance operations it was found that,

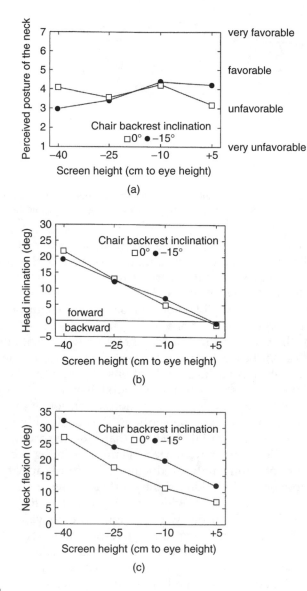

FIGURE 5.11
Touch-typing VDU operation: the eight workstation adjustments tested vs. (a) perceived posture of the neck, (b) head inclination, and (c) neck flexion.

for the working heights that were considered favorable, the head was inclined forward, i.e., average inclinations roughly between 20° and 60°. This range does not match the most favorable head inclinations suggested in the literature, e.g., less than 15° inclined forward (Chaffin 1973). In itself this does not make the role of head inclination as a determinant (or the dominant determinant) of neck load unlikely. It may be that for the experimental

TABLE 5.1

Unfavorable (likely maximum) Neck Flexion vs. Experimental Conditions
(average group scores)

Operation	Average	S.D.	N
Touch-typing VDU operation (chair backrest inclination –15°, visual target height –40 cm)	32.6°	11.3°	8
Pneumatic wrenching (working height –20 cm)	33.6°	4.9°	7
Oxy-gas cutting (working height +20 cm)	27.4°	8.8°	8
Grinding (working height –45 cm)	36.0°	9.7°	7

Note: S.D. = standard deviation. N = number of operators/workers.

Source: Delleman, N.J., 1999. *Working Postures — Prediction and Evaluation*, Soesterberg, the Netherlands: TNO Human Factors.

conditions used (5 min of operation) a working height becomes unfavorable only at a rather high head inclination (it should be noted that for all operations, the lowest working height tested showed the greatest head inclination, and tended to be considered unfavorable for the neck). However, all unfavorable working heights (20 cm below elbow height for pneumatic wrenching, 20 cm above elbow height for oxy-gas cutting, and 45 cm below elbow height for grinding) were explained by the neck flexion measured. At each of these heights an unfavorable maximum neck flexion seemed to have been reached (Table 5.1). Neck flexion also does seem to explain the ambiguous result for a working height of 20 cm above elbow height for oxy-gas cutting, that is, that an equal number of workers considered it favorable and unfavorable. At this height neck flexion was rather small (close to extension), which is known to be a somewhat unfavorable neck position (refer to the touch-typing VDU operation, as described above). From the above it seems that neck flexion/extension is the dominant determinant of neck load, as compared to head inclination.

For particular experimental conditions described above, neck flexion seemed to have reached a maximum (Table 5.1). The results for touch-typing VDU operation, pneumatic wrenching, and grinding are rather close, considering, for example, the different subject groups involved and the differing positions of one of the markers for measurement of trunk posture (i.e., the upper edge of the greater trochanter for touch-typing VDU operation, and the intervertebral disc L5–S1 at the lumbosacral joint for pneumatic wrenching and grinding). A direct comparison of the maximum neck flexion angles found is only possible for oxy-gas cutting and grinding because, in the experiments on these operations, seven of the subjects involved were the same. At maximum, the group averages for these seven subjects were 29.1 for oxy-gas cutting (working height +20 cm) and 36.0 for grinding (working height –45 cm). One reason for the difference seems to be that during oxy-gas cutting at higher working heights (i.e., 10 and 20 cm above elbow height) the possibility of bending the thoracic region of the spine forward is limited, simply due to obstruction by the table used. Because head inclination forward is made possible partly by bending the thoracic spine forward, the maximum neck flexion is not as high as without the limitation.

Looking back on all studies presented above, it is concluded that neck flexion/extension is a determinant of neck load, which should be used as an evaluation criterion for working postures in addition to the traditionally used inclination of the head.

5.3   Recommended Gaze Inclination for VDU Operation

The most favorable gaze inclination for VDU operation has been debated in the literature for several years. Recommendations on gaze inclination roughly range from 15° above to 45° below the horizontal (e.g., Kroemer and Hill 1986a,b, De Wall et al. 1992). In other words, the recommendations vary by as much as 60°. The question is whether this wide range may be narrowed by distinguishing between more and less favorable gaze inclinations. From the current chapter and the two previous chapters we know that various factors concerning the musculoskeletal system of the eyes and the neck play a role in this matter. One may be the inclination of the trunk. In the study on touch-typing VDU operation described above, the most favorable visual target height was 10 cm below eye height (gaze inclination −6°) for an upright backrest (as is generally advised) as well as for a backward inclined backrest (as seen typically in practice; Grandjean et al. 1983, 1984, Ong et al. 1988). Figure 5.11a shows that when gaze is directed more downward (to lower screen positions), an upright trunk is more favorable, while when gaze is directed more upward a backward inclined trunk is more favorable. Eight other studies were found that may distinguish between more and less favorable gaze inclinations.

Berndsen and Delleman (1993) asked 11 touch-typing VDU operators (bank employees) to work for 1 week at each of three visual target heights, i.e., −25, −10, and +5 cm to eye height. Based on workers' perceptions (e.g., localized postural discomfort, perceived posture), the −10 cm adjustment turned out to be most favorable.

Bhatnager et al. (1985) asked subjects to inspect printed circuitboards. Three gaze inclinations to the center of the boards were tested, i.e., +38°, +10°, and −23° to the horizontal (+ = above). Responses were given by pressing either the reject or accept button, i.e., without visual control on the hands. Body part discomfort was lowest at gaze inclination −23°, whereas postural changes (i.e., an indicator of postural fatigue) were least frequent at gaze inclination +10°. It can be concluded that the most favorable gaze inclination is most likely to be between these angles.

Sommerich et al. (1998) studied the effects of gaze inclinations 0° (horizontal), −17.5°, and −35° (i.e., below the horizontal) to the center of the screen.

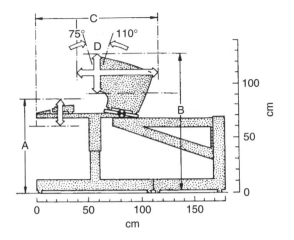

FIGURE 5.12

The easily adjustable experimental VDU workstation used by Grandjean et al. (1983, 1984). The adjustments of keyboard and screen height were controlled with an electrical motor (with push button operation). Screen angle and distance were mechanically controlled by hand. (Reprinted from *Applied Ergonomics*, 15, Grandjean, E., Hünting, W., and Nishiyama, K., Preferred VDT workstation settings, body posture and physical impairments, 99–104, Copyright (1984), with permission from Elsevier Science.)

It was found that computer users, whose primary focus is the screen, prefer gaze inclinations between 0° and –17.5°. This result was confirmed by Psihogios et al. (1998) in a month-long field study.

De Wall et al. (1992) asked subjects to execute computer-aided design (CAD) tasks at three different screen heights relatively close to eye height. Operations were carried out by looking at the screen 90% of the time, i.e., with hardly any visual control on the hands present. From the results it can be concluded that gaze inclinations less than 15° below the horizontal were preferred. On the basis of theoretical biomechanical considerations (a balanced posture of the head, with its center of gravity above its axis of rotation), it was expected that a gaze inclination above the horizontal would be most favorable. However, this was not confirmed by operators' preferences. Apparently, less favorable conditions arise at gaze inclinations above the horizontal (e.g., shortening of the neck and upper back muscles below their optimum, i.e., most favorable, lengths, lengthening of the muscles at the ventral side of the neck beyond their optimum lengths).

Grandjean et al. (1983, 1984) conducted a large field study of the most favorable workstation adjustment for VDU operations. Operators adjusted their workstations according to their preferences (Figure 5.12). The majority of operators were involved in conversational (i.e., interactive) VDU operations for seat-space control at an airline company. For this, according to Läubli (1987), all information is presented on screen (no source documents being used). Therefore, it can be assumed that the screen is their main visual target during a working day. On average, the operators involved in the field study positioned the middle of the screen at a gaze inclination –9° to the horizontal (with 95% of the angles measured between –15° and 0°). Recently, in a similar research approach on office workers, Jaschinski et al. (1998) found that gaze inclinations between horizontal (0°) and 16° below the horizontal are preferred.

Hansson and Attebrant (1986; also Attebrant 1995) tested three relatively low document heights during a word-processing task. The test subjects were sitting with the trunk upright. They were free to look at the manuscript and the keyboard at will. The majority of the subjects were considered to be satisfactory at typewriting and a minority of them (about 20%) showed excellent typewriting skills. Those that were excellent at typewriting looked almost continuously at the manuscript and those who were satisfactory at typewriting looked at the manuscript less. The middle document height was preferred by the majority of subjects. On the basis of the gaze inclinations calculated, the middle and highest document heights can be compared with the visual targets at –40 and –25 cm to eye height, respectively, tested by Delleman (1999, Figure 5.10). The results of workers' perceptions in the latter study indicate that there may be a sort of secondary local optimum at relatively low visual target heights (refer to Figure 5.11a at chair backrest inclination 0°, for comparison with the upright trunk posture in the study by Hansson and Attebrant 1986). Probably, at these heights, the force generated by the lengthening of neck ligaments and passive muscle components reduces the amount of active muscle force needed. Relative to this, higher targets increase the active muscle force needed, whereas lower targets stretch passive neck structures to their maximum. Both effects are considered relatively unfavorable.

The study by Bhatnager et al. (1985) indicated that the most favorable gaze inclination is between –23° and +10° to the horizontal. In the study by Delleman (1999, described above) it was found that gaze inclinations between –18° and +5° to the horizontal induce significantly better perceptions of workers than inclinations outside this range. The studies by Sommerich et al. (1998), Psihogios et al. (1998), De Wall et al. (1992), and Grandjean et al. (1983, 1984) narrow the favorable gaze inclination range to between –15° to the horizontal and the horizontal itself. The study by Grandjean et al. (1983, 1984) and the study by Delleman (1999, described above) indicate that the most favorable gaze inclination is approximately in the middle of this interval, i.e., 6° to 9° below the horizontal. These results lead to the recommendation that for VDU operation without visual control on the hands gaze inclination should be 6° to 9° (range 0° to 15°) below the horizontal, if the chair backrest is adjusted between an upright position and one inclined 15° backward (Figure 5.13). In the case of optometric corrections (e.g., bifocals, multifocals), an ergonomist or an occupational health professional should be consulted for obtaining a favorable working posture (Horgen et al. 1989, 1995), instead of using the recommendation.

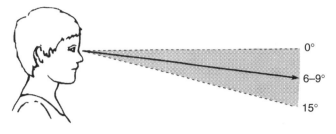

FIGURE 5.13
Recommended gaze inclination (with range) for VDU operation without visual control on the hands, if the chair backrest is adjusted between upright and a 15° backward inclination. In the case of optometric corrections (e.g., bifocals, multifocals), an ergonomist or an occupational health professional should be consulted for obtaining a favorable working posture, instead of using the recommendation.

5.4 Neck Twisting

In the introduction to this chapter it was stated that only a little is known about the effects of neck twisting. This section contains short descriptions of the few studies on this matter.

Wikström (1993) studied the effects of twisted neck and trunk postures (together with the effects of whole-body vibration) on discomfort while driving two forklift trucks and an agricultural tractor. Besides a symmetrical posture (P1), two others were tested: the first with 30° to 50° twisting of the neck and 0° to 5° twisting of the trunk (P2), and the second with 50° to 60° twisting of the neck and 5° to 10° twisting of the trunk (P3). The results showed that the factors of posture and vibration do not interact in relation to discomfort. Discomfort increases with an increasing twist angle and an increasing vibration level. After 90 s in a twisted posture without vibration, the average discomfort in the lower back was between 0.5 and 1.0 on a 10-point rating scale for trunk twist angles of 0° to 5° (P2), and between 1.5 and 2.0 for trunk twist angles of 5° to 10° (P3). Average discomfort in the neck/shoulder was between 1.0 and 2.0 on a 10-point scale for neck twist angles of 30° to 50° (P2), and between 3.0 and 4.0 for neck twist angles of 50° to 60° (P3). The maximum acceptable exposure time (during one day without rest pauses), as estimated by the test subjects, was approximately 6 h at P1, 3 h at P2, and 2 h at P3. According to the author, the 50% reduction of the acceptable exposure time when changing the posture from P1 to P2 was due to neck twisting, while the further reduction when changing from P2 to P3 was due to trunk twisting (although the estimations in terms of hours should be taken with a certain degree of reservation because of the difficulties for subjects to assess for how long they could accept a certain load).

In a field study of VDU and typewriter workplaces, Hünting et al. (1981) disclosed that subjects twisting the neck beyond 20°, when compared to

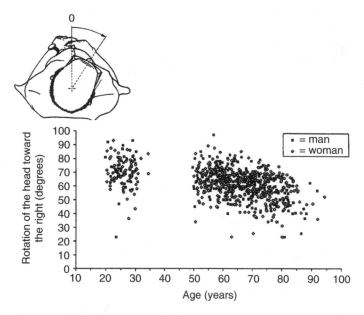

FIGURE 5.14

Maximum rotation of the head vs. the trunk around the vertical axis (neck twisting) for males and females of various ages. (From Steenbekkers, L.P.A. and Van Beijsterveldt, C.E.M., *Design-Relevant Characteristics of Ageing Users*, Delft: Delft University Press, 1998. With permission.)

subjects twisting 20° or less, showed a higher incidence of medical findings in the neck and shoulders, including painfully limited head movement, painful pressure points in the neck muscles, and painful tendon insertions in the shoulder area. Based on a biomechanical analysis, Snijders et al. (1991) showed that twisting beyond 35° causes neck muscle forces and joint reaction forces to increase rapidly.

It is concluded that there is not yet enough data to establish quantitative evaluation criteria for neck twisting. What we do know, however, is that the joint positions toward the end of the range of motion, where a substantial mechanical load is placed on the passive structures such as ligaments, are unfavorable (Harms-Ringdahl and Ekholm 1986). According to the data presented by Steenbekkers and Van Beijsterveldt (1998), limiting neck twisting up to roughly 45° keeps the vast majority of people of working age (less than 65 years old) away from the ends of their ranges of motion (Figure 5.14).

Summary

Studies reporting the effects of head posture (e.g., neck load, health complaints) most often focus on head inclination (i.e., the amount of deviation from the upright posture or the vertical). This measure of musculoskeletal load is to be seen as an equivalent of the force delivered to counteract the gravitational force on the head. In more recent years, attention has been given to neck flexion/extension (i.e., the amount of inclination of the head with respect to the amount of inclination of the trunk). From the literature review reported here it is concluded that neck flexion/extension is a determinant of neck load, and should be used as an evaluation criterion for working postures in addition to the traditionally used inclination of the head. It is recommended to keep neck flexion between 0° and 25°, to avoid unfavorable neck positions for the majority of subjects. A slightly/moderately flexed neck (about 15°) seems to be the most favorable neck position. Only a little is known about the effects of neck lateral flexion and twisting. It is known that people normally do not use lateral flexion during gaze changes. Although there are not yet data for establishing quantitative evaluation criteria for neck twisting, joint positions toward the end of the range of motion, where a substantial mechanical load is on the passive structures such as ligaments, are unfavorable. Neck twisting up to approximately 45° keeps the vast majority of people of working age (less than 65 years old) away from the ends of their ranges of motion.

References

Andersson, G.B.J. and Örtengren, R., 1974. Myoelectric back muscle activity during sitting. *Scandinavian Journal of Rehabilitation Medicine*, Suppl. 3, 73–90.

Andersson, G.B.J., Örtengren, R., Nachemson, R., and Elfström, G., 1974. Lumbar disc pressure and myoelectric back muscle activity during sitting: I. Studies on an experimental chair. *Scandinavian Journal of Rehabilitation Medicine*, 6, 104–114.

Ariëns, G.A.M., 2001. *Work-Related Risk Factors for Neck Pain*, Hoofddorp, the Netherlands: TNO Work and Employment.

Attebrant, M., 1995. Personal communication.

Bendix, T. and Hagberg, M., 1984. Trunk posture and load on the trapezius muscle whilst sitting at sloping desks. *Ergonomics*, 27, 873–882.

Berndsen, M.B. and Delleman, N.J., 1993. Guideline on the viewing angle for touch-typing VDU workers, in *Work with Display Units 92*, Luczak, H., Çakir, A., and Çakir, G., Eds., Amsterdam: North-Holland/Elsevier Science, 476–480.

Bhatnager, V., Drury, C.G., and Schiro, S.G., 1985. Posture, postural discomfort, and performance. *Human Factors*, 27, 189–199.

Chaffin, D.B., 1973. Localized muscle fatigue: definition and measurement. *Journal of Occupational Medicine*, 15, 346–354.

Chaffin, D.B., Andersson, G.B.J., and Martin, B.J., 1999. *Occupational Biomechanics*, 3rd ed., New York: John Wiley & Sons.

Conrady, P., Krueger, H., and Zülch, J., 1987. *Untersuchung der Belastung bei Lupen- und Mikroskopierarbeiten*, Fb 516, Dortmund, Germany: Bundesanstalt für Arbeitsschutz.

Delleman, N.J., 1999. *Working Postures — Prediction and Evaluation*, Soesterberg, the Netherlands: TNO Human Factors.

De Wall, M., Van Riel, M.P.J.M., Aghina, J.C.F.M., Burdorf, A., and Snijders, C.J., 1992. Improving the sitting posture of CAD/CAM workers by increasing VDU monitor working height. *Ergonomics*, 35, 427–436.

Grandjean, E., Nishiyama, K., Hünting, W., and Piderman, M., 1982. A laboratory study on preferred and imposed settings of a VDT workstation. *Behaviour and Information Technology*, 1, 289–304.

Grandjean, E., Hünting, W., and Pidermann, M., 1983. VDT workstation design: preferred settings and their effects. *Human Factors*, 25, 161–175.

Grandjean, E., Hünting, W., and Nishiyama, K., 1984. Preferred VDT workstation settings, body posture and physical impairments. *Applied Ergonomics*, 15, 99–104.

Hansson, J.E. and Attebrant, M., 1986. The effect of table height and table top angle on head position and reading distance, in *Book of Abstracts of the International Scientific Conference Work with Display Units*, Stockholm: Swedish National Board of Occupational Safety and Health, 419–422.

Harms-Ringdahl, K. and Ekholm, J., 1986. Intensity and character of pain and muscular activity levels elicited by maintained extreme flexion position of the lower-cervical-upper-thoracic spine. *Scandinavian Journal of Rehabilitation Medicine*, 18, 117–126.

Harms-Ringdahl, K. and Schüldt, K., 1988. Maximum neck extension strength and relative neck muscular load in different cervical spine positions. *Clinical Biomechanics*, 4, 17–24.

Horgen, G., Aarås, A., Fagerthun, H.E., and Larsen, S.E., 1989. The work posture and the postural load of the neck/shoulder muscles when correcting presbyopia with different types of multifocal lenses on VDU-workers, in *Work with Computers: Organizational, Management, Stress and Health Aspects*, Smith, M.J. and Salvendy, G., Eds., Amsterdam: Elsevier Science, 338–347.

Horgen, G., Aarås, A., Fagerthun, H., and Larsen, S., 1995. Is there a reduction in postural load when wearing progressive lenses during VDT work over a three-month period? *Applied Ergonomics*, 25, 165–171.

Hünting, W., Grandjean, E., and Maeda, K., 1980. Constrained postures in accounting machine operators. *Applied Ergonomics*, 11, 145–149.

Hünting, W., Läubli, T., and Grandjean, E., 1981. Postural and visual loads at VDT workplaces: I. Constrained postures. *Ergonomics*, 24, 917–931.

Jaschinski, W., Heuer, H., and Kylian, H., 1998. Preferred position of visual displays relative to the eyes: a field study of visual strain and individual differences. *Ergonomics*, 41, 1034–1049.

Kilbom, Å., Persson, J., and Jonsson, B., 1986. Risk factors for work-related disorders of the neck and shoulder — with special emphasis on working postures and movements, in *The Ergonomics of Working Postures*, Corlett, E.N., Wilson, J., and Manenica, I., Eds., London: Taylor & Francis, 44–53.

Kroemer, K.H.E. and Hill, S.G., 1986a. Preferred line of sight angle. *Ergonomics*, 29, 1129–1134.

Kroemer, K.H.E. and Hill, S.G., 1986b. Preferred line of sight, in *Book of Abstracts of the International Scientific Conference Work with Display Units*, Stockholm: Swedish National Board of Occupational Safety and Health, 415–418.

Kumar, S., 1994. A computer desk for bifocal lens wearers, with special emphasis on selected telecommunication tasks. *Ergonomics*, 37, 1669–1678.

Läubli, T., 1987. Preferred settings in VDT work: the Zürich experience, in *Work with Display Units 86*, Knave, B. and Widebäck, P.-G., Eds., Amsterdam: North-Holland/Elsevier Science, 249–262.

Lee, K.S., Waikar, A.M., Aghazadeh, F., and Tandon, S., 1986. An electromyographic investigation of neck angles for microscopists, in *Proceedings of the Human Factors Society 30th Annual Meeting*, Santa Monica, CA: Human Factors Society, 548–551.

Lepoutre, F.X., Roger, D., and Loslever, P., 1986. Experimental analysis of a visuo-postural system in an office workstation, in *The Ergonomics of Working Postures*, Corlett, E.N., Wilson, J., and Manenica, I., Eds., London: Taylor & Francis, 363–371.

Nakaseko, M., Morimoto, K., Nishiyama, K., and Tainaka, H., 1993. An image analyzing approach to working postures for screen and keyboard use of visual display units, in *Work with Display Units 92*, Luczak, H., Çakir, A., and Çakir, G., Eds., Amsterdam: North-Holland/Elsevier Science, 235–239.

Ong, C.N., Koh, K., Phoon, W.O., and Low, A., 1988. Anthropometrics and display station preferences of VDU operators. *Ergonomics, 31*, 337–347.

Psihogios, J.P., Sommerich, C.M., Mirka, G.A., and Moon, S.D., 1998. The effects of VDT location on user posture and comfort: a field study, in *Proceedings of the Human Factors and Ergonomics Society 42nd Annual Meeting*, Vol. 2, Santa Monica, CA: Human Factors and Ergonomics Society, 871–875.

Schüldt, K., Ekholm, J., Harms-Ringdahl, K., Németh, G., and Arborelius, U.P., 1986. Effects of changes in sitting posture on static neck and shoulder muscle activity. *Ergonomics, 29*, 1525–1537.

Schüldt, K., Ekholm, J., Harms-Ringdahl, K., Arborelius, U.P., and Németh, G., 1987. Influence of sitting postures on neck and shoulder e.m.g. during arm-hand work movements. *Clinical Biomechanics, 2*, 126–139.

Sommerich, C.M., Joines, S.M.B., and Psihogios, J.P., 1998. Effects of VDT viewing angle on user biomechanics, comfort, and preference, in *Proceedings of the Human Factors and Ergonomics Society 42nd Annual Meeting*, Vol. 2, Santa Monica, CA: Human Factors and Ergonomics Society, 861–865.

Snijders, C.J., Hoek Van Dijke, G.A., and Roosch, E.R., 1991. A biomechanical model for the analysis of the cervical spine in static postures. *Journal of Biomechanics, 24*, 783–792.

Steenbekkers, L.P.A. and van Beijsterveldt, C.E.M., 1998. *Design-Relevant Characteristics of Ageing Users*, Delft, the Netherlands: Delft University Press.

Straumann, D., Haslwanter, Th., Hepp-Reymond, M.-C., and Hepp, K., 1991. Listing's law for eye, head, and arm movements and their synergistic control. *Experimental Brain Research, 86*, 209–215.

Tweed, D. and Vilis, T. 1992. Listing's law for gaze-directing head movements, in *The Head-Neck Sensory Motor System*, Berthoz, A., Graf, W, and Vidal, P.P., Eds., New York: Oxford University Press, 387–391.

Van der Grinten, M.P. and Smitt, P., 1992. Development of a practical method for measuring body part discomfort, in *Proceedings of the Annual International Ergonomics and Safety Conference*, Kumar, S., Ed., London: Taylor & Francis, 311–318.

Wikström, B.-O., 1993. Effects from twisted postures and whole body vibration during driving. *International Journal of Industrial Ergonomics, 12*, 61–75.

6

Trunk

Jaap H. Van Dieën and Maury A. Nussbaum

CONTENTS

Introduction

Low back pain (LBP) is the most prevalent work-related health problem in the industrialized world. The lifetime prevalence is estimated at more than 70% in the general population and as high as 90% in populations exposed to heavy physical loads at work, such as concrete reinforcement workers and nurses. Although most low back complaints recede in about a month, regardless of treatment, recurrence is very common (about 85%), and about 4% of cases become chronic. About 50% of LBP cases result in days lost from work (Van der Hoogen et al. 1997). The direct medical costs and especially the indirect costs, such as workers' compensation and production losses, associated with these complaints are enormous.

Trunk postures are of ergonomic concern chiefly because adverse trunk postures have been shown to be associated with LBP (Hales and Bernard 1996, Burdorf and Sorock 1997, Kuiper et al. 1999). In part because of the limited success of LBP treatment, much emphasis has been placed on ergonomic measures to prevent its first time occurrence (Linton and Van Tulder 2001). Improving workplace design such that unfavorable trunk postures are avoided appears a promising avenue in this respect.

6.1 Definitions and Measurement

In general, posture can be defined as the orientation of body segments in space and in relation to each other. This definition assumes body segments to be rigid links, whose orientation with respect to a neighboring segment is determined by rotations about three axes in one joint. With respect to the trunk, this assumption is clearly not correct. The trunk contains multiple joints, with the total rotations determining the orientation of the trunk as a whole. In many applications, however, trunk posture can be described as if the trunk were a rigid segment. To this end, the orientation of the upper part of the trunk with respect to the pelvis needs to be determined. Reflecting the fact that the motions causing this orientation are in reality not pure rotations, they are here called forward/backward bending, lateral bending, and twisting in accordance with ISO 11226 (ISO 2000) (Figure 6.1).

The effects of a trunk posture are not only determined by these angles between segments, but also by the orientation of the trunk with respect to the gravitational field (Figure 6.2). The angles with respect to the gravitational axis system will be referred to as forward inclination and sideward

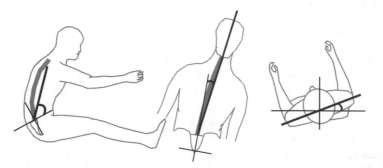

FIGURE 6.1
Forward bending, lateral bending, and torsion or twisting. These classifications of trunk posture are defined in terms of the angle between the trunk and the pelvis and determine the configuration of the spine. Note that the pelvic orientation may be different from vertical.

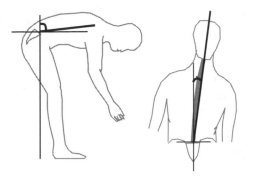

FIGURE 6.2

Forward inclination and sideward inclination. These classifications of trunk posture are defined in terms of the angle between the trunk and the vertical and determine the effect of gravity acting on the upper body.

inclination. Note that flexion and lateral flexion are used in the literature as synonymous with both forward and sideward inclination and of bending and lateral bending. The third degree of freedom in the gravitational system is irrelevant in terms of the effects of posture on an individual. However, when rotating the trunk around a vertical axis while standing and keeping the feet fixed, individuals also rotate the pelvis in the hip joints. Therefore, the overall rotation of the trunk is not equal to twisting.

It is important to note that the above definitions can be unambiguously used only for postures or movements in a primary plane or about a primary axis. The usual way of describing the three-dimensional orientation of a segment is through decomposition of the orientation in three angles about three orthogonal axes, the so-called Euler angles. However, a posture that involves rotations about more than one axis is not uniquely described by three Euler angles. The order in which rotations are performed also determines the resulting posture, or inversely the order of decomposition determines the Euler angles obtained for a certain posture. This is illustrated in Figure 6.3.

In an asymmetric posture, joint rotation about an axis is easily misinterpreted. Consider the following example. The thorax is rotated forward to 90° and subsequently rotated about a horizontal midsagittal axis. The latter movement would be considered lateral bending, with respect to the pelvis axis system (according to the definition given above). With the trunk bent forward less than 90°, the same excursion would be a combination of twisting and lateral bending. However, the effect of the latter excursion on the configuration of the trunk is the same as in twisting when standing upright. This illustrates that, although trunk postures can be clearly defined with respect to a pelvis axis system, the resulting postural angles do not always allow an intuitive interpretation.

A more intuitive description of postures can be obtained by describing the main movement component (usually bending) in the pelvis axis system and the other two components in an axis system connected to the trunk

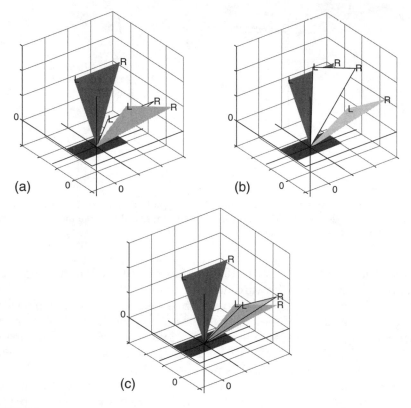

FIGURE 6.3
The dark horizontal plane represents the pelvis, the dark gray triangle represents the trunk in an upright posture. (A) The trunk is first rotated forward 40° (white) and then sideward 20° (light gray). (B) The trunk is first sideward rotated 20° (white) and then forward rotated 40° (light gray). (C) Both final postures, which are not identical. The two primary rotations (20° sideward and 40° forward rotation) thus do not uniquely describe a posture.

(Van Dieën et al. 1998, Gagnon et al. 2000). Measurements of the trunk and pelvis orientations allow for conversions between such axis systems, but classifications based on visual observation may become highly ambiguous because the order of rotations and the axis system used to define angles are usually not made explicit. Consequences of such errors in relation to the evaluation of lifting postures have been discussed by Dempsey and Fathallah (1999). It should also be noted that in the current literature no consensus exists on the axis system to be used to describe trunk posture and loading, which can explain some disparities between studies (Kingma et al. 1999, Plamondon et al. 1999, Gagnon et al. 2000, Van Dieën and Kingma 2001).

Determination of trunk posture requires identification of the position of bony landmarks on the upper trunk or thorax and on the pelvis. Such position identification is often done implicitly, for example, in visual observation of postures. In these cases soft tissue deformation may hamper adequate characterization. An example of this is the impression of a lumbar lordosis in power lifters during lifting, which is actually caused by the prominent contracting gluteus muscle (Cholewicki and McGill 1992). Explicit identification of landmarks is commonly performed using automated video analysis systems along with markers that are usually attached over bony landmarks. On the pelvis a number of bony prominences can be used to this end. On the trunk a marker on the shoulder is often used. Given the considerable range of motion of the shoulder girdle with respect to the thorax, a marker fixed to the thorax is preferable.

Three general methods for determining trunk postures can be differentiated: self-report, observation, and objective measurement. In terms of applicability in ergonomics the three methods have been mentioned here in descending order. Self-reports can be easily administered at relatively low cost, whereas objective measurements require expensive equipment and acquisition of data is time-consuming. Observation can be considered intermediate. In terms of measurement accuracy and validity they have been mentioned in ascending order (Van der Beek and Frings-Dresen 1998). It has been shown that substantial discrepancies between the results of these methods do occur (Burdorf and Laan 1991, Burdorf et al. 1992, Van der Beek et al. 1994, De Looze et al. 1994). Subjective reports of postures are typically obtained using questionnaires and diaries. The results obtained with self-reports, however, are of questionable accuracy and validity (Van der Beek and Frings-Dresen 1998, Li and Buckle 1999). Consequently, observational methods, either direct or video based, have become increasingly popular (Van der Beek and Frings-Dresen 1998, Li and Buckle 1999).

A wide variety of observational methods have become available (for a recent review see Li and Buckle 1999). Nevertheless, the validity of classifications of trunk posture on the basis of observation has been questioned. The validity certainly appears low when subjects change posture relatively frequently (De Looze et al. 1994), due to limitations in the information processing capacity of the observer. Multimoment or time-sampled observations are less demanding for the observer than real-time observations, and consequently the validity of posture classifications is better for these methods (Van der Beek and Frings-Dresen 1998). In addition, validity generally increases when fewer distinct classifications of posture are used, although De Looze et al. (1994) did not find significant differences in validity when classifying trunk forward inclination in 15° or 20° intervals. In addition, the use of coarse categories (e.g., pooling twisting and lateral bending) can in part solve the problem noted above of describing trunk postures in three dimensions, although at the cost of a loss of information. Recently, Paquet et al. (2001) investigated the validity of an observation method based on time sampling with five categories for trunk posture. They found good correspondence to a reference method for the percentage of time spent in a neutral trunk posture, mild forward inclination (>20° and <45°) and extreme trunk forward inclination (>45°). The percentage time in asymmetric postures (sideward inclination or twisting >20°), however, was significantly underestimated. Direct observation has advantages over video-based observation in terms of costs and possibly accuracy, since camera angles may affect classifications (Punnett and Keyserling 1987). On the other hand, video-based observation allows repeated measurements and the use of stills can improve accuracy.

Quantitative measurements of trunk posture can be obtained using several available technologies. Optical methods, such as automated video or film analysis, in principle allow highly accurate three-dimensional determination of trunk posture from marker positions. For laboratory-based measurements, this methodology generally is the first choice; yet applicability outside the laboratory is usually limited. Problems arise in practice because markers can be highly obtrusive, fields of view of cameras are limited, and markers become obscured by objects or body parts.

Several other measurement systems have been developed as alternatives to optical methods (for an overview, see Li and Buckle 1999). Most frequently used are those based on goniometers (e.g., Snijders and Van Riel 1987, Marras et al. 1992). Goniometer-based systems, however, provide only relative rotations of the trunk and pelvis, and do not indicate trunk orientation with respect to the gravitational field. Applicability of electromagnetic tracking devices, which can be used as goniometers (McGill et al. 1997), is severely limited by disturbances caused by metal objects in proximity to the sensors. Recently, inertial sensing (accelerometers, gyroscopes) methods that allow accurate quantification of trunk postures have become available (Baten et al. 1997). Inertial sensing methods can indicate both trunk/pelvis angles and trunk orientation with respect to the vertical. These methods suffer from integration drift, which hampers long-term recording. Goniometric and inertial sensing methods are promising for field use, although the need to attach sensors to the back will continue to limit applicability in some work situations (for example, in seated work). In general, a careful selection of measurement methods is required to fit the aims of the recording and the environment in which recordings are to be made.

6.2   Trunk Anatomy

Although the causes of most cases of LBP are undiagnosed, there is sufficient evidence that the lumbar spinal column and associated soft tissues play an important role in the etiology of the complaint. To understand how trunk posture relates to stress on low back tissues, and ultimately to damage, some understanding of the anatomy of the spine and surrounding musculature is needed.

The spine consists of bony structures called vertebrae, 5 in the lumbar part, 12 in the thoracic part, and 7 in the cervical part. Orientations of the individual lumbar and thoracic vertebrae determine the overall posture of the trunk. In upright stance, the lumbar vertebrae form a curve, concave posteriorly,

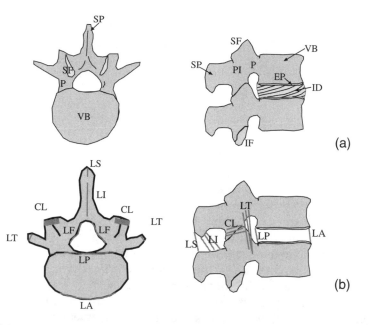

FIGURE 6.4

Schematic overview of motion segment anatomy. The morphology depicted is typical for the lumbar spine; thoracic vertebrae have a slightly different appearance. (a) Bony and cartilaginous structures of the motion segment. VB = vertebral body; SP = spinous process (i.e., the part of the vertebra that can be felt under the skin); SF = superior facet; IF = inferior facet; EP = end plate; ID = intervertebral disc; PI = pars interarticularis; P = pedicle. (b) Ligaments of the motion segment. LS = supraspinous ligament; LI = interspinous ligament; CL = capsular ligaments; LT = transverse ligaments; LP = posterior ligament; LA = anterior ligament; LF = ligamentum flavum.

called lordosis. The thoracic spine forms a convex curve or kyphosis. Although these curves are often referred to as natural, it should be kept in mind that their presence is specific for the standing posture. In trunk forward inclination the lumbar lordosis decreases or disappears, as is the case in sitting and especially slouched sitting. Given the sitting posture typically observed in primates, it can be questioned whether this posture should be considered unnatural.

The general anatomy of motion segments, the fundamental building blocks of the spine, is illustrated in Figure 6.4. A motion segment consists of two vertebrae, connected by pads of soft tissue called intervertebral discs and by a number of fibrous straps called ligaments. These connections allow for a considerable range of motion between two vertebrae. Motions are guided and restricted by two joints on the posterior side of the vertebrae called facet joints or zygapophysial joints. In the lumbar spine, these joints limit torsion and forward shearing of the superior vertebra, as a consequence of the nearly vertical orientation of the joint surfaces at a 45° angle to the frontal plane. The posterior bony parts of the vertebrae also limit extension. The lumbar

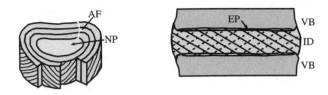

FIGURE 6.5
Schematic illustration of intervertebral disc anatomy. ID = intervertebral disc; VB = vertebral body; EP = end plate; AF = annulus fibrosus; NP = nucleus pulposus.

spine is therefore most flexible in forward and lateral bending. In the lower thoracic spine the facet joints are oriented in the frontal plane, while at higher levels they are more horizontal (and completely horizontal in the cervical spine). As a consequence, the thoracic spine has more mobility in twisting. Although facet joint orientation would allow substantial lateral bending, the range of thoracic movement is limited due to the ribs.

The ligaments have strongly nonlinear material characteristics. In the spine, they therefore resist motions of the vertebrae mainly toward the limits of the range of motion. Posterior ligaments (supraspinous and interspinous ligaments) resist bending, whereas the anterior ligament resists extension. The transverse ligaments resist lateral bending and the capsular ligaments limit torsion.

The intervertebral disc deserves special attention in relation to LBP, because damage to this structure appears to be an important source of pain (Adams and Dolan 1997, Van Dieën et al. 1999). Each disc consists of a gel-like center called the nucleus pulposus that is contained by a ring of fibrous tissue layers, the annulus fibrosus. Discs are bordered on the top and bottom by two plates consisting of bone and cartilage called the end plates (Figure 6.4 and Figure 6.5). The nucleus pulposus has a water content of around 80%, which gives it roughly hydrostatic properties. It therefore distributes forces acting along the axis of the spine evenly over the end plates and also tensions the annulus fibrosus. Fibers in the annulus fibrosus are oriented at an angle of about 60° to the long axis of the spine, and alternate from +60° to −60° between different layers. The annulus is thereby able to resist the hoop stresses caused by the hydrostatic pressure in the nucleus pulposus and also to resist bending and twisting motions of the spine.

The muscles surrounding the spine are illustrated in Figure 6.6. Their main functions can be gleaned from their location with respect to the vertebrae. Muscles posterior to the center of the vertebrae act primarily as trunk extensors. Among these muscles the most important is the erector spinae, a large muscle mass comprising the iliocostalis and longissimus, which in turn comprise many small muscle fascicles that run approximately parallel to the spine. When active unilaterally, the erector spinae mass acts to bend the trunk laterally or to resist a force that would bend the trunk to the opposite side. Other muscles lateral to the center of the vertebrae, especially the oblique abdominal muscles, perform a similar function. The oblique abdominal

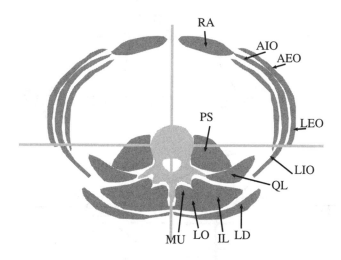

FIGURE 6.6

Schematic overview of the lumbar muscles in a transverse cross-section of the trunk. RA = rectus abdominus; AIO = anterior internal oblique; LIO = lateral internal oblique; AEO = anterior external oblique; LEO = lateral external oblique; MU = multifidus; LO = longissimus; IL = iliocostalis; LD = latissimus dorsi; PS = psoas; QL = quadratus lumborum.

muscles also serve to twist the trunk along its long axis and, when assisted by the rectus abdominus muscle (and usually by gravity), can bend the trunk forward. Because these muscles contribute to moments about more than one axis, in order to obtain a moment about only one axis other resulting moments have to be exerted by other muscles. For example, when the oblique abdominal muscles are recruited to cause pure trunk twisting, the forward and lateral bending components have to be compensated for by the erector spinae.

6.3 Effects of Trunk Posture

The adverse health effects of certain postures are thought to be mainly of mechanical origin. A conceptual model of how maintaining a posture may result in LBP is illustrated in Figure 6.7. Trunk posture is defined, as described above, in terms of the orientation of the trunk in the gravitational field (segment angles) and in terms of thorax orientation with respect to the pelvis (joint angles). Segment angles determine the moment acting about the lumbar spine as a consequence of gravity acting on the upper body, whereas joint angles determine the strain of muscles and other tissues. Tensile strain of tissues and the resulting tissue stress can directly cause tissue damage and discomfort. In addition, the forces produced by stretched structures will produce a moment.

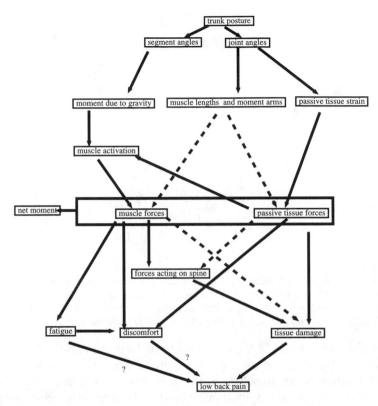

FIGURE 6.7
Conceptual model of the relationship between trunk posture and LBP (for explanation see text). Relationships indicated with dashed lines are not specifically addressed in this chapter.

To maintain a given posture, an individual must equilibrate the sum of this moment, the moment caused by gravity, and the moments due to muscle forces (i.e., their sum equals zero). Muscle moments and passive tissue moments may have the same sign or may counteract each other; their sum is designated the net moment. Moment equilibrium is achieved through modulating muscle activation. In addition to the level of activation of the muscle, the moment it produces depends on its moment arm and length, which are in turn determined by the joint angles. The latter aspect will not be extensively dealt with here. Muscle activation can lead to muscle fatigue, which may contribute to discomfort and potentially to LBP. Muscle and passive tissue forces determine the shear and compression forces acting on the spine, which can cause damage and ultimately LBP. It is uncertain whether direct relationships exist between fatigue and discomfort on the one hand and LBP on the other, but of course these may be relevant outcomes in their own right. Several feedback loops are in reality present but have been omitted from the model. For example, fatigue will affect postures adopted and will affect the distribution or pattern of muscle recruitment.

6.3.1 Trunk Postures and Mechanical Load

The following sections review which structures are loaded most in several trunk postures. In addition, these address the question of whether the level of these loads is high enough to cause clinically relevant damage. Mechanical loads on the low back are only in part determined by postures. External forces acting on the body, for example, when lifting, can contribute substantially. The following discussion refers to the effects of posture exclusively, ignoring other contributions. Moreover, mechanical loads are determined by the three-dimensional posture, because excursions about different axes interact (e.g., Van Dieën 1996, Marras et al. 1998). Nevertheless, for the sake of clarity, effects of bending (or forward inclination), lateral bending (or sideward inclination), and twisting are discussed separately.

6.3.1.1 Tensile Tissue Strain

Tissues in the trunk are generally slack in the neutral upright postures, and any non-neutral posture imposes strain on some tissues. Tensile tissue strain can cause discomfort or damage when it exceeds threshold values or when it is maintained for a prolonged time (Harms-Ringdahl et al. 1983), although information regarding these thresholds and time periods is incomplete.

6.3.1.1.1 Forward Bending

The range of motion of the trunk in the lumbar area in forward bending was reported to be about 55°, or slightly less than 10° of bending per motion segment in one study (Adams and Hutton 1982). Another study (Peach et al. 1998) reported a range of motion of 70°. Age differences may account for these disparate results, as the latter study dealt with college-age subjects. In the thoracic area only limited bending occurs. When the lumbar spine bends forward, passive tissues posterior to the axis of rotation, which is approximately in the center of the intervertebral disc (Pearcy et al. 1984), generate an extension moment as shown in Figure 6.8 (Dolan and Adams 1993). This passive moment is sufficient to carry most of the upper body weight in (near) maximum bending of the spine as is evidenced by the absence of lumbar muscle activity in these postures (the flexion–relaxation phenomenon; Kippers and Parker 1984, McGill and Kippers 1994), although some activity of distant muscles may contribute (Toussaint et al. 1995).

Most of the passive forces will be contributed by passive elongation of muscles and fascia, as the spine with its ligaments and the intervertebral discs can only provide up to 25% of the total passive moment (Adams and Dolan 1991). The exact contribution of passive muscle forces to the extension moment is not known, but in fully bent postures the bellies of the lumbar extensor muscle slips are strained to 1.2 to 1.6 times their length in the upright posture (Macintosh et al. 1993). The relative contribution of the different structures in the spine has been studied in some detail. It appears that the distribution over ligaments and the intervertebral disc depends on

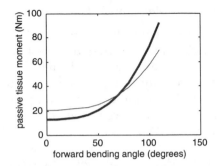

FIGURE 6.8

Passive extension moment as a function of bending angle in males (thin line) and females (thick line), based on regression equations provided by Dolan and Adams (1993).

the bending angle (Adams et al. 1980). The intervertebral disc makes a relatively high contribution in moderate bending, whereas at the end of the range of motion the supraspinous and interspinous ligaments make a larger contribution to the total passive moment. These ligaments are strained to about 1.2 times their rest length at 5° of bending (Panjabi et al. 1982), which is about half the range of motion of a segment.

The supraspinous and interspinous ligaments are also the first to fail when hyperflexion occurs (Adams et al. 1980, 1994). In hyperflexion the posterior part of the disc can also be damaged, leading to herniation of the nucleus pulposus (Adams et al. 1980). It appears, however, that hyperflexion occurs only at about 10° over the *in vivo* range of motion, providing a margin of safety for the spinous tissues (Adams and Hutton 1986). Consequently, it would not seem likely that trunk bending causes tissue strains sufficient to cause damage. With sustained or repeated bending, however, creep will occur (Twomey and Taylor 1982, McGill and Brown 1992, Adams and Dolan 1996). This gradual increase in strain may be a cause of damage to the posterior annulus (Adams and Hutton 1985, Green et al. 1993), and possibly to the posterior and interspinous ligaments (Solomonow et al. 2001).

6.3.1.1.2 Lateral Bending

The range of lumbar trunk motion in lateral bending is about 30° in young subjects and about 20° in subjects older than 65 years (McGill et al. 1999). As in forward bending, passive tissues generate substantial resistive moments (McGill et al. 1994) as shown in Figure 6.9.

Little is known about the contributions of various tissues to passive lateral bending moments. Some studies suggest that the disc contributes only a little (Krismer et al. 2000). The transverse ligaments are strained the most (Panjabi et al. 1982), but their stiffness is unknown. To the authors' knowledge no evidence has been provided that excessive tissue strain in lateral bending can cause clinically relevant damage.

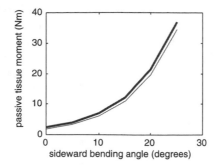

FIGURE 6.9

Passive lateral moment as a function of lateral bending angle in males (thin line) and females (thick line), based on regression equations provided by McGill et al. (1994).

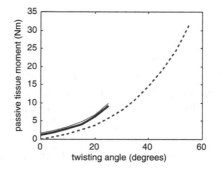

FIGURE 6.10

Passive moment as a function of twisting angle measured between pelvis and sternum (solid lines) and between pelvis and shoulders (dashed lines) in males (thin line) and females (thick line), based on regression equations provided by McGill et al. (1994) and Bodén and Öberg (1998).

6.3.1.1.3 *Twisting*

The twisting range of motion in the lumbar area is limited to about 15° and is not strongly dependent on age (McGill et al. 1999). Because of the considerable twisting mobility in the thoracic area, the range of motion is higher when measured at the level of the shoulders (about 60°; Bodén and Öberg 1998), in line with the definition in Section 6.1, Figure 6.1, or when measured at the level of the sternum (McGill et al. 1994). Passive tissue resistance to twisting is considerable, as shown in Figure 6.10 (McGill et al. 1994, Bodén and Öberg 1998).

In twisted trunk postures, muscle strains will be low as a result of the small moment arms of most muscles about the twisting axis (McGill and Hoodless 1990). In contrast, strains in ligaments of the spine, especially the capsular ligaments, are considerable (Farfan et al. 1970, Panjabi et al. 1982). The intervertebral disc also contributes, due to strain in a portion of the

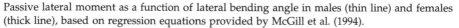

annulus fibers (Haher et al. 1989). Primary resistance to torsion is produced by compression of the facet joint on one side (Adams and Hutton 1981).

There has been considerable debate about whether tissue strains in twisted postures can cause clinically important damage to the spine. Farfan and co-workers (Farfan et al. 1970) have suggested that torsion may cause damage to the annulus fibrosus and as such may be a cause of back pain and disc degeneration. Their arguments have been contested by Adams and Hutton (1981), who showed that the compressed facet joint would probably be the first to fail in torsion, and that extremely high twisting angles would be required to produce disc damage. Tensile strains in the disc that result from twisting depend on the exact location of the axis of rotation, and this axis may be incorrectly imposed in *in vitro* biomechanical experiments (Haher et al. 1989). In addition, it has been shown that frequently repeated twisting can induce damage to the annulus fibrosus even when the angular excursion is small (Liu et al. 1985).

6.3.1.2   Net Moments

Moments required to maintain a certain trunk posture are partially determined by passive tissue resistance. For example, when sitting with a twisted spine, the passive resistance described above would tend to de-rotate the spine, and muscle activity is consequently required to maintain this posture. In many postures, upper body weight contributes substantially and to a much greater degree than the effects of passive tissue resistance. The angle of the trunk with respect to the field of gravity therefore must be identified to obtain meaningful information on mechanical loading. Orientations of the trunk with respect to gravity in the sagittal and frontal planes were defined earlier as forward and sideward inclination, respectively. In forward or sideward inclined postures gravity produces substantial moments about the joints in the lumbar spine. While the moment arms of gravity are close to zero in an upright stance, they increase as a sinusoid of the angle of forward or sideward inclination. As discussed above, these moments are partially counteracted by passive tissues due to bending. Except for extreme postures, muscle forces generate a substantial portion of the counteractive moment. Therefore, moments resulting from gravity are equilibrated by the net effect of all muscles and passive tissues producing moments about the joint considered. Net moments are often used as an indicator of mechanical load, as they reflect the combined load on all these tissues.

Because the net moment is produced by an unknown combination of muscle and passive tissue forces, setting standards with respect to injury thresholds is not possible at this level of analysis. An indication of how load magnitude relates to the capacity of the musculoskeletal system can be obtained by comparison of the net moments during a task with some measure of force or moment generating capacity (i.e., strength). This capacity is most often obtained in isometric maximum voluntary force/moment tests in standardized postures (Garg and Chaffin 1975, Chaffin and Erig 1991,

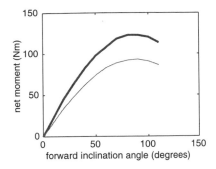

FIGURE 6.11

Net moments on the low back as a function of forward trunk inclination angle for an average male (thin line) and female (thick line).

Gravel et al. 1997). It should be kept in mind that posture in itself will affect strength as a consequence of changes to muscle lengths and moment arms.

It is notable that there can be situations wherein tissues concurrently produce moments about a joint that are opposite in sign, yielding a zero net moment, although considerable mechanical load is still present. In most cases, there is some activity of muscles on the side of a joint opposite to where muscle force is required to produce the moment, probably to guarantee sufficient stability of the joint position. This type of muscle activity is often termed antagonism, and has been shown to occur for several trunk postures and movements (e.g., De Looze et al. 1999).

6.3.1.2.1 Forward Inclination

The net moment on the low back was described earlier as increasing with trunk inclination as a linear function of the sine of the inclination angle. Figure 6.11 gives the net moment as a function of the angle of inclination, estimated for a 50th percentile male and female using a linked segment model (Chaffin and Andersson 1991). Kumar (1996) determined that maximum voluntary extension moments in the upright position averaged 321 Nm for males and 185 Nm for females. The passive tissue moments (Figure 6.8) have the same sign as the moments produced by the extensor muscles. Nevertheless, considerable muscular effort will be required to maintain forward inclined postures, especially in moderate to high levels of forward bending, because the passive moments are typically insufficient to equilibrate gravitational loads.

6.3.1.2.2 Sideward Inclination

As was the case in forward inclination, the net moment in sideward inclination will increase as a linear function of the sine of the inclination angle. Estimates of net moments as a function of bending angle were again made using a linked segment model (Chaffin and Andersson 1991), and are shown in Figure 6.12. Males can on average produce lateral flexion moments of

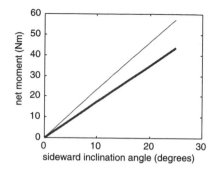

FIGURE 6.12
Net moments on the low back as a function of sideward trunk inclination angle for an average
male (thin line) and female (thick line) subject.

164 Nm; females can produce 109 Nm (Kumar 1996). The passive tissue
moments (Figure 6.9) counteract a large portion of the gravitational moment
and hence assist the moments produced by the muscles. Consequently, only
limited muscular effort is required to maintain laterally flexed postures.

6.3.1.2.3 Twisting

Net moments in twisted postures are zero. Maintaining a twisted posture
requires muscular effort, because the passive tissue moment and muscular
moment are opposite in direction. The muscular moment required to maintain a twisted posture can therefore be estimated from the twisting angle
and is illustrated in Figure 6.10. Using the regression line determined by
Bodén and Öberg (1998), which fits the earlier definition of twisting, the
moment required to maintain a twisting angle of about 50° is estimated at
25 Nm (for males). In males, average maximum twisting strength is 80 Nm
(44 Nm for females). Using these numbers as a reference, maintaining a 50°
twisted posture would require about 30% of an average male's strength,
indicating that keeping such a posture for a prolonged period would be quite
a strenuous activity.

6.3.1.3 Muscle Activation

Measurements of muscle activation through electromyography (EMG) are
often used as indicators of back load. An underlying assumption is that a
straightforward (linear) relationship exists between muscle force and EMG.
Unfortunately this is not the case, because the relationship between muscle
force and activation (or EMG record) is strongly affected by muscle length
and, although less relevant in the context of posture assessment, muscle
contraction velocity. This limitation is especially important in trunk extensor
muscles, which operate over large length ranges and can produce substantial
moments even in the absence of activation (as discussed in Section 6.3.1.1.1).

Nevertheless, if careful calibration procedures are used, trunk muscle EMG measurements can provide an indication of moments acting about the lumbar spine (Van Dieën and Visser 1999, Kingma et al. 2001). Measures of muscle activation can also provide information on how muscles work together, or coordinate, to equilibrate gravitational and passive tissue moments. One important aspect of such coordination is antagonistic muscle activity (opposing the required moment), the presence of which will increase the development of fatigue and the forces acting on the spine.

6.3.1.3.1   Forward Inclination

In forward inclination, the gravitational moment is resisted mainly by the erector spinae muscles. Andersson et al. (1977a) measured erector spinae EMG in several flexed postures, while subjects had their pelvises strapped to a reference frame. In this situation bending angles and inclination angles (as defined above) were equal, but the results cannot be generalized quantitatively to forward inclination in freely adopted postures. Up to about 50° of forward inclination, erector spinae activation increases in both the lumbar and thoracic regions. EMG amplitudes as a function of forward inclination angle can be well described by a linear function of the sine of the angle of inclination. With a further increase in forward inclination, erector spinae muscle activation levels eventually decrease (Kippers and Parker 1984), because of the increasing passive tissue contribution to the moment required (the flexion–relaxation phenomenon, which is illustrated in Figure 6.8). Only minimal antagonistic co-contraction of abdominal muscles is present in unloaded forward inclination (De Looze et al. 1999, 2000).

6.3.1.3.2   Sideward Inclination

Limited data are available on muscle activation when maintaining sideward inclined postures. Lateral bending moments in upright postures, however, are produced mainly by activation of the contralateral latissimus dorsi and erector spinae muscles along with the (lateral parts of the) external oblique abdominal muscles. A small level of coactivation of the muscles on the ipsilateral side has been found (Lavender et al. 1992a,b, Van Dieën and Kingma 1999). These findings on muscle activity can probably be generalized to sideward inclined postures, although activity levels will be relatively low given the substantial passive moment contribution.

6.3.1.3.3   Twisting

Torén (2001) studied muscle activation in twisted postures. Although the normalization procedure used does not allow for quantitative interpretation, the data show that contralateral external oblique and ipsilateral erector spinae muscles likely play a primary role in counteracting the passive tissue moment. This is also consistent with data on twisting moments produced in a neutral posture (Pope et al. 1986, McGill 1991). It can be assumed that the role of the erector spinae muscle is mainly to counteract the bending and lateral bending moments caused by activity of the external oblique. In twisting efforts substantial co-contraction of all trunk muscles is found (Pope et al. 1986, McGill 1991).

6.3.1.4   Spinal Forces and Intra-Discal Pressure

Compression and shear forces acting on the spine cannot be measured directly. Researchers therefore rely on model-based estimates. An indication of compression forces can be obtained from measurements of intra-discal pressure. This is an invasive technique, which is not suitable for routine use. Deformations of the disc occurring in non-neutral postures will also affect the relationship between compression force and intra-discal pressure (Schultz et al. 1979). Model predictions of compression force have been reported to be well correlated with intra-discal pressure measures (Schultz et al. 1982). In addition, different models converge to similar predictions (Hughes et al. 1994, Van Dieën et al. 2000), and models predict directly measurable variables (e.g., moments) fairly well (Granata and Marras 1993, Cholewicki et al. 1995, Nussbaum and Chaffin 1998, Van Dieën et al. 2000). On the basis of these observations, model-based predictions of spine compression are considered sufficiently accurate for comparative use. Predictions of shear forces, in contrast, tend to be very different between models, perhaps not surprisingly given the strong dependency of the predictions on modeling assumptions (Nussbaum et al. 1995, Van Dieën and De Looze 1999). The following discussion is therefore limited to compression forces.

Compression force estimates could in theory be compared to data on the strength of spinal motion segments to derive threshold limit values for trunk posture. Compression strengths of human spinal motion segments range from about 2 to 10 kN (Hansson et al. 1980, Brinckmann et al. 1989). Given limited information on validity of compression predictions, and the fact that motion segment strength data are based on *in vitro* testing, this comparative approach seems unwarranted. Furthermore, the estimates made below show that spine compression resulting only from postural loads is not likely to exceed the compression strength of the spine.

Compression forces were estimated from the net moments in different postures presented earlier (Section 6.3.1.2), using an optimization model described by Van Dieën (Van Dieën 1997, Van Dieën and Kingma 1999, Van Dieën et al. 2000). Monotonic increases of compression with increasing postural deviations are seen in all planes of motion (Figure 6.13). Maximal compression forces are estimated in forward inclination, due to the large net moments occurring in that posture. Results from measurements of intra-discal pressure are available for forward inclination only, but are consistent with these model predictions (Andersson et al. 1977b, Sato et al. 1999, Wilke et al. 1999). From the graphs it is clear that only in forward inclination do compression forces reach levels exceeding 2 kN. However, the compression estimates are based on an average male. Compression strength of the vertebral column of the average male will be around 6.3 kN (Jäger 2001). This

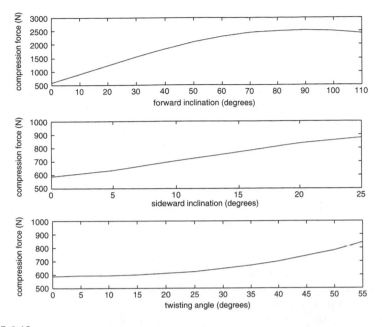

FIGURE 6.13
Estimated spine compression forces (N) in an average male as a function of trunk forward inclination (upper panel), sideward inclination (middle panel), and twisting (lower panel).

strength is lower in some other groups, especially older females, but in these groups compression forces will usually also be lower due to a lower body mass (although shorter muscle lever arms may offset this effect). In the working population in general, spinal compression strength is assumed to be above 3 kN (Waters et al. 1993, Van Dieën and Toussaint 1997, Jäger and Luttmann 1997). However, frequently repeated compression at initially submaximal levels may eventually cause fractures of the vertebral end plates (Hansson et al. 1987, Brinckmann et al. 1988). It is therefore conceivable that frequent forward inclination causes damage even in this population.

6.3.2   Trunk Postures and Discomfort

Subjective perceptions of discomfort resulting from the adoption or maintenance of a specific trunk posture can be measured using one of several existing scales. Among the more commonly used are the various numerical scales (Borg 1970, 1982, Corlett and Bishop 1976), the body mapping of Corlett and Bishop (1976), or a visual analogue scale (e.g., Ulin et al. 1990).

   Studies on the relationship between trunk postures and discomfort have mainly focused on the seated condition. A review of seating research is given in Chapter 7; in this section, the presentation centers on deviations from the upright posture. In contrast to the extensive work on seating, relatively few investigations have examined discomfort in the standing posture. Existing studies can be broadly classified by whether subjective responses were determined when non-neutral trunk postures were held for brief or for more prolonged periods

Discomfort resulting from short-term deviated trunk postures was assessed by Genaidy and Karwowski (1993) and Genaidy et al. (1995). Subjects flexed, extended, laterally bent, and rotated their trunks away from the neutral posture to half or the full range of motion (ROM), and held these positions for 30 or 60 s. Perceived levels of joint discomfort were rated on a 10-point scale (modified from the Corlett and Bishop, 1976, scale), where 0 = none, 5 = moderate, and 10 = extreme. Using a ranking system for assessment of postural deviations corresponding to these discomfort ratings, somewhat inconsistent results were found in terms of relative levels of discomfort resulting from the different postures. More specifically, the studies differed in terms of the rates of discomfort onset associated with each of the deviated postures.

In a more extensive study of this type, Kee and Karwowski (2001) had subjects adopt fixed percentages of their trunk ROM (0, 25, 50, 75, and 100%) that were held for 60 s. The discomfort experienced was quantified using a free modulus method, allowing subjects to freely choose their own range of numbers to reflect discomfort ranging from none to maximal. From these ratings, normalized values were obtained as a function of each individual's minimum and maximum discomfort rating. Increases in discomfort were roughly linear as a function of increasing postural deviation, although with larger increases toward the limit of ROM. Relative discomfort ratings from the different trunk postures were also inconsistent with the earlier work of Genaidy and colleagues noted above (Genaidy et al. 1995). The discrepancies in these studies suggest either that the onset of discomfort associated with deviated trunk postures may not be adequately assessed using short-term trials or that the results may be highly sensitive to the specific procedures and methods employed. While clearly demonstrating that discomfort occurs, the studies provide no clear indication regarding the relative effects of different types of trunk deviation.

Discomfort resulting from more prolonged maintenance of a posture has been investigated during several simulated work activities. Corlett and Manenica (1980) reported results from a study in which subjects performed a tapping task in several postures (with varied working heights and horizontal distances). Higher levels of initial discomfort were found in tasks requiring any forward inclination of the trunk, and in all tasks involving trunk bending a majority of subjects reported back pain at the limit of endurance. Subjects performed a manual tracking task in the study of Boussenna et al. (1982) with the task height at 25, 50, 75, and 100% of shoulder height to their limit of endurance. With increasing trunk inclination, subjects reported a higher level of overall discomfort, using the scales of Corlett and Bishop (1976), and higher final ratings of mid-back and low back discomfort. Similar results were found when subjects performed screw driving tasks for

1 to 2 h at different heights (Ulin et al. 1990); at lower locations, requiring more trunk inclination, higher levels of perceived exertion and discomfort were reported.

Miedema et al. (1997) summarized the results of seven earlier studies (including the first two noted in the previous paragraph), all of which examined static holding in a variety of postures without external loads. Maximal holding times (MHTs) were determined as the duration that the static postures could be held continuously, starting from a rested state. Changes in arm postures were confounded with changes in trunk postures, and thus it is very difficult to specify a clear MHT vs. posture relationship from their data. Nonetheless, postures requiring more forward trunk inclination were generally associated with lower MHTs.

Sudarsan and colleagues (Sudarsan et al. 2001a) have recently reported preliminary results on posture-induced discomfort during static and dynamic forward trunk inclinations and the effects of personal factors (age, gender, and back disability status) and task factors (inclination angle and work–rest cycle). Tasks were performed until subjects reached their maximum pain tolerance or to a limit of 600 s. Subjects with LBP reported generally higher levels of discomfort, particularly during the dynamic tasks. Reported discomfort was also higher for larger duty cycles (80% vs. 60% work), i.e., when the time spent in the inclined posture per cycle increased. Discomfort increased with increasing trunk inclination angle, although the relationship appeared nonlinear, with smaller changes seen at the higher angles.

Divergences in discomfort-based rankings of trunk motions reported suggest that such differences may be minimal or sensitive to specific testing conditions. Studies of longer-term exertions have been limited to forward trunk inclination, but indicate consistently that discomfort increases with increasing angles of postural deviation. As a whole, these studies suggest a monotonic relationship between postural deviation and discomfort. The specific form of the relationship is not clear, however, nor whether the relationship is consistent for different postural deviations (e.g., forward inclination vs. twisting). At present, the evidence does indicate that discomfort related to trunk posture can be minimized by reducing the extent of trunk deviations from the neutral posture.

6.3.3 Trunk Postures and Fatigue

Localized fatigue is a continuous and accumulative process that results from muscular contraction, and can be measured using a wide variety of both subjective (e.g., discomfort) and objective (e.g., strength, EMG) tools. It is important to differentiate the fatigue process, which is ongoing, from endurance, which is a terminal event in which muscle capacity fails to maintain task demands. Given the current complexity involved in objectively monitoring localized fatigue, and disagreement regarding the best methods to quantify fatigue, endurance has been the primary measure of muscular responses to postural deviations. Well-known relationships between endurance time and relative effort level (e.g., percentage of maximum voluntary exertion, or MVE) have been reported for static exertions (Rohmert 1973), while much less is known for intermittent and/or dynamic activities. Furthermore, the use of these relationships to predict fatigue status has been debated both in terms of the accuracy (Van Dieën and Oude Vrielink 1994) and validity (Mathiassen and Winkel 1992).

A small number of studies have determined the effects of different trunk postures on endurance time, which can be interpreted as an indirect indicator of localized fatigue. In Corlett and Manenica (1980), subjects performed a tapping task in several postures with varied working heights and horizontal distances. A tenfold decrease in endurance time was reported between tasks requiring an upright vs. a horizontal trunk. Endurance times decreased substantially with any forward inclination, although only minimal differences were found among several postures requiring 45° to 90° of forward inclination. Similar results were found by Boussenna et al. (1982) and Sudarsan et al. (2001b) regarding a nonlinear decrease in endurance with increasing static forward trunk inclination. Sudarsan et al. (2001b) also observed that endurance times in inclined postures were reduced in individuals with ongoing back pain.

Other studies, while not directly assessing endurance, have either determined the effects of posture on muscle activity or monitored signs of muscle fatigue more specifically. Dolan et al. (1988) showed that lumbar muscle activity was elevated, relative to relaxed standing, in several commonly adopted postures that all resulted in trunk bending (e.g., standing slumped, standing with one leg raised). It can be inferred that these non-neutral postures would also result in more rapid development of muscle fatigue. Asymmetric trunk exertions, particularly involving twisting, have been shown to result in decreased strength (McGill 1992, Van Dieën 1996), higher EMG activity (see Section 6.3.1.3), and more frequent EMG-based signs of muscle fatigue (Kim and Chung 1995, Van Dieën 1996, O'Brien and Potvin 1997).

These reported findings on fatigue and endurance can be adequately explained based on known mechanical and physiological relationships. As the trunk deviates from an upright, neutral posture, increasing muscular activity in the low back is required to equilibrate the increasing spine moments resulting from gravitational loads (as discussed in Section 6.3.1.2). As the moment arms for body segment masses respond as a sine function of the deviation angle, this accounts for the observed nonlinearities in endurance times summarized above. In non-neutral postures, changes in muscle moment arm and muscle length may affect the capacity for moment generation. For a given load, this yields a muscle contraction at another fraction of capacity (%MVE). Because endurance and exertion level (as %MVE) are inversely related, an additional factor may thus contribute to decreased endurance with trunk deviations. The correspondence between the results summarized here and those in the previous section suggests that discomfort related to trunk deviations may be caused in large part by the sequelae of localized fatigue.

In addition to gravitational loads, increasing postural deviations result in passive tissue deformation (of muscles, ligaments, and intervertebral discs, as mentioned in Section 6.3.1.1), which can either contribute to load equilibration (as in forward trunk inclination) or yield additional torques that must be equilibrated (as in trunk twisting). Even small deviations, with small associated muscle forces, may lead to fatigue, because localized blood flow, one factor responsible for fatigue development, is already impaired at very low levels of static muscle contraction. McGill et al. (2000) demonstrated that contractions as low as 2% of MVE resulted in compromised blood flow to the erector spinae muscle mass.

Although fairly limited, the existing evidence indicates that fatigue increases and endurance decreases in a monotonic fashion with increasing deviation from an upright trunk posture. The specific form of the relationship, particularly for deviations in different planes, requires further investigation.

6.3.4 Trunk Postures and Low Back Pain

In a large cross-sectional study on male construction workers, Holmström et al. (1992) studied the relationship between LBP and trunk flexion as assessed by questionnaires. Given the research methodology, flexion probably referred to bending and forward inclination. A significant dependence was seen between the prevalence of LBP and the duration for which flexed postures were adopted. An exposure response relationship between the time spent in stooped postures and LBP risk was also found. Such cross-sectional studies, however, must be interpreted with caution, because they cannot prove causality. The validity of questionnaire-based assessments of posture is also limited (as already noted in Section 6.1).

Two case-control studies on workers in automobile assembly reported high odds ratios for non-neutral trunk postures (Punnett et al. 1991, Norman et al. 1998). Both used video-based observation of postures. Norman et al. found an increased risk for LBP with increased average and peak forward inclination, but not with sideward inclination or twisting angles (with angles defined as in Section 6.1). Punnett et al. studied the time spent in mild flexion (>20°, <45°), severe flexion (>45°), lateral flexion (>20°), and twisting (>20°). Although it was not explicitly stated, it appears that the flexion and lateral flexion angles were defined with respect to the vertical thus referring to inclination, and twisting was defined with respect to the pelvis (as in Section 6.1). They found a significantly higher risk of LBP with increasing time spent in non-neutral trunk postures. In addition, an exposure–response relationship was found for the time spent in non-neutral postures, with an eightfold increase in the risk when the time increased from 0 to 100% of the work cycle.

Finally, two prospective cohort studies on trunk posture and LBP have been conducted (Riihimäki 1985, Hoogendoorn et al. 2000). Riihimäki studied the occurrence of sciatica (LBP with pain radiating to the legs) in a 3-year follow-up of a cohort consisting of over 2000 workers. Postures were assessed using questionnaires. Both twisting and flexion appeared to be associated with an increased risk of sciatica. Hoogendoorn et al. used video-based observations on a cohort of 861 workers, who were followed for 3 years. Postures were classified as neutral (<30° deviation), mild flexion (30° to 60°), extreme flexion (60° to 90°), very extreme flexion (>90°), and twisting (>30°). Again, the frame of reference for these angles was not explicitly stated, but it appears that definitions are such that flexion refers to forward inclination and twisting is defined consistent with the definition given in this chapter (Section 6.1). Low but significant increases in risk of sciatica were found with exposure to non-neutral postures. When flexed postures over 60° were adopted for more than 5% of the work time, the risk increased by 50%. When twisted postures were adopted for more than 10% of the work time, the risk increased by 30%. There were also some indications of an exposure–response relationship, with the risk increasing with longer exposure.

The studies reviewed all show that non-neutral trunk postures constitute a risk factor for LBP. This risk is also consistently shown to increase with increased duration of exposure. Results from the different studies, however, are divergent to a degree that precludes definitive quantitative conclusions. In addition, the data do not allow for sufficient differentiation between postural deviations in different planes or to different angular magnitudes.

6.4 General Evaluation Criteria

The epidemiological data reviewed clearly support the hypothesis that the risk of LBP is related to exposure to non-neutral postures. In addition, exposure–response relationships have been found. This would suggest that setting threshold limit values for exposure to non-neutral postures might help to prevent LBP. However, the large disparities between studies and the arbitrary cutoff points used do not allow this at present.

The short-term effects of trunk posture that have been reviewed (mechanical load, discomfort, and fatigue) might alternatively be used to derive threshold values. For example, assuming that compression-induced damage to the spine is a cause of LBP, the relationship between postural angles and spine compression combined with data on compression strength could be used to this end (Van Dieën et al. 1999). Unfortunately, the etiology of LBP is to a large extent unknown, and as a consequence the selection of short-term effect variables remains somewhat arbitrary. To address the ongoing need for threshold values for use by practitioners, ISO standard 11226 (ISO 2000) has presented data relevant to trunk postures for protection from health risks associated with prolonged (>4 s) static working postures. This standard recommends that asymmetrical trunk postures be avoided, as should trunk inclinations of >60° from vertical. Less deviated trunk postures are deemed acceptable when there is external support. Finally, maximum acceptable holding times are provided for postures between 20° and 60° where there is no external support. These specific recommendations are to be viewed as tentative, as they are mainly based on a limited number of experimental studies of discomfort and fatigue caused by static working postures. It is not clear how effective the recommendations will be in preventing discomfort and fatigue, and certainly not to what extent they will prevent musculoskeletal injury.

Considering the data provided above, it seems unlikely that postural load alone will cause immediate injury. Tissue strains, net moments, and compression forces appear well within the capacity of most individuals. Prolonged submaximal loading of tissues in sustained postures, however, will lead to creep effects (e.g., McGill and Brown 1992). In animal experiments it has been shown that these creep effects may cause muscular cramps, which could lead to pain. Furthermore, reflex control of back muscles appeared affected by creep of spinal ligaments, which might lead to an increased vulnerability to mechanical injury lasting for hours after the exposure (Solomonow et al. 1998).

Static postures may further affect nutrition of the intervertebral disc negatively, whereas movement may promote nutrition (Holm and Nachemson 1983, Van Deursen et al. 2001a,b). However, there is no strong evidence that static postures per se lead to LBP. Most of the data suggest only that non-neutral postures are related to LBP (Burdorf and Sorock 1997, Hoogendoorn et al. 1999, 2000). It may be that the impairment of disc nutrition is stronger when a relatively high intra-discal pressure is maintained for a long time, as would be the case in sustained exposure to non-neutral postures. Finally, non-neutral postures have been shown to require prolonged muscle activity at substantial levels, which does cause muscular fatigue when postures are sustained. Fatigue has been suggested to lead to back pain either directly, as a cause of myalgic pain (Jørgensen 1997), or indirectly, as a cause of mechanical overloading due to reduced motor control (Sparto et al. 1997, Van Dieën et al. 1998).

An alternative approach to setting threshold values would be to define these in terms of the level of a short-term effect variable. They would then have to be based on the relationships between short-term effect variables and LBP risk instead of the relationship between posture and LBP. This might have an advantage in that several important interacting exposure variables are reflected, at the level of such short-term effects. In addition, several short-term effects are important in their own right, such as fatigue and discomfort. When considering threshold values for posture, these would need to be defined in terms of postural angle in combination with duration, duty cycle, and frequency. The latter variables are only beginning to be addressed. In addition, interactions of postural deviations about three axes and interactions with external forces would need to be taken into account. Short-term effect variables might show the integrated effect of all of these exposure variables and their interactions. An encouraging step in this direction is the model developed by Norman et al. (1998), which relates LBP risk to a combination of integrated and peak forces acting on the spine.

On the basis of the present review, we conclude that the state of knowledge of the relationship between posture and short-term effects and between short-term effects and LBP precludes determination of threshold limit values. All the data suggest, however, that negative effects of posture increase more or less monotonically with postural deviation although mostly in a nonlinear way. Finally, the data and methods regarding short-term effect variables provide a means for comparative evaluation of postures in practice and may be used to develop tentative guidelines.

Summary

Adverse trunk postures cause fatigue and discomfort and can cause LBP. The chapter reviews the measurement of trunk posture and its effects on physical loading and health. Trunk postures need to be defined in terms of the orientation of the trunk with respect to the gravitational field (forward or sideward inclined) and in terms of the shape of the trunk (forward bend, sideward bend, or twisted). Measuring trunk postures in these five dimensions is a challenging task. Visual observation can yield valid results only when a fairly coarse categorization of postures is used and when time-sampled observations are made, due to limitations on the information processing capacity of the observer. Interpretation problems often arise when trunk postures involve rotations of the trunk about more than one axis. Objective measurement can provide adequate data at high sampling frequencies, but these techniques are cumbersome and may interfere with task performance. Recent technological developments will make objective measurement of trunk posture in the field possible.

The effects of trunk postures on the musculoskeletal system depend on interactions of the five dimensions in which posture is defined and on inter-actions with any forces exerted on the environment, for example, when holding an object. In general, non-neutral trunk postures cause more mechanical load, more muscle fatigue, more discomfort, and more health risk. Short-term effects increase monotonously when looking at any one dimension at a time. This suggests that the same may hold for health risks.

The state of knowledge precludes determination of threshold limit values. However, data regarding the short-term effects of adopting certain postures can provide a means for comparative evaluation of postures in practice and may be used to develop tentative guidelines.

References

Adams, M.A. and Dolan, P., 1991. A technique for quantifying the bending moment acting on the lumbar spine *in vivo*. *Journal of Biomechanics* 24, 117–126.

Adams, M.A. and Dolan, P., 1996. Time-dependent changes in the lumbar spine's resistance to bending. *Clinical Biomechanics* 11, 194–200.

Adams, M.A. and Dolan, P., 1997. Could sudden increases in physical activity cause degeneration of intervertebral discs? *Lancet* 350, 734–735.

Adams, M.A. and Hutton, W.C., 1981. The relevance of torsion to the mechanical derangement of the lumbar spine. *Spine* 6, 241–248.

Adams, M.A. and Hutton, W.C., 1982. Prolapsed intervertebral disc: a hyperflexion injury. *Spine* 7, 184–191.

Adams, M.A. and Hutton, W.C., 1985. Gradual disc prolapse. *Spine* 10, 524–531.

Adams, M.A. and Hutton, W.C., 1986. Has the lumbar spine a margin of safety in forward bending? *Clinical Biomechanics* 1, 3–6.

Adams, M.A., Hutton, W.C., and Stott, J.R.R., 1980. The resistance to flexion of the lumbar intervertebral joint. *Spine* 5, 245–253.

Adams, M.A., Green, T.P., and Dolan, P., 1994. The strength in anterior bending of lumbar intervertebral discs. *Spine* 19, 2197–2203.

Andersson, G.B.J., Örtengren, R., and Herberts, P., 1977a. Quantitative electromyographic studies of back muscle activity related to posture and loading. *Orthopedic Clinics of North America* 8, 85–95.

Andersson, G.B.J., Örtengren, R., and Nachemson, A., 1977b. Intradiscal pressure, intra-abdominal pressure and myoelectric back muscle activity related to posture and loading. *Clinical Orthopaedics and Related Research* 129, 156–164.

Baten, C.T.M., Oosterhoff, P., Luinge, H., Veltink, P.H., Van Dieën, J.H., Dolan, P., and Hermens, H.J., 1997. Quantitative assessment of mechanical low back load in the field: validation in asymmetric lifting, in *Proceedings of the 13th Triennial Congress of the International Ergonomics Association*, Seppälä, P., Luoparjärvi, T., Nygård, C.-H., and Mattila, M., Eds., Vol. 4, Helsinki: Finnish Institute of Occupational Health, 488–490.

Bodén, A. and Öberg, K., 1998. Torque resistance of the passive tissues of the trunk at axial rotation. *Applied Ergonomics* 29, 111–118.

Borg, G., 1970. Perceived exertion as an indicator of somatic stress. *Scandinavian Journal of Rehabilitation Medicine* 2, 92–98.

Borg, G.A.V., 1982. Psychophysical bases of perceived exertion. *Medicine and Science in Sports and Exercise* 14, 377–381.

Boussenna, M., Corlett, E.N., and Pheasant, S.T., 1982. The relation between discomfort and postural loading at the joints. *Ergonomics* 25, 315–322.

Brinckmann, P., Biggeman, M., and Hilweg, D., 1988. Fatigue fracture of human lumbar vertebrae. *Clinical Biomechanics* 3, 1–27.

Brinckmann, P., Biggeman, M., and Hilweg, D., 1989. Prediction of the compressive strength of human lumbar vertebrae. *Clinical Biomechanics* 4, 1–27.

Burdorf, A. and Laan, J., 1991. Comparison of three methods for the assessment of postural load on the back. *Scandinavian Journal of Work, Environment & Health* 17, 425–429.

Burdorf, A. and Sorock, G., 1997. Positive and negative evidence of risk factors for back disorders. *Scandinavian Journal of Work, Environment & Health* 23, 243–256.

Burdorf, A., Derksen, J., Naaktgeboren, B., and Van Riel, M., 1992. Measurement of trunk bending during work by direct observation and continuous measurement. *Applied Ergonomics* 23, 263–267.

Chaffin, D.B. and Andersson, G.B.J., 1991. *Occupational Biomechanics*, 2nd ed., New York: Wiley.

Chaffin, D.B. and Erig, M., 1991. Three-dimensional biomechanical static strength prediction model sensitivity to postural and anthropometric inaccuracies. *IIE Transactions* 23, 215–227.

Cholewicki, J. and McGill, S.M., 1992. Lumbar posterior ligament involvement during extremely heavy lifts estimated from fluoroscopic measurements. *Journal of Biomechanics* 25, 17–28.

Cholewicki, J., McGill, S.M., and Norman, R.W., 1995. Comparison of muscle forces and joint load from an optimization and EMG assisted lumbar spine model: towards development of a hybrid approach. *Journal of Biomechanics* 28, 321–331.

Corlett, E.N. and Bishop, R.P., 1976. A technique for assessing postural discomfort. *Ergonomics* 19, 175–182.

Corlett, E.N., and Manenica, I., 1980. The effect and measurement of working postures. *Applied Ergonomics* 11, 7–16.

De Looze, M.P., Toussaint, H.M., Ensink, J., Mangnus, C., and Van der Beek, A., 1994. The validity of visual observations to register postural aspects of a manual materials handling job. *Ergonomics* 37, 1335–1343.

De Looze, M.P., Groen, H., Horemans, H., Kingma, I., and Van Dieën, J.H., 1999. Abdominal muscles contribute in a minor way to peak spinal compression in lifting. *Journal of Biomechanics* 32, 655–662.

De Looze, M.P., Boeken-Kruger, M.C., Steenhuizen, S., Baten, C.T.M., Kingma, I., and Van Dieën, J.H., 2000. Trunk muscle activation and low back loading in lifting in the absence of load knowledge. *Ergonomics* 43, 333–344.

Dempsey, P.G. and Fathallah, F.A., 1999. Application issues and theoretical concerns regarding the 1991 NIOSH equation asymmetry multiplier. *International Journal of Industrial Ergonomics* 23, 181–191.

Dolan, P. and Adams, M.A., 1993. Influence of lumbar and hip mobility on the bending stresses acting on the lumbar spine. *Clinical Biomechanics* 8, 185–192.

Dolan, P., Adams, M.A., and Hutton, W.C., 1988. Commonly adopted postures and their effect on the lumbar spine. *Spine* 13, 197–201.

Farfan, H.F., Cossette, J.W., Robertson, G.H., Wells, R.V., and Kraus, H., 1970. The effect of torsion on the lumbar intervertebral joints: the role of torsion in the production of disc degeneration. *Journal of Bone and Joint Surgery* 52A, 468–497.

Gagnon, M., Larrivé, A., and Desjardins, P., 2000. Strategies of load tilts and shoulder positioning in asymmetrical lifting. A concomitant evaluation of the reference systems of axes. *Clinical Biomechanics* 15, 478–488.

Garg, A. and Chaffin, D.B., 1975. A biomechanical computerized simulation of human strength. *A.I.I.E. Transactions* 7, 1–15.

Genaidy, A.M. and Karwowski, W., 1993. The effects of neutral posture deviations on perceived joint discomfort ratings in sitting and standing postures. *Ergonomics* 36, 785–792.

Genaidy, A., Barkawi, H., and Christensen, D., 1995. Ranking of static non-neutral postures around the joints of the upper extremity and the spine. *Ergonomics* 38, 1851–1858.

Granata, K.P. and Marras, W.S., 1993. An EMG assisted model of loads on the lumbar spine during asymmetric trunk extensions. *Journal of Biomechanics* 26, 1429–1438.

Gravel, D., Gagnon, M., Plamondon, A., and Desjardins, P., 1997. Development and application of predictive equations of maximal static moments generated by the trunk musculature. *Clinical Biomechanics* 12, 314–324.

Green, T.P., Adams, M.A., and Dolan, P., 1993. Tensile properties of the annulus fibrosus. II. Ultimate tensile strength and fatigue life. *European Spine Journal* 2, 209–214.

Haher, T.R., Fehny, W., Baruch, H., Devlin, V., Welin, D., O'Brien, M., Ahmad, J., Valenza, J., and Parish, S., 1989. The contribution of the three columns of the spine to rotational stability: a biomechanical model. *Spine* 14, 663–673.

Hales, T.R. and Bernard, B.P., 1996. Epidemiology of work-related musculoskeletal disorders. *Orthopedic Clinics of North America* 27, 679–709.

Hansson, T.H., Roos, B., and Nachemson, A., 1980. The bone mineral content and ultimate compressive strength in lumbar vertebrae. *Spine* 5, 46–55.

Hansson, T.H., Keller., T.S., and Spengler, D.M., 1987. Mechanical behaviour of the human lumbar spine II. Fatigue strength during dynamic compressive loading. *Journal of Orthopaedic Research* 5, 479–487.

Harms-Ringdahl, K., Brodin, H., Eklund, L., and Borg, G., 1983. Discomfort and pain from loaded passive joint structures. *Scandinavian Journal of Rehabilitation Medicine* 15, 205–211.

Holm, S. and Nachemson, A., 1983. Variations in the nutrition of canine intervertebral discs induced by motion. *Spine* 8, 866–874.

Holmström, E.B., Lindell, J., and Moritz, U., 1992. Low back and neck/shoulder pain in construction workers: occupational workload and psychosocial risk factors. Part 1: Relationship to low back pain. *Spine* 17, 663–671.

Hoogendoorn, W.E., Van Poppel, M.N.M., Bongers, P.M., Koes, B.W., and Bouter, L.M., 1999. Physical load during work and leisure time as risk factors for back pain. *Scandinavian Journal of Work, Environment & Health* 25, 387–403.

Hoogendoorn, W.E., Bongers, P.M., De Vet, H.C., Douwes, M., Koes, B.W., Miedema, M.C., Ariëns, G.A., and Bouter, L.M., 2000. Flexion and rotation of the trunk and lifting at work are risk factors for low back pain: results of a prospective cohort study. *Spine* 25, 3087–3092.

Hughes, R.E., Chaffin, D.B., Lavender, S.A., and Andersson, G.B.J., 1994. Evaluation of muscle force prediction models of the lumbar trunk using surface electromyography. *Journal of Orthopaedic Research* 12, 689–698.

ISO, 2000, *ISO 11226 Ergonomics — Evaluation of Static Working Postures*, Geneva: International Organization for Standardization.

Jäger, M., 2001. Belastung und Belastbarkeit der Lendenwirbelsäule im Berufsalltag, Dortmund, VDI-Verlag.

Jäger, M. and Luttmann, A., 1997. Assessment of low-back load during manual materials handling, in Seppälä, P., Luoparjärvi, T., Nygård, C.-H., and Mattila, M., Eds., *Proceedings of the 13th Triennial Congress of the International Ergonomics Association*. Vol. 4, Helsinki: Finnish Institute of Occupational Health, 171–173.

Jørgensen, K., 1997. Human trunk extensor muscles. Physiology and ergonomics. *Acta Orthopaedica Scandinavica* 160, Suppl. 637, 1–58.

Kee, D. and Karwowski, W., 2001. The boundaries for joint angles of isocomfort for sitting and standing males based on perceived comfort of static joint postures. *Ergonomics* 44, 614–648.

Kim, S.H., and Chung, M.K., 1995. Effects of posture, weight and frequency on trunk muscular-activity and fatigue during repetitive lifting tasks. *Ergonomics* 38, 853–863.

Kingma, I., Van Dieën, J.H., De Looze, M.P., Toussaint, H.M., Dolan, P., and Baten, C.T.M., 1999. On the use of axis systems in quantification of lumbar loading during asymmetric lifting. *Journal of Biomechanics* 32, 637–638.

Kingma, I., Baten, C.T.M., Dolan, P., Adams, M.A., Toussaint, H.M., Van Dieën, J.H., and De Looze, M.P., 2001. Lumbar loading during lifting: a comparative study of three measurement techniques. *Journal of Electromyography and Kinesiology* 11, 337–345.

Kippers, V. and Parker, A. W., 1984. Posture related to myoelectric silence of erectores spinae during trunk flexion. *Spine* 9, 740–745.

Krismer, M., Haid, C., Behensky, H., Kapfinger, P., Landauer, F., and Rachbauer, F., 2000. Motion in lumbar functional spine units during side bending and axial rotation moments depending on the degree of degeneration. *Spine* 25, 2020–2027.

Kuiper, J., Burdorf, A., Verbeek, J.H.A.M., Frings-Dresen, M.H.W., Van der Beek, A.J., and Viikari-Juntura, E.R.A., 1999. Epidemiologic evidence on manual materials handling as a risk factor for back disorders: a systematic review. *International Journal of Industrial Ergonomics* 24, 389–404.

Kumar, S., 1996. Isolated planar trunk strengths measurement in normals. 3. Results and database. *International Journal of Industrial Ergonomics* 17, 103–111.

Lavender, S.A., Tsuang, Y.H., Andersson, G.B.J., Hafezi, A., and Shin, C.C., 1992a. Trunk muscle co-contraction: the effects of moment direction and moment magnitude. *Journal of Orthopaedic Research* 10, 691–700.

Lavender, S.A., Tsuang, Y.H., Hafezi, A., Andersson, G.B.J., Chaffin, D.B., and Hughes, R.E., 1992b. Coactivation of the trunk muscles during asymmetric loading of the torso. *Human Factors* 34, 239–247.

Li, G.Y. and Buckle, P., 1999. Current techniques for assessing physical exposure to work-related musculoskeletal risks, with emphasis on posture-based methods. *Ergonomics* 42, 674–695.

Linton, S.J. and Van Tulder, M.W., 2001. Preventive interventions for back and neck pain problems: what is the evidence? *Spine* 26, 778–787.

Liu, Y.K., Goel, V.K., Dejong, A., Njus, G., Nishiyama, K., and Buckwalter, J., 1985. Torsional fatigue of the lumbar intervertebral joints. *Spine* 10, 894–900.

Macintosh, J.E., Bogduk, N., and Pearcy, M.J., 1993. The effects of flexion on the geometry and actions of the lumbar erector spinae. *Spine* 18, 884–893.

Marras, W.S., Fathallah, F.A., and Miller, R.J., 1992. Accuracy of a three-dimensional lumbar motion monitor for recording dynamic trunk motion characteristics. *International Journal of Industrial Ergonomics* 9, 75–85.

Marras, W.S., Davis, K.G., and Granata, K.P., 1998. Trunk muscle activities during asymmetric twisting motions. *Journal of Electromyography and Kinesiology* 8, 247–256.

Mathiassen, S.E. and Winkel, J., 1992. Can occupational guidelines for work-rest schedules be based on endurance time data? *Ergonomics* 35, 253–259.

McGill, S.M., 1991. Electromyographic activity of the abdominal and low back musculature during the generation of isometric and dynamic axial trunk torque: implications for lumbar mechanics. *Journal of Orthopaedic Research* 9, 91–103.

McGill, S.M., 1992. The influence of lordosis on axial trunk torque and trunk muscle myoelectric activity. *Spine* 17, 1187–1193.

McGill, S.M. and Brown, S., 1992. Creep response of the lumbar spine to prolonged full flexion. *Clinical Biomechanics* 7, 43–46.

McGill, S.M. and Hoodless, K., 1990. Measured and modelled static and dynamic axial trunk torsion during twisting in males and females. *Journal of Biomedical Engineering* 12, 403–409.

McGill, S.M. and Kippers, V., 1994. Transfer of loads between lumbar tissues during the flexion-relaxation phenomenon. *Spine* 19, 2190–2196.

McGill, S.M., Seguin, J., and Bennett, G., 1994. Passive stiffness of the lumbar torso in flexion, extension, lateral bending, and axial rotation. *Spine* 19, 696–704.

McGill, S.M., Cholewicki, J., and Peach, J.P., 1997. Methodological considerations for using inductive sensors (3space isotrak) to monitor 3-D orthopaedic joint motion. *Clinical Biomechanics* 12, 190–194.

McGill, S.M., Yingling, V.R., and Peach, J.P., 1999. Three-dimensional kinematics and trunk muscle myoelectric activity in the elderly spine — a database compared to young people. *Clinical Biomechanics* 14, 389–395.

McGill, S.M., Hughson, R.L., and Parks, K., 2000. Lumbar erector spinae oxygenation during prolonged contractions: implications for prolonged work. *Ergonomics* 43, 486–493.

Miedema, M.C., Douwes, M., and Dul, J., 1997. Recommended maximum holding times for prevention of discomfort of static standing postures. *International Journal of Industrial Ergonomics* 19, 9–18.

Norman, R., Wells, R., Neumann, P., Frank, J., Shannon, H., and Kerr, M., 1998. A comparison of peak vs cumulative physical work exposure risk factors for the reporting of low back pain in the automotive industry. *Clinical Biomechanics* 13, 561–573.

Nussbaum, M.A. and Chaffin, D.B., 1998. Lumbar muscle force estimation using a subject-invariant 5-parameter EMG-based model. *Journal of Biomechanics* 31, 667–672.

Nussbaum, M.A., Chaffin, D.B., and Rechtien, C.J., 1995. Muscle lines-of-action affect predicted forces in optimization-based spine muscle modeling. *Journal of Biomechanics* 28, 401–409.

O'Brien, P.R. and Potvin, J.R., 1997. Fatigue-related EMG responses of trunk muscles to a prolonged, isometric twist exertion. *Clinical Biomechanics* 12, 306–313.

Panjabi, M.M., Goel., V.K., and Takata, K., 1982. Physiologic strains on the lumbar spinal ligaments. *Spine* 7, 192–203.

Paquet, V.L., Punnett, L., and Buchholz, B., 2001. Validity of fixed-interval observations for postural assessment in construction work. *Applied Ergonomics* 32, 215–224.

Peach, J.P., Sutarno, C.G., and McGill, S.M., 1998. Three-dimensional kinematics and trunk muscle myoelectric activity in the young lumbar spine: a database. *Archives of Physical Medicine and Rehabilitation* 79, 663–669.

Pearcy, M., Portek, I., and Shepherd, J., 1984. Three-dimensional X-ray analysis of normal movement in the lumbar spine. *Spine* 9, 294–297.

Plamondon, A., Gagnon, M., and Gravel, D., 1999. Comments on "Asymmetric low back loading in asymmetric lifting movements is not prevented by pelvic twist." *Journal of Biomechanics* 32, 635.

Pope, M.H., Andersson, G.B.J., Broman, H., Svensson, M., and Zetterberg, C., 1986. Electromyographic studies of the lumbar trunk musculature during the development of axial torques. *Journal of Orthopaedic Research* 4, 288–297.

Punnett, L. and Keyserling, W.M., 1987. Exposure to ergonomic stressors in the garment industry: application and critique of job-site work analysis methods. *Ergonomics* 30, 1099–1116.

Punnett, L., Fine, L.J., Keyserling, W.M., Herrin, G.D., and Chaffin, D.B., 1991. Back disorders and nonneutral trunk postures of automobile assembly workers. *Scandinavian Journal of Work, Environment & Health* 17, 337–346.

Riihimäki, H., 1985. Back pain and heavy physical work: a comparative study of concrete reinforcement workers and maintenance house painters. *British Journal of Industrial Medicine* 42, 226–232.

Rohmert, W., 1973. Problems in determining rest allowances. Part 1. Use of modern methods to evaluate stress and strain in static muscular work. *Ergonomics* 4, 91–95.

Sato, K., Kikuchi, S., and Yonezawa, T., 1999. *In vivo* intradiscal pressure measurement in healthy individuals and in patients with ongoing back problems. *Spine* 24, 2468–2474.

Schultz, A., Andersson, G., Örtengren, R., Haderspeck, K., and Nachemson, A., 1982. Loads on the lumbar spine. Validation of a biomechanical analysis by measurements of intradiscal pressure and myoelectric signals. *Journal of Bone and Joint Surgery* 64A, 713–720.

Schultz, A.B., Warwick, D.N., Berkson, M.H., and Nachemson, A.L., 1979. Mechanical properties of human lumbar spine motion segments, Part I. Responses in flexion, extension, lateral bending and torsion. *Journal of Biomechanical Engineering* 101, 46–52.

Snijders, C.J. and Van Riel, M.P.J.M., 1987. Continuous measurements of spine movements in normal working situations over periods of 8 hours or more. *Ergonomics* 30, 639–653.

Solomonow, M., Zhou, B.H., Harris, M., Lu, Y., and Baratta, R.V., 1998. The ligamento-muscular stabilizing system of the spine. *Spine* 23, 2552–2562.

Solomonow, M., Eversull, E., Zhou, B.H., Baratta, R.V., and Zhu, M.P., 2001. Neuromuscular neutral zones associated with viscoelastic hysteresis during cyclic lumbar flexion. *Spine* 26, E314–E324.

Sparto, P.J., Parnianpour, M., Reinsel, T.E., and Simon, S., 1997. The effect of fatigue on multijoint kinematics, coordination, and postural stability during a repetitive lifting test. *Journal of Orthopaedic and Sports Physical Therapy* 25, 3–12.

Sudarsan, S.P., Keyserling, W.M., Martin, B.J., and Haig, A.J., 2001a. Assessment of posture-induced discomfort during static and dynamic trunk flexion, in *Proceedings of the Rehabilitation Society of North America (RESNA) Annual Meeting*, Reno, NV, 233–235.

Sudarsan, S.P., Keyserling, W.M., Martin, B.J., and Haig, A.J., 2001b. Effects of low back disability status on postural endurance time during static trunk postures, in *Proceedings of the Rehabilitation Society of North America (RESNA) Annual Meeting*, Reno, NV, 230–232.

Torén, A., 2001. Muscle activity and range of motion during active trunk rotation in a sitting posture. *Applied Ergonomics* 32, 583–591.

Toussaint, H.M., De Winter, A.F., De Haas, Y., De Looze, M.P., Van Dieën, J.H., and Kingma, I., 1995. Flexion relaxation during lifting: implications for torque production by muscle activity and tissue strain at the lumbo-sacral joint. *Journal of Biomechanics* 28, 199–210.

Twomey, L.T. and Taylor, J.F., 1982. Flexion creep deformation and hysteresis in the lumbar vertebral column. *Spine* 7, 116–122.

Ulin, S.S., Ways, C.M., Armstrong, T.J., and Snook, S.H., 1990. Perceived exertion and discomfort versus work height with a pistol-shaped screwdriver. *American Industrial Hygiene Association Journal* 21, 588–594.

Van der Beek, A.J. and Frings-Dresen, M.H.W., 1998. Assessment of mechanical exposure in ergonomic epidemiology. *Occupational and Environmental Medicine* 55, 291–299.

Van der Beek, A.J., Braam, I.T.J., Douwes, M., Bongers, P.M., Frings-Dresen, M.H.W., Verbeek, J.H.A.M., and Luyts, S., 1994. Validity of a diary estimating exposure to tasks, activities, and postures of the trunk. *International Archives of Occupational Environment and Health* 66, 173–178.

Van der Hoogen, J.M.M., Koes, B.W., Devillé, W., Van Eijk, J.T.M., and Bouter, L.M., 1997. The prognosis of low back pain in general practice. *Spine* 22, 1515–1521.

Van Deursen, D.L., Snijders, C.J., Van Dieën, J.H., Kingma, I., and Van Deursen, L.L.J.M., 2001a. Passive vertebral rotation causes instantaneous depressurization of the nucleus pulposus. *Journal of Biomechanics* 34, 405–408.

Van Deursen, D.L., Snijders, C.J., Kingma, I., and Van Dieën, J.H., 2001b. Torsion induced changes in stress distribution in porcine intervertebral discs *in vitro*. *Spine* 26, 2582–2586.

Van Dieën, J.H., 1996. Asymmetry of erector spinae muscle-activity in twisted postures and consistency of muscle activation patterns across subjects. *Spine* 21, 2651–2661.

Van Dieën, J.H., 1997. Are recruitment patterns of the trunk musculature compatible with a synergy based on the maximization of endurance? *Journal of Biomechanics* 30, 1095–1100.

Van Dieën, J.H. and De Looze, M.P., 1999. Sensitivity of single-equivalent trunk extensor muscle models to anatomical and functional assumptions. *Journal of Biomechanics* 32, 195–198.

Van Dieën, J.H. and Kingma, I., 1999. Total trunk muscle force and spinal compression are lower in asymmetric moments as compared to pure extension moments. *Journal of Biomechanics* 32, 655–662.

Van Dieën, J.H. and Kingma, I., 2001. Reporting net moments about the lumbar spine. *Clinical Biomechanics* 16, 348–349.

Van Dieën, J.H. and Oude Vrielink, H.H.E., 1994. The use of the relation between relative force and endurance time. *Ergonomics* 37, 231–243.

Van Dieën, J.H. and Toussaint, H.M., 1997. Evaluation of the probability of spinal damage caused by sustained cyclic compression loading. *Human Factors* 39, 469–480.

Van Dieën, J.H. and Visser, B., 1999. Estimating net lumbar moments from EMG data. The validity of calibration procedures. *Journal of Electromyography and Kinesiology* 9, 309–315.

Van Dieën, J.H., Van der Burg, P., Raaijmakers, T.A.J., and Toussaint, H.M., 1998. Effects of repetitive lifting on the kinematics, inadequate anticipatory control or adaptive changes? *Journal of Motor Behavior* 30, 20–32.

Van Dieën, J.H., Weinans, H., and Toussaint, H.M., 1999. Fractures of the lumbar vertebral endplate in the etiology of low back pain. A hypothesis on the causative role of spinal compression in a specific low back pain. *Medical Hypotheses* 53, 246–252.

Van Dieën, J.H., Hoozemans, M.J.M., Van der Burg, J.C.E., Jansen, J.P., Kingma, I., and Kuijer, P.P.F.M., 2000. The importance of antagonistic co-contraction of trunk muscles for spinal loads during lifting and pulling tasks: implications for modeling approaches, in *Proceedings of the IEA 2000/HFES 2000 Congress*, Santa Monica, CA: Human Factors and Ergonomics Society, 617–620.

Waters, T.R., Putz-Anderson, V., Garg, A., and Fine, L.J., 1993. Revised NIOSH equation for the design and evaluation of manual lifting tasks. *Ergonomics* 36, 749–776.

Wilke, H.J., Neef, P., Caimi, M., Hoogland, T., and Claes, L.E., 1999. New *in vivo* measurements of pressures in the intervertebral disc in daily life. *Spine* 24, 755–762.

7

Pelvis

CONTENTS

Introduction

This chapter contains three sections on the pelvis and its neighboring musculoskeletal segments, the spine and the lower extremities. The first section discusses sitting behavior and factors influencing this and introduces methods of assessing sitting posture, including observational tools and posture classification techniques. The second section reports research concerning various aspects of seating concepts and arrangements, also addressing new seating concepts such as the knee support chair, the sit-stand workplace, and dynamic sitting. The third section presents an overview of the general anatomy of the pelvis and its linkages to adjoining segments, discusses experimental data on the relationship of spinal, pelvic, and lower extremity postures, and provides a summary of the ergonomic implications of this knowledge.

7.1   Sitting Behavior

Antonio Grieco, Giovanni Molteni, and Giovanni De Vito

After this book was compiled, we were saddened to hear of the death of one of our authors, Professor Antonio Grieco. His major contribution to developing occupational health research is recognized internationally and he will be greatly missed by colleagues worldwide.—*The editors*

Introduction

Sitting behavior can be defined as the postures and positions of body segments held by a subject when sitting, that is the person's postural range and movement frequencies. Of the many parameters that could influence sitting behavior, the perception of comfort, due to task characteristics, workstation layout, anthropometric dimensions, and training, seems to be the most important determinant of the behavior of the person doing the sitting.

The general use of a chair as a working tool is relatively recent and not even common in some countries of the world; in fact, many ancient societies had no knowledge of seats of any kind. Today millions of people in non-Western countries work and rest in postures such as deep squatting, cross-legged sitting, long sitting, and sitting on the heels with the knees resting on the floor (Bridger 1991b). Seats originated, at least in part, as status symbols since only the chief had the right to be raised by a seat — hence the gradual development of ceremonial stools, which indicated status by their size and decoration. This status function has persisted to the present day as in many companies there is a different type of chair corresponding to each salary level.

At the beginning of the 20th century it was gradually realized that well-being and efficiency could be improved and fatigue reduced if people could sit at their work. The reason is physiological: standing requires considerable static muscular effort to maintain a fixed position, but this is greatly reduced when sitting. Today more than 70% of workers in developed countries have sedentary jobs (Reinecke et al. 1992) and concerns about musculoskeletal problems have led to the increasing application of ergonomics in changing the workplaces to provide good sitting postures.

In general the attention of ergonomics research has been focused on defining "the most correct" sitting position, even though the most important problem at the workplace is not the posture in itself but for how long it is maintained (Laville 1980). It can be observed that workplace design should concentrate on the need for operators to vary their sitting behavior according to job requirements rather than on finding a single, so-called optimal position (Cantoni et al. 1984, Grieco 1986).

7.1.1 Measuring Sitting Behavior

Postural analysis and its evaluation rely on data from laboratory studies, which have used the techniques of intervertebral disc pressure measurement, electromyography, radiography, observation and recording of positions of body segments, and subjective assessment. Observational tools such as the posture classification techniques that are available are often not ideal for measuring seated activities (Graf et al. 1995) because they were designed as multipurpose analysis systems, basically developed for assessing heavy, manual work. Branton and Grayson (1967) designed a tool for measuring sitting behavior by means of observations guided by the following considerations. Because in most other respects human behavior tends to avoid unpleasant and obviously harmful situations, it may be assumed that the sitter is not actively seeking discomfort. Left to themselves, sitters will neither deliberately nor unconsciously take up uncomfortable postures but seek to minimize discomfort as far as the situation permits. If, then, a sufficiently

Name	Surname		Age	Weight	Height		
Trunk				**Trunk**		**Legs**	**Time**

Trunk			Trunk		Trunk	Legs	Time
	[1] TRUNK IN FRONT OF THE SWITCH BOARD		[1] OCCUPYING ALL THE SEAT	[1] UPRIGHT	[1] SUPPORTED	[1] AT 90°	Min. ...
					[2] UNSUPPORTED	[2] CROSSED	Sec. ...
[1] UNTWISTED	TRUNK AT SIDE OF SWITCH BOARD	[1] R [3] L			[3] LAT.BENT R	[3] FEET UNDER THE SEAT	
					[4] LAT.BENT L		
[1] TWISTED	LEGS AT SIDE AND TRUNK IN FRONT OF THE SWITH BOARD	[4] R [5] L	[1] OCCUPYING A SMALL PART OF THE SEAT	[5] KYPHOSIS	[6] LAT.BENT R	[4] EXTENDED KNEES	
					[7] LAT.BENT L		
	LEGS IN FRONT AND TRUNK AT SIDE OF THE SWITCH BOARD	[6] R [7] L		[8] BENT	[9] LAT.BENT R	[5] STRADDLED	
					[10] LAT.BENT L		
					[10] LAT.BENT L		
			11 SLIDING POSITION				

R = Right
L = Left

FIGURE 7.1

Cantoni et al.'s (1984) observational tool. (From Cantoni, S. et al., in *Ergonomics and Health in Modern Offices*, Grandjean, E., Ed., London: Taylor & Francis, 1984, 455–464. With permission from Taylor & Francis (http://www.tandf.co.uk).)

large number of people are observed in a particular seat, a naturalistic range of postures will be found by which the sitters achieve minimal discomfort in that seat. Branton and Grayson's method describes each posture as represented by a set of four figures. The first refers to the position of the head (free of support, against headrest, against side-wing, supported by hands), the second to the trunk (free from backrest, against backrest, lounging–slumped back), the third to the arms (free from armrest, on armrest), and the last to the legs (free — both feet on floor, crossed at knee, crossed at ankle, stretched forward). Branton and Grayson also suggested video recording and observing the sitting behavior with freeze frames at 160:1 real time in order to bring out the dynamics of the sitting situation.

Cantoni et al. (1984) proposed an observational tool (shown in Figure 7.1) that evaluated the position of the trunk in the horizontal, sagittal, and frontal planes, the way of sitting on the seat, and the positions of the legs. The authors emphasized the need to measure postural fixity by recording every change of posture during the observation time, the time spent in different situations of lumbar load, and the frequency of posture change.

Some years later, Graf et al. (1995) designed the posture classification system shown in Figure 7.2 specifically for seated activity, taking into account the need to measure postural fixity as suggested by Cantoni et al. (1984). The classification system was based on a matrix containing 34 positions when shoulders were parallel to the hip and 34 when the body was twisted. Using

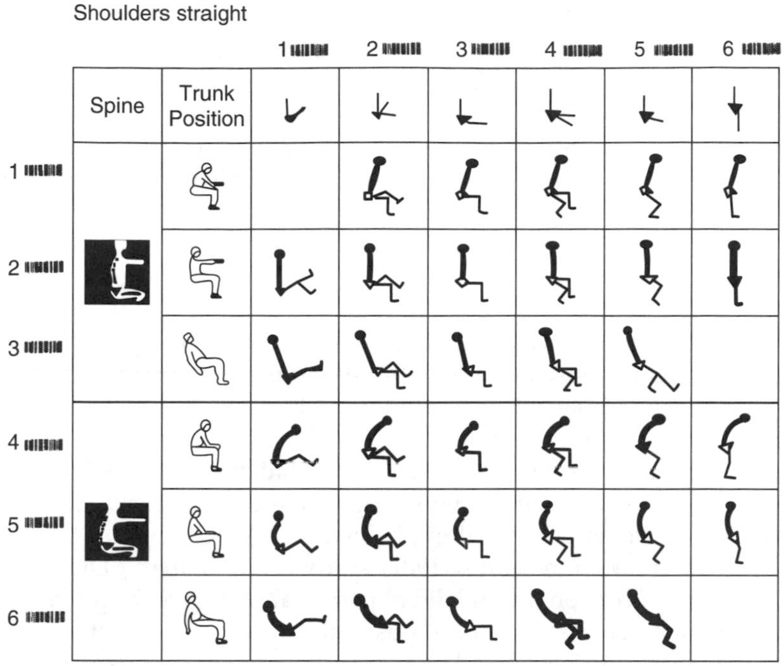

FIGURE 7.2

Posture classification system. Side one of the sitting position classification matrix. The reverse side of the matrix is used when the shoulders are not parallel to the hip (the body is twisted). Leg positions are numbered from 1 to 6 (columns). Data can be recorded manually, using column and row numbers, or by the barcodes shown for automatic data entry. (Reprinted from *International Journal of Industrial Ergonomics*, 15, Graf, M. et al., An assessment of seated activity and postures at five workplaces, 81–91, Copyright (1995), with permission from Elsevier Science.)

this system, the subjects' body and leg positions were recorded once per minute for 1 or 2 h. Only the last 50 observations were analyzed to avoid the worst of the complications arising from the subjects' awareness of being observed. The posture of the subjects' shoulder was first recorded, whether bent sideward or twisted relative to the hip. The next decision was whether the lumbar spine was in a lordotic or kyphotic curve. The observers where instructed to score a kyphotic posture only when the spine was markedly curved forward in the lumbar region. A distinction was then made between whether the upper trunk was inclined forward relative to the hip, was over the hip, or was inclined backward. Finally, the leg position could be recorded. Nevertheless, leg position and back curve are often difficult to record due to viewing difficulties.

The three classification systems presented above are the most validated found in the literature. The last system of Graf et al. (1995) can be considered the most fully developed as it considers all possible aspects of sitting behavior and for this reason, in the authors' opinion, should be the appropriate one to use in studies today.

7.1.2   Influence of Task on Sitting Posture and Behavior

It seems obvious that task requirements would significantly affect sitting behavior, but few studies have been made to document this (Graf et al. 1995). Nevertheless most of these studies definitely indicate the task as the most powerful determinant of sitting position. In effect, observational analysis of sitting behavior should be integrated with a detailed description of the workplace and task demands. Important parameters concern the characteristics of the task itself defined by mean duration of specific task components: for example, manual actions, use of telephone, conversation, data entry, computer-aided design/computer-aided manufacturing (CAD/CAM) work, software programming, accounting machine operation, and paperwork. Furthermore, data entry activities would be better described by means of keying speed, because muscular effort, related to high speed, limits the freedom to select a number of body positions.

Because the visual display unit (VDU) workstation is one of the most common types of workplaces requiring sitting, some observational studies have been carried out to assess the sitting behavior of VDU operators. The most striking result from Grandjean's (1987) study showed that VDU operators moved only very occasionally and did not noticeably change the main postural elements (defined as trunk posture, positions of forearms, wrists, and legs), which were obviously determined by the positions of the keyboard and screen.

Figure 7.3 shows the rather normal distribution of trunk postures observed. The majority of subjects preferred a trunk flexion posture within

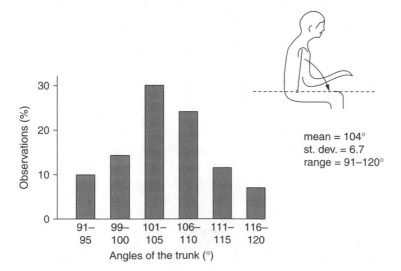

mean = 104°
st. dev. = 6.7
range = 91–120°

FIGURE 7.3
Distribution of trunk postures in 59 VDU operators (236 observations) while performing their usual daily jobs. Trunk posture is assessed as the angle of a line to the horizontal between the hip and shoulder points. (From Grandjean, E., Ed., *Ergonomics in Computerized Offices*, London: Taylor & Francis, 1987, 445–455. With permission from Taylor & Francis (http://www.tandf.co.uk).)

the range of trunk–thigh angles between 100° and 110°, while only 10% demonstrated an upright trunk posture. Moreover, 80% of the subjects rested their forearms or wrists on armrests, where provided. If no special support was provided, about 50% of the subjects rested their forearms and wrists on the desk surface in front of the keyboard. These data seemed to confirm the general impression gained when observing the sitting postures of many VDU operators in offices: most of them lean back and often stretch out their legs. They seem to put up with having to bend the head forward and lift their arms. This posture is very similar to that of car drivers.

To describe the effects of different tasks on sitting behavior, a strict description of task demands should also take into account the crucial importance of visual distance, since it has a strong influence on posture. As seen in operators performing paperwork, leaning forward of the trunk is required for writing tasks, whether during general office work or during attendance at lectures. Accurate manual work characterized by high visual load shows the same leaning forward associated with a shortening of the visual distance, which is normally obtained by increasing the desk height. On the other hand, working with a VDU requires an upright or forwardly inclined trunk position when keying is performed at high speed, while conversation and VDU tasks at low keying speed leave much more room for the operator to choose more comfortable sitting positions, such as leaning backward with the back supported by a backrest so that less muscular effort is required. In that case the visual distance required is often achieved by bending the neck forward.

The Branton and Grayson (1967) method was designed to analyze sitting behavior on a specific seat. Nevertheless, its use has been evaluated mainly on train seats and the results have shown different behaviors, which were mainly related to the height of the subjects. Furthermore, sitting on a train seat is influenced by instability due to jolts of the carriage and therefore the applicability of the results of this study to office work is questionable.

Cantoni et al. (1984) investigated the effect on sitting behavior following the transformation of workplaces of the National Telephone Company from the traditional electromechanical switchboard to an ergonomically designed VDU-operated switchboard. Four subjects were observed independently by four experts using the observational tool described in Figure 7.1, and the results are shown in Figure 7.4. This shows the average proportion of time for which different postures were held during the working day (supported back, upright back, kyphosis, bent back, break) and the load on the lumbar spine (at the L_3–L_4 disc) at the two workplaces. The average decrease in lumbar load from the changes in postures in the new VDU workstation was highly significant, demonstrating the positive effect of ergonomics workplace redesign. Nevertheless, the number of changes of posture per hour was reduced with the new VDU-operated switchboard, leading to a significant increase in postural fixity (as shown in Figure 7.5). Cantoni et al. (1984) concluded that the induced fixity could have been compensated for by introducing a 15-min break every 2 h.

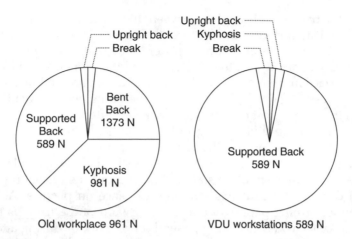

FIGURE 7.4
Average proportion of time for which different postures were held in the study of Cantoni et al. (1984). Loads on the lumbar spine (at the L_3–L_4 disc) are also indicated for the different postures adopted at the two workplaces.

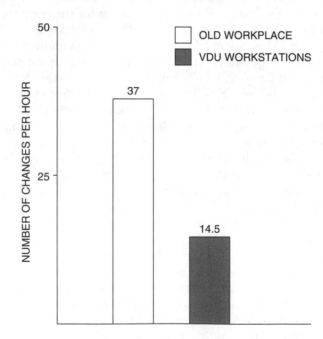

FIGURE 7.5
Number of changes of posture per hour at the two workplaces studied by Cantoni et al. (1984).

In a study comparing the effects on posture of different types of task (both VDU tasks and others), Graf et al. (1995) observed the frequencies of various sitting positions adopted by operators in a field study of five different

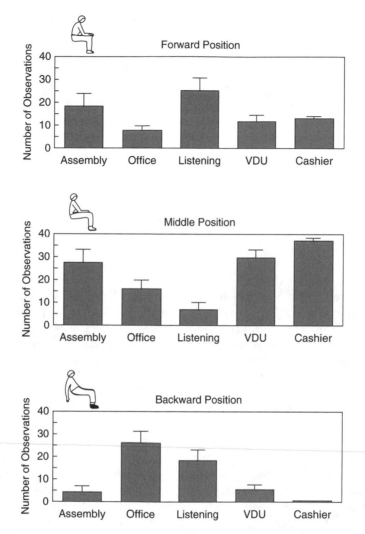

FIGURE 7.6
Group means frequencies and standard errors of trunk positions during five different tasks. (Reprinted from *International Journal of Industrial Ergonomics*, 15, Graf, M. et al., An assessment of seated activity and postures at five workplaces, 81–91, Copyright (1995), with permission from Elsevier Science.)

occupational tasks: assembly (including welding onto electronic circuit boards), general office work (at VDU, telephone calls, reading paper), listening during lectures, VDU work (including programming), and the work of a supermarket cashier. The results are shown in Figure 7.6.

The assembly workers, VDU workers, and cashiers were found to sit principally in the forward or middle sitting positions illustrated in Figure 7.6, rarely leaning backward, whereas the general office workers leaned backward more often than forward. General office workers used their sitting position options more than the other workers, and 80% of the general office

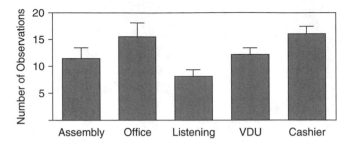

FIGURE 7.7
Group means frequencies and standard errors of changes in posture in five different tasks. (Reprinted from *International Journal of Industrial Ergonomics*, 15, Graf, M. et al., An assessment of seated activity and postures at five workplaces, 81–91, Copyright (1995), with permission from Elsevier Science.)

observations showed kyphotic positions. The lecture attendees rarely adopted the middle position and mainly had the upper torso supported by the desk or leaning on the backrest, reducing the load on their spine and muscles. A common work posture, named "coat-hanger," was seen adopted by the general office group and was described as sitting on the front of the chair but leaning the shoulder on the backrest, with the back falling in the hollow space at the base of the backrest. The cashiers were most often observed in twisted positions, although twisting was also found rather often at the other work tasks (15 to 20%).

Type and characteristics of manual work also provide a very important descriptor of sitting behavior. In fact, some special work situations, such as manual work requiring freedom of movement or physical effort, call for an upright trunk position with elbows down and forearms horizontal. The old mechanical typewriters, for example, which required a significant key depression force, were more easily operated in such a posture, while modern keyboards are mainly operated by finger movements with hardly any assistance of the forearms. These conditions may explain why the VDU operators in modern offices working at low keying speed prefer to lean back, keep the upper arm slightly forward with the wrists on a support and adjust the keyboard to a rather high level.

Figure 7.7 describes the data from Graf et al. (1995) concerning the frequencies of changes in posture. In fact the office workers and cashiers changed position most frequently, while the listeners had the lowest frequency of changes.

A component of sitting behavior is leg crossing, that is, putting a knee or an ankle on the knee of the opposite leg. Graf et al. (1991) observed VDU workers crossing their knee frequently (30%) but rarely observed this in assembly workers and lecture attendants. Graf et al. (1995) showed this option as impossible for cashiers due to limited legroom. Nevertheless, even if leg crossing is often seen at the workplace, field studies seldom report data on this topic, perhaps due to viewing difficulties. Reasons for leg crossing are alternate loading and unloading of soft tissues and muscles, stability of the lower limbs in prolonged sitting, and cultural reasons (for example, related to short skirts). Leg crossing is, however, coupled with an unfavorable effect of hindrance of blood circulation in the lower legs. Takishita et al. (1991) showed that leg crossing markedly increased blood pressure in a sitting position in patients with orthostatic hypotension but not in normal subjects, while Snijders at al. (1995) found reduced tension of abdominal muscles when sitting on a firm seat with the use of backrest, and suggested that leg crossing is physiologically valuable and helpful as an optional posture.

Nevertheless, in the authors' opinion, it seems likely that the search for a comfortable posture in which to perform a particular task is based on the following ranking of adjustments: first comes the visual distance, then the reaching of the best angle between elbow and desk surface, and third the seat pan height, but this appears to be a topic that has received little research attention.

The literature mainly reports cross-sectional studies comparing sitting behavior in different tasks, but when the aim of the study is the comparison between the effects of modifications to the workstation layout, it becomes crucial to compare the effects in the same tasks, and a control group should be chosen in order to assess the possible bias driven by the modification itself.

On account of the important role of perception of comfort, every observational analysis should be coupled with administered or self-administered questionnaires to collect subjective ratings of sitting discomfort. The questionnaire should also contain body diagrams for identifying the areas of discomfort.

7.1.3 Need for Variation in Posture during Work

Derived from the studies on postural fixity, the idea has therefore developed that no seated position should be maintained for a prolonged period, and therefore the optimum seating behavior involves regular changes in position (Graf et al. 1991, 1993, Grieco and Molteni 1999). This is the principle of "dynamic sitting," which refers to achieving regular changes in posture, also through job enlargement, active breaks, and design of tasks or layout of equipment to enforce some standing and walking. However, the term *posture variation* should probably be preferred, as it better describes optimum sitting behavior.

Because no position should be held continually, there is not a single ideal sitting position. In fact the degree of postural fixity, that is, lack of movement, is itself further suspected to be a cause for musculoskeletal disorders (Grieco 1986). The movement factor is important when the task to be performed requires extended periods of sitting, as in most occupations today. The workstation characteristics should therefore be designed to give the maximum comfort in different positions.

7.1.4 Workstation Layout

Sitting behavior is influenced by perception of comfort, task characteristics, workstation layout, anthropometric dimensions, and training.

As described above, "dynamic sitting," or better "postural variation," should be the main objective of workstation design. In fact, modern workers spend more and more time, both at work and at leisure, in fixed positions, and in this respect postural fixity can in itself be considered as a risk factor, because correct intervertebral disc nutrition mainly depends on alternating hydrostatic pressures, above and below a critical hydrostatic pressure value.

The workstation and its components should have the correct dimensions and be easily adjustable to meet the anthropometric needs of a wide range of users. Normal design criteria should accommodate at least 90% of the potential users, but there is still a need to provide workstation modifications to help the 10% of the population with body dimensions falling outside this range. Of the three furniture elements of the workstation (desk, seat, footrest), at least two should be height-adjustable. The best solution is an adjustable chair and an adjustable desk; an economically satisfactory solution may, however, be an adjustable chair and an adjustable footrest with a fixed height desk. Even with this, the positions of other critical items (and particularly keyboards and VDUs) should be adjustable both vertically and horizontally.

Desk depth is a very important factor quite often overlooked; for example, a shallow depth of desk hinders operators from positioning the VDU at an adequate visual distance in front of them. Moving the VDU to the left or right side of the desk can frustrate all the efforts aimed at favoring a good posture. In this respect, flat screen monitors will be helpful for solving borderline situations.

The chair is the most important workplace component for the designer's attention and should have the following characteristics:

- Multiple adjustments and proper shapes and profiles, to ensure adaptability and comfort, solidity, and stability
- Ease of adjustment while seated
- Maintenance over time

A good chair, in fact, is one that supports the body in multiple desirable positions. Nevertheless, operators need training on "body awareness" (that is, the feeling of the body position), training on the correct distances/dimensions within the workstation, and above all training on the use of the "high-tech" chair, because training in proper seat adjustments is a central component for ensuring that users are accommodated (Helander et al. 1995). In fact, Henriques (1985) conducted a study on furniture adjustments and pointed out that only 5% of the users had voluntarily adjusted their furniture.

Summary

Sitting behavior can be defined as the postures and positions of body segments held by a person when sitting, that is, the person's postural range and movement frequencies within their repertoire of sitting behavior, which is influenced by their perception of comfort, task characteristics, workstation layout, anthropometric dimensions, and training.

Observational analysis, by means of field studies, is needed to describe and analyze sitting behavior, and it should be integrated with a detailed description of the workplace and task demands, including visual distance and keying speed. Comfort perception can be assessed by means of questionnaires and subjective rating of sitting discomfort, together with body diagrams for identifying areas of discomfort.

It would seem obvious that task requirements significantly affect sitting behavior, but only a few studies have been conducted to document how sitting behavior is affected by task demands. Manual work, requiring freedom of movement or physical effort, calls for an upright trunk position with elbows down and forearms horizontal, while the VDU operators in modern offices, at least when working at a low keying speed, prefer to lean back, keep the upper arm slightly forward with the wrists on a support, and adjust the keyboard to a rather high level. In fact, the latter condition gives the operators much more freedom to vary their sitting behavior.

Leg crossing has been observed in VDU workers and rarely in other tasks. Reasons for leg crossing are alternate loading and unloading of soft tissues and muscles, stability of the lower limbs in prolonged sitting, and cultural reasons. When sitting on a firm seat with the use of backrest, it has been suggested that this optional posture is physiologically valuable and helpful.

However, a basic principle of "posture variation" is that no sitting position should be maintained for a prolonged period, and therefore optimum seating behavior involves regular changes in position. This should be the main objective of workstation design. In fact, a prolonged fixed posture can in itself be considered as a risk factor for musculoskeletal disorders.

The layout and dimensions of workplaces should be designed to provide not only physiologically comfortable postures but also body movement for people with a wide variety of body types and dimensions.

Of the three furniture elements of the workstation (desk, seat, footrest), at least two should be height-adjustable. The best solution is an adjustable chair and an adjustable desk; an economically satisfactory solution, however, may be an adjustable chair and an adjustable footrest with a fixed-height desk. It is important that the surface of the footrest be large enough, in order not to constrain working posture. A footrest may also be necessary when the adjustment ranges of the workstation do not accommodate a small user.

In conclusion, different types of training and practical interventions to encourage people to move around and change sitting positions have been demonstrated to have positive influence in introducing the concept of "posture variation."

7.2   Seating Concepts

Tom Bendix and Bob Bridger

Introduction

The huge interest in chair design and sitting posture during the 1970s and 1980s was stimulated largely by the study of Magora (1972). He had demonstrated that there was a U-shaped relation between duration of sitting and low back pain (LBP) (Figure 7.8a). This study led to the concept that long-term sitting is a risk factor for LBP, and probably that it is a main explanation of the increase in sick leave due to LBP. Later studies, however, have shown that sitting is not an important risk factor for LBP (Hoogendoorn et al. 1999, Hartvigsen et al. 2000). Some of the mistakes seem to have been due to the "healthy worker effect" (Hartvigsen et al. 2001): if people are followed over years, some people change from heavy work to sedentary work; those with LBP for whatever reason typically choose lighter jobs that are normally more sedentary (Figure 7.8b). Seated work is predominantly a problem for those people who already have problems with LBP.

Moreover, for the general population — including many with actual back pain — there is the question of whether ergonomic intervention actually prevents/reduces back pain. Seemingly, the effect of such intervention is less than usually believed. At least one large study questions the efficacy of

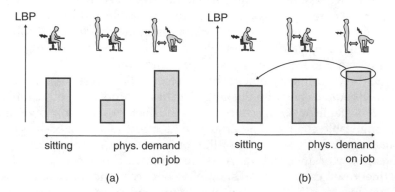

(a)   (b)

FIGURE 7.8

Magora (1972) found almost as much LBP among sedentary workers as among workers in physically demanding jobs (a). Later studies (e.g., Hartvigsen et al. 2000) indicate that those with LBP select lighter work that is often more sedentary (b).

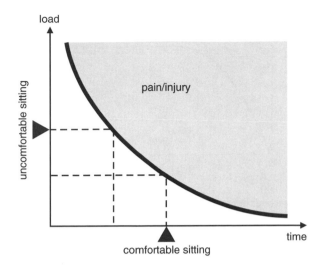

FIGURE 7.9
One reason that sitting postures seemingly have very little influence on LBP may be the "buffer" effect as described further in the text.

primary ergonomic intervention in postal workers (Daltroy 1997), but issues other than sitting were also addressed in that study.

The minor correlation between sitting at work and LBP may in part be explained by the "buffer effect," as illustrated in Figure 7.9. If an uncomfortable posture in an awkwardly adjusted workstation is adopted (left arrow), the subject may change posture after a period indicated by the upper dotted line when the area of pain/injury is reached. If the ergonomic adjustment leads to a more comfortable posture (lower horizontal dotted line), corresponding to smaller loads on various structures, then the person will instead tend to sit in a more constrained posture for a longer period of time and by another route again reach the area of pain/injury (Winkel 1987).

It is likely, however, that, in general, improvements of a very poorly adjusted workplace may imply a certain health benefit (Verbeek 1991) (Figure 7.10, left), whereas further "improvements" (Figure 7.10, right) are absorbed by the buffer effects described in Figure 7.9. Arguments on the less convincing influence of different sitting workplace arrangements will be given below.

Regarding neck and shoulder pain, there seems to be a stronger association with sitting than for LBP. However, here it is more difficult to distinguish between the effect of the task itself as opposed to the influence of different workstation adjustments. At least some studies, however, indicate that workplace adjustment does play a role. Luopajärvi (1987) found that focusing on instruction of the workers to adjust the seated workplace and encouraging them to use it actually reduced "tension–neck syndrome," as compared to a control group. The paper does not, however, describe which adjustment is advocated, and also the paper assessed the effect only immediately after

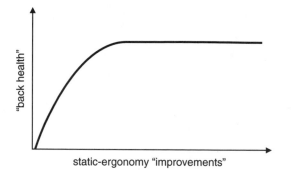

FIGURE 7.10
It is likely that only very poorly adjusted workstations benefit from ergonomic improvements regarding LBP-health issues.

the intervention period. Moreover, in such studies it may be difficult to determine whether any beneficial effects are due to improved ergonomics or are merely artifacts of the experimental procedure. It should be noted, however, that the evidence for the famous Hawthorne effect has itself been called into question on many occasions (e.g., Wickström and Bendix 2000), as has the influence of the placebo effect (Hrobjartsson and Gotzsche 2001).

7.2.1 Optimizing Workplace Adjustments in Sedentary Work

The optimal sitting-workplace adjustment should be based on the following:

- Health of the worker
- Subjective preference
- Biomechanics of sitting

Because the health of the worker is almost impossible to document, subjective preference must be given a high priority, with the hope that it will fit with the biomechanical concepts. However, subjective preferences are also influenced by habits, which may not always coincide with proper health influence or biomechanical optimality.

Working postures are typically the result of an interaction between three classes of variables:

1. The anthropometry of the worker
2. The requirements of the task (principally the visual and manual requirements)
3. The design and dimensions of the workspace

Thus, workplace optimization in sedentary work needs to take a more holistic approach and not be restricted to the selection of an "ideal" chair. Rather, the design and layout of the desk, the positioning of work objects in relation to the seat, and the task demands all have to be considered as well.

It should also be kept in mind that different tasks may call for different adjustments, seen from all three of the above perspectives. This section focuses on tasks such as handwriting and reading. Workplace design for VDU adjustment is discussed in Chapters 4 and 5.

7.2.2 Seating Arrangements

Many different seating concepts have been proposed over the years. Interested readers can be referred to Mandal (1985), Schoberth (1962), and Åkerblom (1948). For a discussion of some cross-cultural aspects of sitting posture see Bridger (1991b, 1995). In the following sections, various seating concepts are described, with the emphasis on the static as well as the dynamic biomechanical aspects.

In Schoberth's X-ray investigation from 1962, he demonstrated that, even in an upright sitting position without a backrest, 16% of the children had a kyphotic spinal posture (Schoberth 1962). On the other hand, in a relaxed, hunched posture, 11% maintained a lordosis (see also Section 7.3). Therefore, the expressions "increased lordosis/reduced kyphosis," and vice versa, is used below.

7.2.2.1 Seat Inclination

Traditionally, seats were horizontal. Based on the mathematical force parallelogram-based arguments of Strasser (1913, cited in Åkerblom 1948), it was argued that the backrest would be utilized more if the seat was slightly backwardly inclined. This was the main reason for making chairs with seats slanted backward. Such an inclination may be appropriate for relaxed sitting, but does not seem to be optimal for work at a table.

In the 1970s Mandal suggested an alternative approach, arguing that chairs should be higher and that the seats should slant forward to enable a more open trunk–thigh angle. It was argued that this would result in a more lordotic lumbar posture with less constraint of the trunk, because the backrest would be used less (Mandal 1982). An increase of the seat height has to follow a forward slant of the seat, if a supported forward/downward inclination of the thighs is wanted. Bridger (1988) observed that subjects participating in a laboratory study comparing forward sloping and conventional chairs (Figure 7.11) alternate between an upright "erect" and an upright "slumped" sitting posture by "rocking" backward and forward over the tuberosities. The mean pelvic tilt and lumbar angles when sitting on the forward-sloping chairs tended more toward anterior tilt and lumbar lordosis than on conventional chairs. Bridger et al. (1989a) showed that both the forward-sloping seat and the downward inclination of the thighs tend to reduce lumbar flexion when people sit on chairs without backrests. It seems

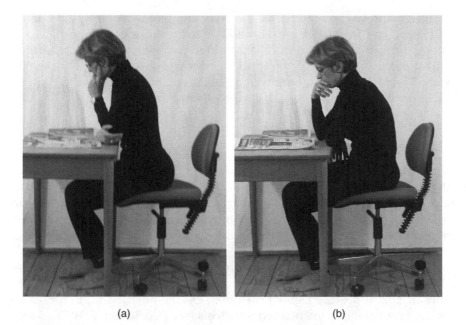

(a)   (b)

FIGURE 7.11
"Rocking" over the ischial tuberosities (a) Erect sitting posture with lumbar lordosis. (b) Erect "slumped" posture with posterior pelvic tilt and lumbar flattening.

that the effect of the downward inclination of the thighs is to maintain an equilibrium position of the pelvic musculature that is similar to that found in standing, with some of the lumbar lordosis retained. The effect of the sloping seat is to cause the pelvis to "rock" forward on the ischial tuberosities, resulting in a reduction in lumbar flexion. About 20 years earlier, Branton (1969) carried out a field study of the sitting postures of train passengers and observed how they tended to "rock" backward and forward over the tuberosities. Branton pointed out that, in the sitting posture, the weight of the trunk is borne by only two points of support — the ischial tuberosities — and is unstable in the sagittal plane.

The instability of the pelvis as a base of support in sitting has led to two, quite different approaches to the design of seats.

1. Design concepts that attempt to stabilize the pelvis by means of contoured seating and snugly fitting backrests and lumbar supports. This design philosophy is the basis of much modern office chair design. The pelvis is held in an anteriorly tilted position by a lumbar support to assist the weakened hip flexors, and the seat contouring prevents the sitter from sliding forward. Postural change is accomplished by means of a tilt mechanism on the backrest.
2. Design concepts that free the pelvis to "rock" about a neutral position such as the neutral positions in Figure 7.23 (see p. 174). These

FIGURE 7.12
The three types of chairs tested in a study by Bendix (1984) comparing backward inclined, forward inclined, and tiltable seats.

designs include chairs with forward-sloping seats, tiltable seats, and even giant, inflated physiotherapy balls (Bridger et al. 2000). This design philosophy is based on the idea that, first, the body posture should be correct with a trunk–thigh angle of at least 105° (Keegan 1953). This will automatically place the pelvis in a neutral position and therefore provide the basis for a neutral spinal posture around which task movements can take place.

Various studies have not clarified what is expected to be the optimal seat height/inclination (Bendix 1984, 1987). The three types of chair adjustment shown in Figure 7.12 were tested. With a forward tilted seat, as compared to a backward tilt, the lumbar spine moves on average into a less kyphotic/ more lordotic posture, with less utilization of the backrest. Seen in isolation, the reduced kyphosis may be beneficial according to disc-pressure measurements (Andersson and Örtengren 1974), although the direct interpretation of such a measure has been questioned (Wilke et al. 2001). Moreover, the health-related value of disc pressure has been questioned as well, because discal flow/nutrition may be much more important (Holm and Nachemson 1983, Horner and Urban 2001) than the pressure within it, especially if the pressure is in the middle of its entire range.

Regarding less utilization of the backrest, a possible health value of a back support has not really been documented, but the evident reduction in spinal load when such as support is used seems to be a source of reduced fatigue in the back. It is very likely that in case of hip joint pain, a more open trunk-hip angle is wanted, as obtained in a chair with a high and forwardly inclined seat. In conclusion, it is not clear what should generally be preferred seen from a biomechanical point of view, if there is any general preference at all: a slightly backwardly or forwardly inclined seat.

Seat height is further discussed in the Section 7.2.2.3. For tiltability, see Sections 7.2.2.4, 7.2.2.5, and 7.2.2.7.

FIGURE 7.13
Three backrest arrangements tested for their influence on lumbar curvature, showing that a backrest generally facilitates kyphosis. (From Bendix, T. et al., *Ergonomics*, 39, 533–542, 1996.)

7.2.2.2   Influence of a Backrest

It is generally believed that a backrest facilitates lordosis. It is evident that if a person sits on a chair without a backrest, and somebody pushes a backrest toward the lower back, the lumbar curve shifts into a posture with less kyphosis/more lordosis. However, in practice, the backrest may facilitate the forward movement of the buttocks and kyphosis of the lumbar spine to stabilize the trunk against the backrest. This question was addressed in a study where the three types of chairs were compared (Bendix et al. 1996). This facilitation of kyphosis is seen for reading and handwriting tasks. Only for assembly work did the prominent backrest (Figure 7.13, right) increase the lordosis as compared to sitting without a backrest.

A clinical study of people with LBP considered to be of discogenic origin without actual herniation has, however, demonstrated that a portable lumbar roll reduced back and leg pain, and shifted the remaining pain from irradiating leg pain to be localized more centrally in the spine, a "centralization of the pain" (Williams et al. 1991). In accordance to the principles of the McKenzie concept — that lordosis may push nuclear material in a posterior annular fissure away from the peripheral part of the intervertebral disc toward its center (Donelson et al. 1997) — it is likely that it is a lordosis that creates such an effect from the use of a lumbar roll. However, whether these people actually did obtain a lordosis with the lumbar roll was not recorded. It may be that in the case of a *loose* lumbar roll, as opposed to the *stable* backrests in Figure 7.13, a lordotic posture is actually adopted.

7.2.2.3   Seat Height

Mandal once demonstrated that people spontaneously prefer a very high seat. However, the evidence came from a study using a height-adjustable chair without a backrest (Mandal 1982). A similar study where the subjects had the option of utilizing a backrest (Bendix and Bloch 1986) showed a preference for much lower seats — about 3 to 5 cm above popliteal height.

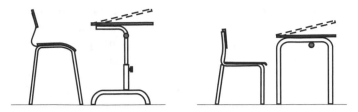

FIGURE 7.14
The high and forwardly inclined seated chair (left) was preferred for mixed schoolwork (hand-writing–reading–listening) by schoolchildren over traditional chairs (right). (From Aagaard-Hansen, J. and Storr-Paulsen, A., *Ergonomics*, 38, 1025–1035, 1995.)

It is very likely that the difference across the studies can be explained as follows: if there is no backrest, subjects will prefer a seat height where they "miss" the backrest as little as possible; the higher the seat, the less a backrest is needed as the sitting posture tends to approximate a standing posture. If a backrest is available, users will prefer a seat height where they gain advantage from it, i.e., a relatively lower seat height where they can recline against the backrest without fear of sliding off the seat.

For school children, there seems to be a clear preference for a relatively high and forwardly inclined seat (Figure 7.14, left) over the conventional low adjustment (Figure 7.14, right) (Aagaard-Hansen and Storr-Paulsen 1995), but it is uncertain how far that implies any health benefit. It is unclear why children preferred the high furniture. Practical constraints meant that only very high or rather low chairs were tested. Whether something in between would have been chosen if available, and if upholstered, should be addressed. In the height-preference studies mentioned above, it was recorded that children generally prefer relatively higher adjustments than adults (Bendix and Bloch 1986). But whether this is because children are accustomed to, for them, relatively high furniture, or whether it is an indication of this being the optimal adjustment, which is then "destroyed" by the relative low adjustments that adult people adapt themselves to, cannot be substantiated.

7.2.2.4 Work Surface Height and Inclination

The absolute table height is actually in itself not influencing the posture of the user much. Rather it is the "seat-to-work surface height difference" that is relevant. This will influence the inclination of the trunk. In a study by Bendix and Bloch (1986), the subjective preference, using a chair with a tiltable seat and a backrest, when doing miscellaneous paperwork, was a work surface height of about 3 to 5 cm above elbow level (defined as the height of the elbow, while sitting upright with the shoulder relaxed, the upper arm vertical, and the elbows at 90°).

In the study on the school furniture (Figure 7.14), at the conventional low seat adjustment the horizontal work surface was compared with a sloped work surface. The latter was clearly preferred. In another study, different work surfaces with the surface horizontal or sloped at 22° or 45° were tested regarding neck-muscle tension, head position, trunk posture, and subjective acceptability (Bendix and Hagberg 1984). Neck tension and flexion reduced with increasing inclination. Also a change in lumbar curvature toward increased lordosis/reduced kyphosis was obtained with increasing desktop inclination, and the entire trunk became more vertical, indicating a more open hip–trunk angle. Correspondingly a steep inclination was preferred, but only for predominantly reading tasks. For handwriting the horizontal surface was preferred, mainly because the users felt that blood left the forearm when sloped upward during the writing task. Thus, for deskwork it may be recommended that a separate sloping desk is used for reading tasks. That leaves the rest of the table for handwriting.

Bridger (1988) obtained similar results as Bendix and Hagberg (1984). In Bridger's study subjects carried out a writing task for 20 minutes at each of four workstations combining a conventional chair or a chair with a forward sloping seat, a horizontal work surface or a tilted work surface. It was found that seat tilt influenced mainly the posture of the lower limbs, pelvis, and lumbar spine (with the forward sloping seat resulting in a more open trunk–thigh angle, more anterior tilt of the pelvis, and less trunk flexion). The tilted work surface caused the trunk to be more erect and there was less flexion in the neck. The most erect postures were found when the chair with the sloping seat and the tilting work surface were combined, with the effects of the two design features being additive.

7.2.2.5   Knee Support Chair

Knee support sitting stems from an old African way of sitting and was taken into the modern office by the Norwegian furniture company Stokke, and named Balans Chair. The idea is that the trunk–thigh angle can be even more open than with just a usual forwardly inclined seat and, according to the advertisements, maintain lordosis as in the standing position.

A kinematic study showed that the lumbar spine actually is more lordotic/less kyphotic than when sitting on a traditional office chair (Bendix et al. 1988a), but also that the reason for this change toward lordosis is mostly explained by the lack of a backrest rather than by the relatively steep inclination of the thighs. Moreover, the lumbar curvature adopted during sitting on a knee support chair is much less lordotic than in standing (see also Section 7.3). Also, on this chair, the basic biomechanical principles exist that in standing the trunk rests on the hip joints, whereas in sitting the pelvis rests on the ischial tuberosities. Therefore, the pelvis must rotate backward when sitting down, and consequently the lordosis shifts into/toward kyphosis (see Section 7.3).

The knee support chair "improves" lumbar spine posture, according to traditional thinking, but seems not to be suited for whole-day use because a backrest is missing.

From the discussion thus far it can be concluded that seat height and inclination predominantly influence the posture of the pelvis and lumbar spine; the seat-to-table height difference influences mainly trunk inclination, whereas work surface inclination has the highest potential for influencing head, neck, and trunk inclination, as well as the lumbar curvature. The latter is also influenced by a backrest, but in the direction of increased kyphosis. Seat tilt creates a bottom-up effect, with biomechanical adjustments starting from below, as illustrated in Figure 7.18 in Section 7.3. Conversely, work surface inclination creates a top-down effect.

7.2.2.6   Sit-Stand Workplace

With the idea of inducing postural variability, a sit-stand workplace has been introduced. It seems evident that the option of changing working position is comfortable, but a systematic evaluation was not convincing in terms of inducing health effects on the neck–shoulder regions (Winkel and Oxenburgh 1990). In the lumbar spine more variation in the muscle activity was recorded, but neither the spontaneous utilization of the optional change in position nor a possible health effect on LBP has been tested.

7.2.2.7   Dynamics of Sitting

How far do different sitting workplace adjustments facilitate movement while sitting, as opposed to constrained sitting? The intention with a tiltable seat (Figure 7.12, middle seat) is to facilitate trunk movements and to let them take place using an axis under the seat, thus saving stress on the lumbar spine, especially during forward reaching tasks. However, studies could not demonstrate any differences either in movement frequency across the three chairs on Figure 7.12 (Jensen and Bendix 1992) or in trunk flexion or inclination when comparing forward reaching movements (Bendix et al. 1988b).

There are other chairs on the market that are more unstable — aiming to facilitate dynamic sitting (e.g., the seat pivoting on one point under the seat instead of having a transverse axis as on the usual tiltable office chair, as depicted in Figure 7.12) — but to the authors' knowledge, the effects have not been studied. However, sitting behavior on such a multidynamic chair (with seat inclination and backrest inclination following buttock/upper leg posture and trunk posture, respectively) shows that other chairs with a relatively fixed relationship between seat inclination and backrest inclination (i.e., "fixed chair," "synchro chair," "inclination chair," and "tip-up chair") provide proper body support less well (Delleman and Miedema 1998).

Different continuous passive movement (CPM) chairs exist, e.g., one with a pump connected to an air bag placed on the backrest (Reinecke et al. 1994) (Figure 7.15). Probably this is particularly successful in a car seat, compared with (for example) office seating. Evaluative studies are ongoing (Hazard 2001).

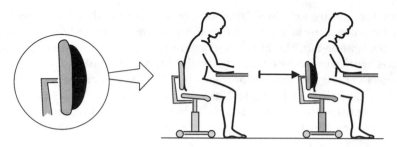

FIGURE 7.15
A CPM backrest, moving the lumbar spine slowly and cyclically.

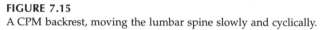

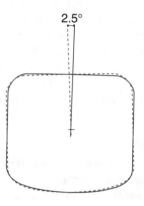

FIGURE 7.16
A seat (seen from above) rotated passively and cyclically 1.25° in each direction across the midsagittal line has been able to demonstrate a significant reduction in sitting-related low back pain. (From Van Deursen, L.L. et al., *European Spine Journal*, 8, 187–193, 1999.)

Another utilization of CPM is a seat that is slowly, passively rotated to each side, with the motion induced by a motor (Van Deursen et al. 1999), as illustrated in Figure 7.16 (refer also to Figure 7.3). A controlled study showed that this reduced pain perception significantly in chronic nonspecific LBP patients, who had LBP problems during sitting, when sitting uninterrupted for an hour.

A final consideration in the design of chairs is the provision of the swivel facility. It seems to be the case that, in most forms of postural adaptation to work demands, the greatest changes in posture occur at the distal rather than the proximal segments. For example, when reaching for an object, we tend to extend the arms first, rather than the trunk. The pelvis seems to be an exception to this rule. When reaching for objects at different horizontal locations, the tendency is for the pelvis to rotate around the vertical axis (see Delleman 1999, for further discussion). Why this happens is not clear, although it may be to minimize the effects of torsion on the lumbar motion segments. Whatever the reason, it makes sense to incorporate some kind of swivel facility into chair seats so that these rotational movements can be accomplished with minimum effort.

Summary

The influence of workplace adjustments on sitting posture seems reasonably well established. However, the various existing studies of seating concepts presented above do not provide a basis for a clear recommendation on one optimal seating concept, in relation to possible influence on health concerns. One reason for this may be that, according to the more recent epidemiological studies, sitting actually does not constitute the health hazard as was previously believed. Thus, taking the epidemiological data into account, it is likely that it does not matter much how a chair is constructed and adjusted, as long as it can be adjusted to fit the worker in relation to the rest of the workstation and the requirements of the task.

The biomechanical influences of various seating concepts are as follows:

- The higher the seat, the more the seat must be inclined forward to conform to the thighs.
- A backrest facilitates in general a kyphosis, probably because the user intuitively pushes the lower back against it to obtain stability.
- Probably for this last-mentioned reason, the knee support chair is less well rated by most users, although it could seem to facilitate biomechanically advantageous postures.
- Children prefer relatively higher furniture adjustments than adults, but the implications of this are unclear.
- The seat heights normally used do not influence the number or quantity of movements that the user performs during sitting.
- There are some indications that CPM seats may offer advantages in certain situations, for example, during car driving.
- The sit-stand workplace may be comfortable, but how much the users actually utilize it or, if they do, whether it has any health benefit has not been validated.

Does the theoretically less constraint of the trunk (backrest less used) — which in some studies, however, could not be recorded — and reduced kyphosis with a relatively high and forwardly inclined seat of a conventional chair or a knee support chair, have advantages over a more stabilized posture induced by better utilization of a backrest in a slightly lower chair? Only a slanted desktop seems to provide benefits at the biomechanical level, with regard to personal preferences and, to some extent, to pain perception.

However, as long as a certain sitting workplace adjustment is not documented as being the optimal in terms of positive influence on health, it is suggested to allow people themselves to arrange their sitting workplace, when they start a new job, provided they really have a chance and equipment to do so, i.e., a few different adjustable chairs and a table adjustable in height and in inclination should be available, along with time and encouragement to figure out which adjustment they really prefer. It is unclear if such a procedure improves health, but at least it gives the user a "feeling of ownership of the idea." As a result of our poor understanding of health issues and sitting, people with actual back pain aggravated by sitting should have the opportunity of trying different concepts and selecting the one that is most comfortable for them.

7.3   Pelvis and Neighboring Segments

Bob Bridger and Tom Bendix

Introduction

It is a fact that the spine rests on top of the pelvis. It is unavoidable, therefore, that anything that affects the posture or stability of the pelvis of an upright human will also affect the spine. This applies to many postures and movements, although the standing and sitting postures are of most interest to ergonomists.

In this section, an overview is given of the general anatomy of the pelvis and its linkages to adjoining segments. Next, the experimental data that illustrate the relationship among pelvic posture, spinal posture, and related structures are reviewed. Finally, the ergonomic implications of this knowledge is summarized.

What Is the Pelvis? The pelvis is a ring-shaped structure made up of three bones, the sacrum and the two innominate bones. The sacrum lies below the lumbar spine and consists of a number of fused vertebrae. The three pelvic bones are held together in a ring shape by ligaments (Figure 7.17). The innominate bones are made from the fusion of three other bones: the ilium, the ischium, and the pubis. The pubis lies at the anterior part of the pelvis. It joins the other bones, completing the ring shape and acting like a strut to prevent the pelvis from collapsing under weight bearing (Tile 1984). The posterior structures of the pelvis, the sacrum, and the ilia carry out the actual weight-bearing function.

Function of the Pelvis. The pelvis can be likened to an arch, which transfers the load of super-incumbent body parts to the femoral heads in standing and to the ischial tuberosities in sitting. When it is viewed from the rear, the sacrum resembles the keystone of the arch. The load from above is transmitted through the innominate bones to the femoral heads. The heavier the load, the more tightly the sacrum is held in place. However, when viewed from

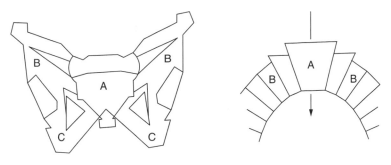

FIGURE 7.17
The pelvis as an arch (from the rear) with (A) the sacrum, (B) the innominate bones, (C) the ischia (see Bridger 1995 for a more detailed discussion). (Redrawn from Tile 1984.)

above, the sacrum has the wrong shape for a keystone — it tends to slide forward, out of the arch. Under weight bearing, strong ligaments resist the tendency for the sacrum to slide forward anteriorly. It is these posterior sacroiliac ligaments that stabilize the joint in the transverse plane. DonTigny (1985) has pointed out that standing postures in which the person has to bend forward slightly from the hip (e.g., when washing dishes at a sink) increase the tendency for the sacrum to be anteriorly displaced, thereby increasing the tension in the sacroiliac ligaments. Small displacements of the sacrum can occur causing soft tissues to be "pinched" and this can cause pain — pain that can be mistaken for LBP.

7.3.1 The Pelvis and Related Structures

In the standing posture, the curvature of the spine depends on both *structural* and *functional* factors and there are large differences between individuals. Structurally, it depends on the angle of the articular facet of the sacrum within the sacroiliac joint. Kapandji (1992) describes "dynamic" and "static" types of individuals. In the former, the sacrum tends to the horizontal and the upper surface of the first sacral vertebra (S1) is angled downward. In the latter, the sacrum is more vertical and the upper surface of S1 tends to the horizontal. Functionally, lumbar curvature is influenced by the tilt of the pelvis and this is influenced by the equilibrium position of the various muscle groups that cross the pelvis. This was first discussed in the context of postural faults by Kendall et al. (1971) who distinguished between a "swayback" posture (with shortened hip flexors and erectores spinae muscles and weakened abdominal and hip extensors) and a "flatback" posture caused by weakness in the hip flexors and erectores spinae. In the former, the pelvis is held in an anterior tilt and, in the latter, the pelvis tilts rearward. These postural faults were held to be caused by muscle imbalances (Figure 7.18).

In those cases where the flatback posture is caused by a backward tilt of the pelvis, it is likely to be a consequence of tight hamstrings. Using an analogy with a boat at anchor (Figure 7.18, right), the tilt of the pelvis is due

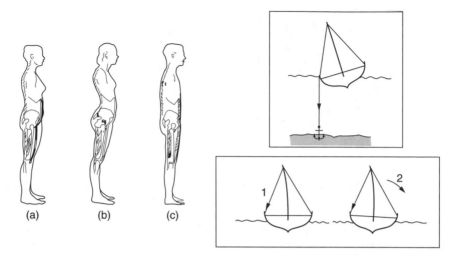

FIGURE 7.18

Standing posture and postural faults. (a) Normal standing, (b) "swayback" posture with excessive lumbar lordosis and anterior pelvic tilt, (c) "flatback" posture. (Redrawn from Kendall et al. 1971.) At the right side of the figure the pelvis is likened to a boat, anchored at sea. A pull from below will tilt the entire structure, whereas a pull from above the pelvis (boat) will only bend the spine (mast).

to the pull in the hamstrings and other muscles attached to the legs, in the same way that the tilt of a boat is caused by tension in the lines attached to the anchor. In humans, the mast (spine) is flexible and curves reflexively to resist the tilting to keep the head erect.

7.3.1.1   Pelvic Tilt and Lumbar Lordosis

The tilt of the pelvis may be measured using an anthropometric caliper with integral inclinometer. The points of the caliper are placed on the anterior and posterior superior iliac spines (Figure 7.19). In the standing position, with the pelvis held in an anteriorly tilted position, the spinal column rests on the sacrum. As shown in Figure 7.20, the superior surface of the sacrum slopes forward. Mandal (1985) describes this as the "base angle" of the lumbar spine — the angle between the plane through the L5-S1 interspace and the horizontal. The greater the base angle, the more pronounced the lumbar lordosis. According to Schoberth (1962), "We have never seen a lumbar curve where the base angle was less than 18 degrees. Conversely, a rounded back has never been seen where the base angle was more than 10 degrees."

That the degree of lumbar lordosis depends on the base angle is illustrated in Figure 7.21 (from Bridger 1985). These data were obtained from radiographs of the spines of five volunteers in three postures: standing and two sitting postures. For each successive triplet of vertebrae from S1 to L1, the geometric centers of the images of each vertebra were established and radii

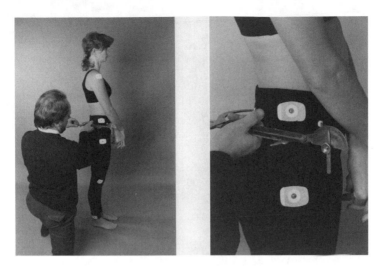

FIGURE 7.19
Measurement of pelvic tilt using a caliper with inclinometer attached. The points of the caliper are placed over the anterior and posterior superior iliac spines.

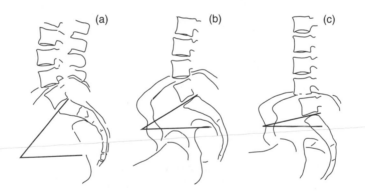

FIGURE 7.20
Relationship between base angle of the lumbar spine and lumbar curvature. (a) Standing, (b) sitting on a chair with a forward sloping seat, (c) sitting on a chair with a flat seat.

of curvature calculated. As can be seen, in all cases, the greatest amount of curvature occurs in the articulation of S1, L5, and L4, diminishing rapidly through triplets L5, L4, L3 and L4, L3, L2 before appearing to reverse sign at L2.

7.3.1.2 Lumbar Spine, Sacrum, Pelvis, and Lower Limbs as a Linkage

Together, these structures form a coordinated articular system. In the standing position, the line of gravity of the superincumbent body parts tends to fall posteriorly to the center of the hip joint (Kapandji 1992). With the ground reaction force passing through the center of the joint, a force couple is

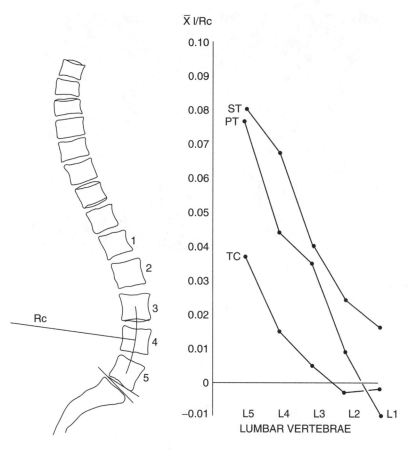

FIGURE 7.21
Mean inverse radii of curvature $(1/R_c)$ for five successive lumbar vertebral triplets. The larger
the reciprocal of the radius, the greater the curvature. (Redrawn from Bridger 1985.) The data
indicate that most of the curvature is at the base of the lumbar spine. ST: radii of curvature in
standing, PT: radii of curvature when sitting on a forward sloping chair, TC: radii of curvature
when sitting on a conventional chair.

produced that causes a tendency for the pelvis to rotate backward. This is
prevented by the iliofemoral ligament and by the iliopsoas muscle, which is
in a lengthened and stretched position in standing.

Humans have evolved to stand and walk on two legs. The infant crawls
on all fours and has a primary C-shaped curve along the length of its spine.
Standing, and then walking, is achieved partly by extending the hip joint so
that the trunk rotates around the femur toward the vertical, and partly by
extending the lumbar spine to bring the trunk into its final vertical position.
The remaining hip extension is necessary for walking to allow the trailing
leg to lag behind the body during the stance phase of gait.

We can imagine the sagittally viewed pelvis as being fixed on the hip joint
by muscles and ligaments (Figure 7.22) that prevent the upper body from
"jackknifing" over the legs either anteriorly or posteriorly. The back extensor

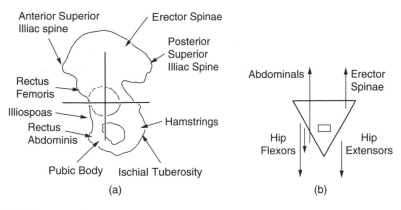

FIGURE 7.22
(a) The muscular attachments of the pelvis. (b) Imbalances in muscle length or tone, for whatever reason, can create changes in the tilt of the pelvis.

muscles, hip flexors, and extensors all play a role in stabilizing the pelvis in the standing position. The abdominal muscles are relaxed in normal standing because of the natural tendency for the pelvis to rotate backward anyway.

A complex system of reflexes controls the posture of the whole body such that changes in pelvic tilt can be compensated for by changes in other structures, notably the lumbar spine. These postural reflexes act to maintain the torso and head in an erect position with respect to gravity. As soon as the pelvis tilts, for any reason, compensatory flexion or extension of the lumbar spine occurs. This system of reflexes accounts for much of the behavior of the lower limbs, pelvis, and spine as a linkage and explains why different spinal postures are found in different upright body positions.

7.3.1.3 Pelvic Tilt in Different Body Positions

It was Keegan (1953) who first pointed out that the main difference between the seated work position and standing was the loss of the lumbar lordosis when sitting. Keegan held that in order to adopt the sitting position, the hip joints have to flex, causing the hamstring muscles to lengthen and tighten, pulling at their pelvic attachments and causing the pelvis to tilt backward. The result was a loss of the normal lumbar lordosis. This finding has subsequently been confirmed by many independent investigators (e.g., Bendix and Biering-Sorensen 1983, Brunswic 1984, Bridger 1988, Link et al. 1990, Bridger et al. 1992, Lord et al. 1997). It should be noted that, while Keegan (1993) and Lord et al. (1997) used radiographic methods, the other groups used bony landmarks at the skin surface to estimate the location and alignment of the underlying structures. The latter method is acceptable when the aim is to compare different body positions but the data should be interpreted with caution when making absolute statements about posture.

It seems to be true that the pelvis is in an anterior position in standing and there is a distinct lumbar lordosis, which is lost when sitting down

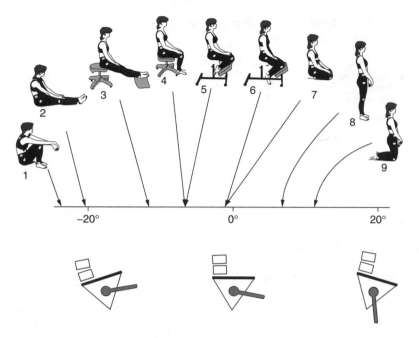

FIGURE 7.23
Angles of pelvic tilt in different body positions.

TABLE 7.1

Mean (and standard deviation) Angles of Pelvic Tilt and Lumbar Angle (degrees) in the Nine Body Positions Shown in Figure 7.23

Body Position	Males ($N = 25$)		Females ($N = 25$)	
	Pelvic Tilt	Lumbar Angle	Pelvic Tilt	Lumbar Angle
1	−25.2(6.8)	−15.4(6.6)	−21.8(5.2)	−8.3(4.6)
2	−23.5(7.3)	−15.1(6.8)	−17.6(5.2)	−5.8(5.8)
3	−13.9(6.7)	−7.9(6.6)	−10.1(4.7)	−0.7(6.6)
4	−12.3(7.6)	−5.6(6.6)	−9.7(5.2)	1.8(5.7)
5	−8.1(6.9)	0.0(6.6)	−5.1(4.6)	2.2(7.0)
6	−8.6(8.2)	−2.5(8.6)	−5.2(5.5)	3.3(6.8)
7	−1.6(6.7)	7.1(6.7)	−1.2(4.6)	8.1(8.8)
8	7.4(4.5)	18.7(4.8)	6.3(5.0)	19.9(7.9)
9	13.4(4.8)	25.5(8.4)	10.0(6.3)	22.9(9.9)

Source: Bridger, R.S. et al., *International Journal of Industrial Ergonomics*, 9, 235–244, 1992.

because the lumbar spine flexes to stop the head and torso from jackknifing backward over the legs as the pelvis tips backward. Bridger et al. (1992) investigated lumbar curvature and pelvic tilt in the body positions depicted in Figure 7.23. Table 7.1 summarizes the angles of pelvic tilt and lumbar lordosis in the different body positions in Figure 7.23. Positive integers represent lumbar lordosis and anterior pelvic tilt. As can be seen, the body positions fall roughly into three categories. Positions 8 and 9 are characterized by anterior pelvic tilt and marked lumbar lordosis. Positions 1, 2, 3, and 4 are characterized by negative pelvic tilt and a distinct loss of lordosis (negative lumbar angles). Positions 5, 6, and 7 have slight backward tilt of the pelvis and lumbar angles around zero.

7.3.1.4 Effect of the Lower Limbs on Pelvic Tilt and Lumbar Angle

As can be seen from Figure 7.23, with the torso and head erect, the posture of the pelvis and lumbar spine is largely influenced by the position of the lower limbs. Flexion at the hip and knee changes the lengths of the muscles around the pelvis and alters the equilibrium position by altering the balance of forces around the pelvis according to the length–tension relationship. In position 9 of Figure 7.23, for example, the knees are flexed by approximately 90°, shortening and therefore weakening the hamstring muscles, causing the pelvis to tilt anteriorly. The lumbar spine extends to keep the torso and head erect, resulting in an increase in lumbar lordosis.

Link et al. (1990) investigated indices of the relationship between muscle length and reduction in lumbar curvature from standing to sitting. Indices of hamstring and hip flexor muscle length are obtained by measuring the range of motion, in degrees, of the hip joint in flexion and extension, respectively (Figure 7.24). For measurement of hamstring and iliopsoas length, the knee is extended (A and B) and for rectus femoris length it is flexed (C). These angular measures are referred to as "lengths" because it is the passive length of the muscle, as it reaches its elastic limit, that prevents further rotation of the hip joint (see Bridger et al. 1989b, for a more detailed description of these hip mobility indices and the methods used to obtain them). Two chairs were investigated by Link et al. (1990): a conventional chair and a chair with a forward sloping seat. Hip flexor length was the only variable that related significantly to the change in lumbar curvature when subjects sat down. Subjects with short hip flexors experienced the largest change in lumbar curvature when sitting down on either of the chairs. This was interpreted as follows. In standing, those with short hip flexors have greater lumbar curvature because the iliopsoas muscle is "tight" and pulls on the pelvis, holding it in an anteriorly tilted position. As soon as the hips flex, the iliopsoas muscle shortens, allowing the pelvis to tilt rearward (due to the action of the force couple described above). Thus, those who have the most lumbar curvature in standing have the most to lose when they sit down.

These findings, and their interpretation, are supported quite independently by the findings of Bridger et al. (1992). For body positions 1, 3, 4, 5, and 6 in Figure 7.23, the hip flexor length was the best predictor of the change in pelvic tilt from that in standing. This indicates that, as soon as a standing person flexes the hip joints to sit down, the hip flexors (iliopsoas and rectus femoris) slacken off almost immediately and the anteriorly tilted pelvis tilts backward, flattening the lumbar curve. This happens long before the hips have been flexed sufficiently to stretch the hamstrings. Thus, contrary to the

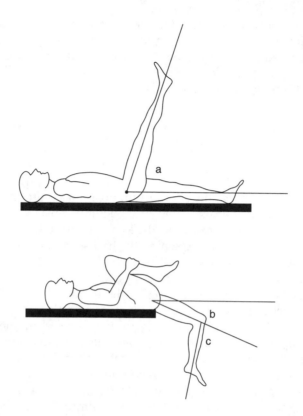

FIGURE 7.24
Joint ranges measured to obtain angular indices of hip muscle lengths. For measurement of hamstring and iliopsoas length, the knee is extended (a and b) and for rectus femoris length it is flexed (c).

standing position where a "flat" back may be due to tight hamstrings, normal individuals develop a flat back when they sit down because of the shortening and weakening of the hip flexors, which begins to happen almost as soon as the hips begin to flex.

Furthermore, in many sitting postures experienced in everyday life, the knees as well as the hips are flexed. Because the hamstrings cross both the hip and knee joints, this knee flexion will counteract the effect of hip flexion on the hamstrings and, to a certain extent, will prevent the hamstrings from becoming stretched. The hip flexors, however, cross only the hip joint, and their length is unaffected by the posture of the knees. Thus, the process of postural adaptation to sitting depends on two factors, the shortening and weakening of the hip flexors, which depend only on the amount of hip flexion, and the lengthening and stretching of the hamstrings, which depend on both hip and knee flexion.

Position 2 in Figure 7.23 is an exception to this, however. Here, hamstring length accounted for more of the variability in change in pelvic tilt. This is not surprising, as the knees are extended causing maximum hamstring stretch in the seated position and causing a large posterior pelvic tilt in subjects with short hamstrings. In postures 1 and 6, both muscle groups were associated with change in pelvic tilt. For females, the hip flexor index was the best predictor of change in pelvic tilt in most of the postures.

In general, loss of lordosis and change in pelvic tilt depend more on the extent to which the body position requires the hip to be flexed than on the angle of knee flexion. Both Brunswic (1984) and Eklund and Liew (1991) found that hip flexion is more powerful than knee flexion in its effect on spinal posture (up to eight times more powerful). Figure 7.25 summarizes the relationship between lumbar curvature, hip flexion, and knee flexion in different sitting positions.

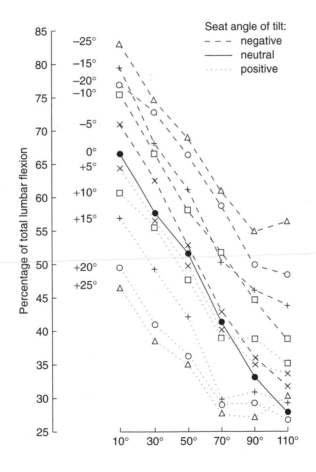

FIGURE 7.25

Relationship between lumbar flexion (as a percentage of total flexion) and knee flexion (degrees) on seats with angles of tilt ranging from 25° backward ("negative tilt") to 25° forward ("positive tilt"). (From Brunswic, M., *Physiotherapy*, 70, 39–43, 1984. With permission.)

TABLE 7.2

Ranges of Motion (degrees) of the Pelvis
(sagittal tilting) in the Nine Body Positions
Shown in Figure 7.23

Body Position	Males (N = 25)	Females (N = 25)
1	8.0	13.5
2	8.4	16.0
3	18.2	23.9
4	18.9	25.6
5	24.0	28.0
6	23.4	28.3
7	25.4	32.2
8	26.7	25.1
9	18.1	22.9

Source: Bridger, R.S., 1991a. Interrelationships be-
tween Spinal Curves, Pelvic Tilt and Hip Mobility and
Their Implications for Workspace Design, Ph.D. thesis,
University of Cape Town.

7.3.1.5   Pelvic Range of Movement in Different Body Positions

As has been shown, the lumbar spine, pelvis, and hip joints form a co-
articulated linkage and can move in relation to each other and the rest of
the body. Humans can arch their backs or tilt the pelvis forward or backward.
When these volitional movements cease, the joints return to a resting position
in which the forces acting on them are in equilibrium. Lumbar and pelvic
angles in these resting positions were measured by Bridger (1991a) in the
body positions in Figure 7.23. However, in each position, there is a range of
movement (ROM) of the spine and pelvis. Table 7.2 gives the range of pelvic
tilt in each of the body positions in Figure 7.23. The data were obtained by
having subjects carry out maximum volitional anterior and posterior pelvic
tilts in each body position.

As can be seen, the body positions differ not only in terms of the absolute
lumbar and pelvic angles (Table 7.1) but also in the degree of postural
constraint, with females tending to have greater ROM in all body positions.
Postures 1 and 2 are the most constrained and postures 6, 7, and 8 are the
least constrained. Interestingly, for both males and females, significant associations were found between the ability to tilt the pelvis forward from rest
in positions 2, 3, and 5 and the index of hamstring length. Individuals with
short hamstrings had a lesser ability to tilt the pelvis forward — as might
be expected, because the knees are extended in these positions and the
hamstring muscles are likely to be stretched already and will therefore limit
further forward movement.

These findings provide support for the recommendation that is often made
by ergonomists that the knees should be flexed when a seated person leans
forward. In constrained seated postures, the flexion strain on the lumbar spine

will increase as the person leans forward, as the hamstrings prevent the pelvis from doing so. The findings also suggest that kneeling or semikneeling seated postures are optimal for trunk posture in seated work, partly because excessive loss of lumbar lordosis is avoided and partly because the pelvis is less constrained. Movements of the trunk can therefore take place by tilting the pelvis forward on the hip joints, rather than by flexing the lumbar spine.

7.3.1.6   Lumbar Spine Range of Motion in Different Body Positions

Eklund and Liew (1991) and Bridger et al. (1992) captured data on the range of flexion/extension of the lumbar spine. This enables lumbar curves in any particular body position to be "normalized" (Figure 7.25). According to Bridger et al. the standing lumbar lordosis is located at about 65 to 70% of the range of motion, where 0% is full flexion and 100% is full extension of the lumbar spine. This estimate is somewhat lower than that of Eklund and Liew, possibly due to a more severe method of attaining full extension. However, it seems that standing is a relatively extended position and any workplace factors that increase the standing lordosis (such as having to adopt position 9 in Figure 7.23 for long periods) will move the spine to the extremes of its joint range. This is commonly held to cause stress and discomfort (Van Wely 1970, Eklund and Liew 1991) and will shift more of the compressive load of the upper body onto the facet joints of the posterior spinal column, a possible factor in the degeneration of these joints (Adams and Hutton 1980).

The only sitting postures in Figure 7.23 that are close to the midpoint of the lumbar ROM are 5, 6, and 7. Postures 1, 2, and 3 are not recommended for any type of sustained work as they are close to the fully flexed position. Furthermore, these postures are constrained posteriorly such that, even with conscious effort, it is not possible for subjects to tilt the pelvis forward far enough to the middle of the lumbar ROM. Indeed, even subjects using conventional chairs were unable to reproduce their normal standing lumbar lordosis by means of a maximal volitional anterior pelvic tilt maneuver.

7.3.2   Ergonomic Implications

Evidence is mounting that both postural constraint and extreme lumbar postures contribute to the prevalence of back problems and the degeneration of spinal structures such as the intervertebral discs (Keyserling et al. 1988). Several ergonomic interventions are available to improve the posture of the pelvis and reduce postural constraint.

7.3.2.1   Sitting Posture: Improving the Posture of the Pelvis and Related Segments

Bendix et al. (1985, 1986) evaluated an office chair with a seat that could tilt from 5° backward to 10° forward over a transverse axis. The purpose of the seat was to prevent postural fixity by providing users with the ability to "rock" backward and forward at will. Subjects were found to prefer the tiltable seat and to use the flexibility provided to change posture — tilts exceeding 2° were observed each minute, on average, with as many as ten smaller tilts each minute. The seat was moved more frequently and with a greater range when the seat height was 6 cm above the subjects' popliteal height compared to 1 cm below. Foot swelling was found to be greater with the higher seats, whether tiltable or fixed forward sloping, but there were no effects of seat type on lumbar muscle activity. However, Van Dieën et al. (2001) found greater stature gains (indicative of lower trunk loading) when subjects sat on dynamic chairs compared to a chair with a fixed seat and backrest. Van Deursen et al. (2000) designed a dynamic office chair in which rotations about an axis perpendicular to the seat were applied by a motor at an amplitude of 0.6° and a frequency of 0.08 Hz. The application of these gentle twisting motions was found to result in increases in spinal length over a 1-h period of sitting, significantly more so than when subjects sat in a static control chair.

These findings support the hypothesis that rotation applied to the vertebrae via the pelvis during sitting reduces pressure in the nucleus pulposus and allows fluid to enter, increasing disc thickness and improving the nutritional status of the disc.

When provided with additional workstation flexibility in the form of a tilting seat, subjects do use it and appear to prefer it to similar chairs with fixed seats. Interestingly, in the 1985 and 1986 studies by Bendix et al., the tilt facility was used less frequently and the muscular load was higher in a typing task than in the subjects' ordinary deskwork (not involving typing). This indicates that tasks involving typing are constraining and should be combined with other work.

Udo et al. (1999) compared a tilting seat with a fixed seat during the performance of a word processing task. The tilting frequencies were much lower than in Bendix et al.'s (1985, 1986) experiments (25 per hour) but a reduction in back discomfort was reported when the tilting seat was used. Greater low back electromyographic (EMG) activity was recorded with the tilting seat, suggesting that the reduction in pain was bought about by the reduction of static back muscle activity. As with Bendix et al.'s experiments, no differences in lower leg swelling were found. However, Stranden (2000) did find that tilting seats produced a significant reduction in leg swelling (although the difference was small, of the order 1% of calf volume, which may explain why previous researchers found no difference). According to Stranden, the key to venous pump activation is leg movement. Plantarflexion, in particular, is immediately effective and can be elicited by reclining on a tilting seat or by resting the feet on *dynamic footrests* that are pivoted to allow a "treadle pumping" kind of action.

7.3.2.2   Standing Posture: Improving the Posture of the Pelvis and Related Segments

Several researchers have investigated the effects of footrests/footrails and toe space in standing.

7.3.2.2.1   Footrests and Footrails

Placing one foot up on a footrest is thought to "release" the ipsilateral iliopsoas muscle and allow the pelvis to tip backward slightly, thereby preventing excessive lumbar lordosis and additional compression of the facet joints. Bridger and Orkin (1992) determined the effect of a footrest on pelvic angle in standing. The footrest raised the resting foot 250 mm above the level of the floor and resulted in a net posterior rotation of the pelvis of 4° to 6° (up to 12° in some subjects, similar to the pelvic adaptation to a high saddle seat according to the data of Bendix and Hagberg 1984). Whistance et al. (1995) confirmed this finding: use of a footrail reduced anterior pelvic tilt, straightened the supporting leg, and increased the plantarflexion of the supporting foot. The footrest would appear to be a valid way of reducing lumbopelvic constraint in standing workers and help prevent discomfort in the lumbopelvic region. Rys and Konz (1994) have reviewed recent research on the ergonomics of standing. Use of a 100-mm foot platform by subjects was perceived as more comfortable than normal standing in 9 of 12 body regions, including the neck. Use of either a flat or a 15° tilted platform was perceived to be better than a simple footrail, but all standing aids were preferred to standing on a bare floor. During a 2-h period of standing, subjects placed one foot on the platform 83% of the time, switching their foot from the platform to the floor every 90 s on average. The freedom to stand with one foot forward and elevated seems to be an important feature of a well-designed standing workplace.

7.3.2.2.2   Toe Space

Panels or obstructions in front of benches cause users to stand farther away from the work surface (Figure 7.26). Fox and Jones (1967) observed that, after having to lean forward to work for many years, dentists did so by arching the back in the thoracic region or by flexing the lumbar spine or both. Constrained standing for many years seems to cause people to use the pelvis as if it were part of the legs and to bend forward from the segments above — using the lumbar spine as a false joint to lean forward. Whistance et al. (1995) found that a lack of toe space does indeed cause people to deviate from the neutral standing position, leaning forward by means of a combination of pelvic tilting and lumbar flexion and placing more stress on the spine. The provision of toe space can prevent this (Rys and Konz 1994, Whistance et al. 1995).

FIGURE 7.26
Experimental investigation of standing posture — lack of toe space. (From Whistance, R.S. et al. 1995. *Ergonomics,* 38, 2485–2503. With permission from Taylor & Francis (http://www.tandf.co.uk).)

7.3.2.3   Reclining and Lying Down: Improving the Posture of the Pelvis and Related Segments

Few jobs encourage people to lie down when they are working! Nevertheless, extreme postures and postural constraint can occur when lying down, particularly when lying in the supine and prone positions. When lying supine on a hard surface, additional hip extension is needed before the surface supports the weight of the legs. The legs then act as a lever to tilt the pelvis anteriorly and exaggerate the lumbar lordosis. Clinicians often recommend the adoption of the "Fowler" position to prevent this from happening: the trunk is raised 10° to 15° and a pillow placed under the knees to flex the hip joints. Lying flat on one's back is not recommended as a working position, although car mechanics sometimes lie on boards when working underneath cars. If such a posture is unavoidable, an ergonomic recommendation would be that there should be sufficient space above the worker to enable the knees to be flexed. In prone lying, pillows are often placed under the upper body, causing increased lordosis, best prevented by not lying or sleeping in this position at all.

7.3.3   Final Remarks

It is often suggested that lumbar spine degeneration is a disease of Western society caused by an "unnatural lifestyle." Several cross-cultural investigations of the incidence of spinal pathology have been carried out. Fahrni and Trueman (1965) compared the incidence of disc space narrowing and pathological change in the spines of North Americans and Europeans with that in a jungle-dwelling tribe in India. In all groups, lumbar degeneration increased with age, as might be expected, but the trends were steeper for the Western group. Disc space narrowing was much less marked in the Indians and was hardly age related at all.

Fahrni and Trueman, finding that the Indians in their study habitually squatted and avoided extreme lumbar postures, suggested that excessive lumbar lordosis caused by the prevailing postural repertoire might be the cause of accelerated disc degeneration in Western society. Although habitual daily activities influence the incidence of disease, there is no reason to think that the mechanical strain on the spine is greater in people who stand compared to people who sit — what evidence there is suggests the opposite. In addition, not all cross-cultural studies show spinal pathology to be a disease of industrialized countries (e.g., Tower and Pratt 1990). Anderson (1992) reports studies of central Asian peasants and of subsistence farmers in the remote highlands of Papua New Guinea that show there to be as high a prevalence of mechanical back pain among these communities as there is in the industrially developed world.

Postural behavior and motor habits develop through childhood and are influenced by custom and the particular environment. Milne and Mireau (1979) found that hamstring length decreased throughout childhood and adolescence, but there was a marked drop around the age of 6 years, which they attributed to increased chair sitting when the children started school. Preschool children spend long periods sitting on the floor in postures that require the joints to be held in a wide range of quasi-static postures. Fisk and Baigent (1981) found that a spinal pathology in young people (Scheuermann's disease) was related to extreme hamstring shortness. They postulated that the hamstring shortness restricted movements of the hip, thereby placing a greater postural adaptation load on the spine in movements requiring flexion of the trunk.

Although a great deal is known about the requirements for workspace design in the Western post-industrial milieu, it is clear that much more research is required to elucidate the relationship between the postural behaviors and motor habits that people in a particular culture acquire over the course of their lives and the incidence of discomfort and disease. Whether current design practice is optimal is not known, particularly for implementation in developing countries. Further research in this area will undoubtedly lead to a better understanding of the true limitations and possibilities of human anatomy.

Summary

The evidence confirms that the pelvis is, indeed, the foundation on which the spine stands and that pelvic constraint and posture have a profound influence on the spine. For optimal pelvic posture and minimal pelvic constraint a number of ergonomic recommendations can be made for the design of fixed workstations:

- For seated work, design around a trunk–thigh angle of at least 105°
- Provide seats with a forward tilt facility and backrests with lumbar supports
- For standing workers, provide a footrest or footrail
- Avoid static postures that constrain the pelvis

Although it is rarely possible for practitioners to measure the pelvic tilt of workers, estimates can be made by observing the sagittal angle of the trouser seams at the level of the pelvis. In standing, with the pelvis in an anteriorly tilted position, the seam is approximately vertical. In slumped sitting, it points backward by up to 45°. When the lumbar support of an office chair is used correctly, the seam is more vertical, indicating that the support is preventing posterior pelvic tilt.

References

Aagaard-Hansen, J. and Storr-Paulsen, A., 1995. A comparative study of three different kinds of school furniture. *Ergonomics*, 38, 1025–1035.

Adams, M.A. and Hutton, W.C., 1980. The effect of posture on the role of the apophyseal joints in resisting intervertebral compressive forces. *Journal of Bone and Joint Surgery*, 2B, 358–362.

Åkerblom, B., 1948. Standing and Sitting Posture, Thesis, Akademiske Bokhandeln, Stockholm.

Anderson, R., 1992. The back pain of bus drivers. *Spine*, 17, 1481–1489.

Andersson, G. and Örtengren, R., 1974. Lumbar disc pressure and myoelectric activity during sitting. II. Studies on an office chair. *Scandinavian Journal of Rehabilitation Medicine*, 6, 115–121.

Bendix, T., 1984. Seated trunk posture at various seat inclinations, seat height and table heights. *Human Factors*, 26, 695–703.

Bendix, T., 1987. Adjustment of the seated workplace — with special reference to heights and inclinations of seat and table. *Danish Medical Bulletin*, 34, 125–139.

Bendix, T. and Biering-Sorensen, F., 1983. Posture of the trunk when sitting on forward-sloping seats. *Scandinavian Journal of Rehabilitation Medicine*, 15, 197–203.

Bendix, T. and Bloch, I., 1986. How high should a seated workplace with a tiltable chair be adjusted? *Applied Ergonomics*, 17, 127–135.

Bendix, T. and Hagberg, M., 1984. Trunk posture and load on the trapezius muscle whilst sitting at sloping desks. *Ergonomics*, 27, 873–882.

Bendix, T., Winkel, J., and Jessen, F., 1985. Comparison of office chairs with fixed forwards or backwards inclining, or tiltable seats. *European Journal of Applied Physiology*, 54, 378–385.

Bendix, T., Jessen, Bo., and Winkel, J., 1986. An evaluation of a tiltable office chair with respect to seat height, backrest position and task. *European Journal of Applied Physiology*, 55, 30–36.

Bendix, A.F., Jensen, C.V., and Bendix, T., 1988a. Posture, acceptability and energy consumption on a tiltable and a knee-support chair. *Clinical Biomechanics*, 37, 66–73.

Bendix, T., Jessen, F., and Krohn, L., 1988b. Biomechanics of forward-reaching movements while sitting on fixed forward- or backward-inclining or tiltable seats. *Spine*, 13, 193–196.

Bendix, T., Poulsen, V., Klausen, K., and Jensen, C.V., 1996. What does a backrest actually do to the lumbar spine? *Ergonomics*, 39, 533–542.

Branton, P., 1969. Behaviour, body mechanics and discomfort. *Ergonomics*, 12, 316–327.

Branton, P. and Grayson, G., 1967. An evaluation of train seats by observation of sitting behaviour. *Ergonomics*, 10(1), 35–51.

Bridger, R.S., 1985. A preliminary study of chairs with forward sloping seats, and sitting postures. *South African Journal of Physiotherapy*, 41(3), 74–77.

Bridger, R.S., 1988, Postural adaptations to a sloping chair and worksurface. *Human Factors*, 30, 237–247.

Bridger, R.S., 1991a. Interrelationships between Spinal Curves, Pelvic Tilt and Hip Mobility and Their Implications for Workspace Design, Ph.D. thesis, University of Cape Town.

Bridger, R.S., 1991b. Some fundamental aspects of posture related to ergonomics. *International Journal of Industrial Ergonomics*, 8(1), 3–15.

Bridger, R.S., 1995. *Introduction to Ergonomics*. New York: McGraw-Hill.

Bridger, R.S. and Orkin, D., 1992. Effect of a footrest on standing posture. *Ergonomics South Africa*, 4, 42–48.

Bridger, R.S., Von Eisenhart-Rothe, C., and Henneberg, M., 1989a. The effects of seat slope and hip flexion on spinal angles in sitting. *Human Factors*, 31, 679–688.

Bridger, R.S., Wilkinson, D., and Van Houweninge, T., 1989b. Hip joint mobility and spinal angles in standing and in different sitting postures. *Human Factors*, 31, 229–241.

Bridger, R.S., Orkin, D., and Henneberg, M., 1992. A quantitative investigation of lumbar and pelvic postures in standing and sitting: interrelationships with body position and hip muscle length. *International Journal of Industrial Ergonomics*, 9, 235–244.

Bridger, R.S., Kloote, C., Rowlands, B., and Fourie, G., 2000. Palliative interventions for sedentary low back pain: the kneeling chair, the physiotherapy ball and conventional ergonomics compared, in *Proceedings of IEA/HFES2000*, San Diego, CA: Human Factors and Ergonomics Society and the International Ergonomics Association, 5-87–5-90.

Brunswic, M., 1984. Ergonomics of seat design. *Physiotherapy*, 70, 39–43.

Cantoni, S., Colombini, D., Occhipinti, E., Grieco, A., Frigo, C., and Pedotti, A., 1984. Posture analysis and evaluation at the old and new work place of the telephone company, in *Ergonomics and Health in Modern Offices*, Grandjean, E., Ed., London: Taylor & Francis, 455–464.

Daltroy, L.H., 1997. A controlled trial of an educational program to prevent low back injuries. *New England Journal of Medicine*, 337, 322–328.

Delleman, N.J., 1999. Working Postures: Prediction and Evaluation, Ph.D. thesis, Vrije University, Amsterdam.

Delleman, N. and Miedema, M., 1998. Multi-dynamic office chairs — natural sitting behaviour with proper body support, in *PREMUS-ISEOH '98, 3rd International Scientific Conference on Prevention of Work-Related Muscoloskeletal Disorders*, Helsinki: Finnish Institute of Occupational Health, 152.

Donelson, R., Aprill, C.N., Medcalf, R., and Grant, W., 1997. A prospective study of centralization of lumbar and referred pain: a predictor of symptomatic discs and annular competency? *Spine*, 22, 1115–1122.

DonTigny, R.L., 1985. Function and pathomechanics of the sacro-iliac joint. *Physical Therapy*, 65, 35–44.

Eklund, J. and Liew, M., 1991. Evaluation of seating: the influence of hip and knee angles on spinal posture. *International Journal of Industrial Ergonomics*, 8, 67–73.

Fahrni, W.H. and Trueman, G.E., 1965. Comparative radiological study of the spines of a primitive population with North Americans and northern Europeans. *Journal of Bone and Joint Surgery*, 47-B, 552–555.

Fisk, J.W. and Baigent, M.L., 1981. Hamstring tightness and Scheuermann's disease. *American Journal of Physical Medicine*, 60, 122–125.

Fox, J.G. and Jones, J.M., 1967. Occupational stress in dental practice. *British Dental Journal*, 123, 465–473.

Graf, M., Guggenbühl, U., and Krueger, H., 1991. Movement dynamics of sitting behaviour during different activities, in *Proceedings of the 11th Congress of the International Ergonomics Association*, Quéinnec, Y., and Daniellou, F., Eds., London: Taylor & Francis, 15–17.

Graf, M., Guggenbühl, U., and Krueger, H., 1993. Investigation on the effects of seat shape and slope on posture, comfort and back muscle activity. *International Journal of Industrial Ergonomics*, 12, 91–103.

Graf, M., Guggenbühl, U., and Krueger, H., 1995. An assessment of seated activity and postures at five workplaces. *International Journal of Industrial Ergonomics*, 15, 81–90.

Grandjean, E., 1987. Postural problems in office machine work stations (introductory paper), in *Ergonomics in Computerized Offices*, Grandjean, E., Ed., London: Taylor & Francis, 445–455.

Grieco, A., 1986. Sitting posture: an old problem and a new one. *Ergonomics*, 29, 345–362.

Grieco, A. and Molteni, G., 1999. Seating and posture in VDU work, in *The Occupational Ergonomics Handbook*, Karwowski, W. and Marras, W., Eds., Boca Raton, FL: CRC Press, 1779–1791.

Hartvigsen, J., Leboeuf-Yde, C., Lings, S., and Corder, E.H., 2000. Is sitting-while-at-work associated with low back pain? A systematic, critical literature review, *Scandinavian Journal of Public Health*, 28, 230–239.

Hartvigsen, J., Bakketeig, L.S., Leboeuf-Yde, C., Engberg, M., and Lauritzen, T., 2001. The association between physical workload and low back pain clouded by the "healthy worker" effect: population-based cross-sectional and 5-year prospective questionnaire study. *Spine*, 26, 1788–1793.

Hazard, R., 2001. Personal communication.

Helander, M.G., Zhang, L., and Michel, D., 1995. Ergonomics of ergonomic chairs: a study of adjustability features. *Ergonomics*, 38(10), 2007–2078.

Henriques, V.E., 1985. Ergonomics is good for what you need. *The Office*, 62–64.

Holm, S. and Nachemson, A., 1983, Variations in the nutrition of the canine intervertebral disc induced by motion. *Spine*, 8, 866–874.

Hoogendoorn, W.E., Van Poppel, M.N., Bongers, P.M., Koes, B.W., and Bouter, L.M., 1999. Physical load during work and leisure time as risk factors for back pain. *Scandinavian Journal of Work, Environment & Health*, 25, 387–403.

Horner, H.A. and Urban, J.P., 2001. 2001 Volvo Award Winner in Basic Science Studies: effect of nutrient supply on the viability of cells from the nucleus pulposus of the intervertebral disc. *Spine*, 26, 2543–2549.

Hrobjartsson, A. and Gotzsche, P.C., 2001. Is the placebo powerless? An analysis of clinical trials comparing placebo with no treatment. *New England Journal of Medicine*, 344, 1594–1602.

Jensen, C.V. and Bendix, T., 1992. Spontaneous movements with various seated-workplace adjustments. *Clinical Biomechanics*, 7, 87–90.

Kapandji, I.A., 1992. *The Physiology of the Joints*. Vol. 3, *The Trunk and Vertebral Column*. London: Churchill Livingstone.

Keegan, J.J. 1953. Alterations of the lumbar curve related to posture and seating. *Journal of Bone and Joint Surgery*, 35A, 589–603.

Kendall, H.O., Kendall, F.P., and Boynton, D.A., 1971. *Muscles: Testing and Function*, Baltimore: Williams & Wilkins.

Keyserling, W.M., Punnett, L., and Fine, L.J., 1988. Trunk posture and back pain: identification and control of occupational risk factors. *Applied Industrial Hygiene*, 3, 87–92.

Kim, J.Y., Stuart-Buttle, C., and Marras, W.S., 1994. The effects of mats on back and leg fatigue. *Applied Ergonomics*, 25, 29–34.

Laville, A., 1980. Postural reactions related to activities on VDU. Postural problem, section 5, in *Ergonomics Aspects of VDU*, Grandjean, E. and Vigliani, E., Eds., London: Taylor & Francis, 167–174.

Link, C.S., Nicholson, G.G., Shaddeau, S.A., Birch, R., and Gossman, M., 1990. Lumbar curvature in standing and sitting in two types of chairs: relationship of hamstring and hip flexor muscle length. *Physical Therapy*, 70, 611–618.

Lord, M.J., Small, J.M., Jocylance, M., Dinsay, R.N., and Watkins, R.G., 1997. Lumbar lordosis: effects of standing and sitting. *Spine*, 22, 2571–2574.

Luopajärvi, T., 1987. Prevention of work-related neck and upper-limb disorders, in *Work-Related Musculo-Skeletal Disorders, Proceedings of an International Symposium*, Osterholtz, U., Karmaus, W., Hullmann, B., and Ritz, B., Eds., Hamburg: University Hospital Eppendorf, Institute of Medical Sociology, 420–434.

Magora, A., 1972. Investigation between low back pain and occupation. 3. Physical requirements: sitting, standing and weight lifting. *Industrial Medicine*, 41, 5–9.

Mandal, Å.C., 1985. *The Seated Man: Homo sedens*. Klampenborg, Denmark: Dafnia Publications.

Mandal, C., 1982. The correct height of school furniture. *Human Factors*, 24, 257–269.

Milne, R.A. and Mireau, D.R., 1979. Hamstring distensibility in the general population: relationship to pelvic tilt and low back stress. *Journal of Manipulative and Physiological Therapeutics*, 2, 146–150.

Reinecke, S.M., Hazard, R.G., Coleman, K., and Pope, M.H., 1992. A continuous passive lumbar motion device to relieve back pain in prolonged sitting, in *Advances in Industrial Ergonomics and Safety IV*, Kumar, S., Ed., London: Taylor & Francis, 971–976.

Reinecke, S.M., Hazard, R.G., and Coleman, K., 1994. Continuous passive motion in seating: a new strategy against low back pain. *Journal of Spinal Disorders*, 7, 29–35.

Rys, M. and Konz, S., 1994. Standing. *Ergonomics*, 37, 677–687.

Schoberth, H., 1962. *Sitzhaltung, Sitzschaden, Sitzmöbel* [Seating: Posture, Injuries and Furniture], Berlin: Springer Verlag.

Snijders, C.J., Slagter, A.H.E., Van Strik, R., Vleeming, A., Stoeckart, R., and Stam, H.J., 1995. Why leg crossing? *Spine*, 20, 1989–1993.

Stranden, E., 2000. Dynamic leg volume changes when sitting in a locked and free-floating tilt office chair. *Ergonomics*, 43, 421–433.

Takishita, S., Touma, T., Kawazoe, N., Muratani, H., and Fukiyama, K., 1991. Usefulness of leg-crossing for maintained blood pressure in a sitting position in patients with orthostatic hypotension; Case reports. *Angiology*, 42, 421–415.

Tile, M., 1984. *Fracture of the Pelvis and Acetabulum.* Baltimore: Williams & Wilkins.

Tower, S.S. and Pratt, W.B., 1990. Spondylolysis and associated spondylolisthesis in Eskimo and Athabascan populations. *Clinical Orthopaedics and Related Research,* 250, 171–175.

Udo, H., Fujimora, M., and Yoshinaga, F., 1999. The effect of a tilting seat on back, lower back and legs during sitting work. *Industrial Health,* 37, 369–381.

Van Deursen, D.L., Goossens, R.H.M., Evers, J.J.M., Van der Helm, F.C.T., and Van Deursen, L.L.J.M., 2000. Length of the spine while sitting on a new concept for an office chair. *Applied Ergonomics,* 31, 95–98.

Van Deursen, L.L., Patijn, J., Durinck, J.R., Brouwer, R., Erven-Sommers, J.R., and Vortman, B.J., 1999. Sitting and low back pain: the positive effect of rotary dynamic stimuli during prolonged sitting. *European Spine Journal,* 8, 187–193.

Van Dieën, J.H., De Looze, M.P., and Hermans, V., 2001. Effects of dynamic office chairs on trunk kinematics, trunk extensor EMG and spinal shrinkage. *Ergonomics,* 44, 739–750.

Van Wely, P., 1970. Design and disease. *Applied Ergonomics,* 1, 262–269.

Verbeek, J.H., 1991. The use of adjustable furniture: evaluation of an instruction programme for office workers. *Applied Ergonomics,* 22, 179–184.

Whistance, R.S., Adams, L.P., Van Geems, B.A., and Bridger, R.S., 1995. Postural adaptations to workbench modification in standing workers. *Ergonomics,* 38, 2485–2503.

Wickström, G. and Bendix, T., 2000. The "Hawthorne Effect" — what did the original Hawthorne studies actually show? *Scandinavian Journal of Work, Environment & Health,* 26, 363–367.

Wilke, H., Neef, P., Hinz, B., Seidel, H., and Claes, L., 2001. Intradiscal pressure together with anthropometric data — a data set for the validation of models. *Clinical Biomechanics,* 16(Suppl. 1), S111–126.

Williams, M.M., Hawley, J.A., McKenzie, R.A., and Van Wijmen, P.M., 1991. A comparison of the effects of two sitting postures on back and referred pain. *Spine,* 16, 1185–1191.

Winkel, J., 1987. On the significance of physical activity in sedentary work, in *Work with Display Units,* Knave, B. and Widebäck, B.-G., Eds., Amsterdam: North Holland, Elsevier Science.

Winkel, J. and Oxenburgh, M., 1990. Towards optimizing physical activity in VDT/office work, in *Promoting Health and Productivity in the Computerized Office: Models of Successful Ergonomic Interventions,* Sauter, S., Dainoff, M., and Smith, M., Eds., London: Taylor & Francis, 94–117.

8

Leg and Foot

CONTENTS

Introduction

This chapter contains three sections on the lower extremities. The first section begins with a brief general overview of studies on pedal operation (which is one of the main work applications in which evaluation of leg posture is required), and then focuses on the use of leg-operated pedals in a sitting posture, particularly on the kinematics and kinetics of force exertion, strength data, perception of pedal forces, preferred postures and movements, and optimum layout of pedals.

The second section reports on the effects of flooring during prolonged standing, and in particular on workers' health (in relation to causes of pain in the lower extremities and low back), body part discomfort, overall tiredness, localized muscle fatigue (usually monitored by electromyographic [EMG] recordings), body sway, and effects on cardiovascular function (leg swelling and skin temperature).

The third section describes the basic postures and dynamic postural responses of the body that occur during walking. These are relevant to study of slips and falls, the biomechanics of slips, and the environmental factors affecting postures, in particular slip resistance of the shoe–floor interface and its measurement.

8.1   Pedal Operation

Xuguang Wang and Margaret I. Bullock

Introduction

Two categories of pedals are distinguished according to the mode of operation: ankle- and leg-operated pedals. The first requires a rotation of the foot about a fixed point such as the supported heel point. The second requires a thrusting action of the whole lower limb. The focus of this section is on the operation of these two kinds of pedals.

Foot controls often restrict the posture of the user even more than hand controls, and inappropriate pedal design may contribute to muscle fatigue and cause discomfort for users. Surprisingly, very few studies of the comfort of pedal operation have been reported (see Haslegrave 1995, for a review). Kroemer (1971) conducted an extensive review of the investigations on pedal operations, and also conducted an experiment showing that the subject could perform the foot control task with considerable accuracy and very quickly. However, most of the studies reviewed by Kroemer (1971) concerned hinged pedals with a pivot near the pedal surface. The review focused mainly on the influence of pivot location on the speed of activation. It also focused on static forces applicable to pedals. Pedal sizes and relative positions between pedals, especially between brake and accelerator pedals, were also studied by several investigators (Johansson and Rumar 1971, Drury 1975, Snyder 1976, Glass and Suggs 1977, Morrison et al. 1986, Hoffman 1991). Most of these studies aimed at reducing the braking reaction. For the same purpose, a combined accelerator–brake pedal was also studied (Konz et al. 1971, Poock et al. 1973). As Casey and Rogers (1987) rightly pointed out, pedal design recommendations should not be evaluated from a single criterion of reducing reaction time; other considerations such as anthropometry and control modulation must also be taken into account.

It is believed that good pedal design needs to take into account not only the operator's strength capability but also biomechanics and kinematics of exertion. In the following sections, to begin with, pedal force control during pedal operation is analyzed. Then, the results of the kinematic analysis of the lower limb movement for leg-operated pedal operations are presented, mainly based on a study by Wang et al. (2000). Afterward, preferred lower limb positions for driving are reviewed, and optimum layout of automobile pedals is discussed. In the case of the seated operator with no backrest, the results from an extensive study by Bullock (1991) for determining the pedal location and path are reviewed. Finally, static foot strength is presented as well as the perception of pedal force.

The scope of Section 8.1 is limited to pedal operation in a seated position, more specifically for driving. Moreover, the emphasis is on biomechanical analysis. For those who are interested in general guidelines and recommendations for the design, selection, and location of foot controls, the work of Bullinger et al. (1991) presents a systematic approach to optimize the design variables based on ergonomic and safety considerations.

8.1.1 Force Control When Depressing a Pedal

8.1.1.1 Ankle-Operated Pedals

An ankle-operated pedal is designed so that the pedal can be depressed through foot rotation about the heel, which is supported by the floor. It is

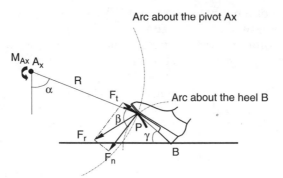

FIGURE 8.1
A biomechanical model for force control when depressing an ankle-operated pedal.

generally considered as a suitable choice only when the required pedal force is not high.

A biomechanical analysis of the control of pedal force is helpful to understand pedal operation and its design. As shown in Figure 8.1, the pedal force F_r can be decomposed into two components, one perpendicular (F_n) and the other tangent (F_t) to the pedal arc. The static equilibrium equation (Equation 8.1) can be written as follows:

$$M_{Ax} = F_n R \sin\beta \pm F_t R \cos\beta \qquad (8.1)$$

where M_{Ax} is the pedal resistance moment around the pivot Ax, R is the distance between Ax and the contact point P and β the angle between the direction of F_n and $\overrightarrow{PAx}$. The second term of the right side of Equation 8.1 is the contribution of the foot–pedal friction. Its sign depends on the direction of foot pedal relative movement, which is directly related to the direction of foot movement (upward or downward). For a pendant pedal such as the one shown in Figure 8.1, foot pedal friction disappears only when the pivot Ax, contact point P and heel B are aligned. Non-alignment of these three points causes sliding of the foot on the pedal surface. If the coefficient of friction between shoe and pedal surface is denoted as μ, then Equation 8.1 can be written

$$M_{Ax} = F_n R(\sin\beta \pm \mu \cos\beta) \qquad (8.2)$$

From Equations 8.1 and 8.2, it can be seen that the pedal force F_r and its direction are completely determined by three factors: pedal resistance, its location with respect to the foot, and foot–pedal friction. Different shoes and pedal surfaces change the foot–pedal friction. This may affect the perception of the force required for a pedal operation, implying that the pedal surface should be appropriately selected for an ankle-operated pedal. A low coefficient of friction is generally suggested to avoid the impression of the shoe sticking on the pedal (Bullinger et al. 1991). To reduce the amplitude of foot slip movement on the pedal surface, this simple biomechanical analysis (illustrated in Figure 8.1) also suggests that the pedal pivot Ax, the foot–pedal contact point P (assumed to be at the ball of the foot) and the heel B should be as closely aligned as possible. In addition, pedal travel length should be small (although the effect of this on sensitivity of control must also be considered).

Figure 8.1 shows only the pendant type of pedal, for which the pedal pivot is above the floor and farther from the heel than the pedal surface itself. There exists another type of ankle-operated pedal, "organ-type," for which the pivot is close to the heel. The general suggestions for pedal design from the above biomechanical analysis also apply to organ-type pedals. The pedal pivot should be as close as possible to the heel to reduce the relative sliding movement of the foot on the pedal. A more detailed analysis of this kind of pedal can be found in Bullinger et al. (1991), who made a systematic analysis of pivot location on pedal operation. In addition to the argument for reducing the effect of friction, they also favored the close alignment of the axis of the foot movement and the pedal pivot because the relationship between foot rotation γ and pedal travel angle α becomes linear under this condition (see Figure 8.1 for angle definitions). A linear relationship between foot movement and pedal rotation is required for sensitive control operation.

8.1.1.2   Leg-Operated Pedals

A leg-operated pedal is designed so that the pedal is depressed through a thrusting action of the leg involving the movement of the whole lower limb without the heel fixed on the floor. Because the muscles of the whole lower limb are involved in such an action, a leg-operated pedal is generally considered a suitable choice when the required pedal force is high. The static equilibrium equation (Equation 8.1) holds also for this type of pedal. Compared to an ankle-operated pedal with the heel fixed to the floor, the kinematics of a leg-operated pedal exertion does not force the foot (shoe) to slide on the pedal surface. Although the tangent force component F_t is still provided by the shoe–pedal friction, its amplitude cannot be simply determined by μF_n. The shoe may roll over the pedal surface. To avoid the slipping effect, a certain level of shoe–pedal friction is required.

One can easily verify from Equation 8.1 that the moment produced by the tangent force F_t becomes negative when the foot–pedal contact point P is outside the arc $P_s P_1$ (Figure 8.2), where P_s is the intersection between $\overrightarrow{OAx}$ and the pedal circle and P_1 is the tangent point from Ax to the pedal circle. O and Ax are, respectively, the pedal circle center and the pedal pivot. From a biomechanical point of view, a negative moment means that it cancels out one part of the muscle force to overcome the pedal resistance. As a consequence, the pedal arc should be situated between P_s and P_1.

For an ankle-operated pedal, its pedal force F_r is solely imposed by the pedal design characteristics. However, for a leg-operated pedal, its pedal

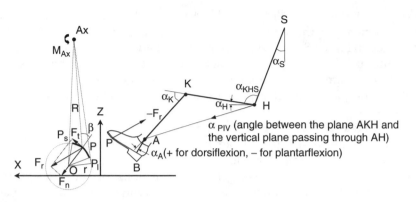

FIGURE 8.2
A biomechanical model for force control when depressing a leg-operated pedal.

force F_r depends also on the direction of the force exerted (or chosen) by the operator. The investigation of the control of pedal force direction may reveal some insight into motor control strategies. Wang et al. (2000) tested an intuitive hypothesis stating that pedal force direction is controlled such that the minimum joint muscle moments are applied, as described by the following equation:

$$Minimize\ f = k_A * \left\| \mathbf{M}_A \right\|^q + k_K * \left\| \mathbf{M}_K \right\|^q + k_H * \left\| \mathbf{M}_H \right\|^q \tag{8.3}$$

where M_A, M_K, and M_H are the joint moments at the ankle, knee, and hip and k_A, k_K, and k_H the associated weighting coefficients, which are inversely proportional to each maximum joint strength. The parameter q is defined as the best fit to the experimental data. Wang et al. (2000) showed that the force direction predicted by minimizing joint muscle moments agrees with their experimental data. Both simulation and experiments showed that pedal force direction is strongly dependent on the direction $\overrightarrow{HP}$ connecting the hip joint H and the foot–pedal contact point P. This suggests that the pedal thrusting direction should be specified by pedal designers to follow the direction $\overrightarrow{HP}$ to minimize joint muscle work during a pedal operation. It should be noted that the direction $\overrightarrow{HP}$ is mainly dependent on seat height.

8.1.2 Kinematics of Lower Limb Movement during a Pedal Operation by Leg Thrusting

For an ankle-operated pedal, lower limb position is completely imposed by pedal position because the heel is fixed to the floor. However, for a leg-operated pedal without the heel fixed on the floor, there is at least one degree of redundancy, given that the lower limb has at least three degrees of freedom

TABLE 8.1

Experimental Conditions Investigated in Wang et al.'s (2000) Study of Clutch Operation

Parameters That Could Be Adjusted by Subjects	Controlled Parameters	Subject Stature (Mean ± SD)	Experimental Apparatus
Clutch pedal position both in longitudinal (X) and vertical (Z) directions	Seat height (200–400 mm) Pedal travel length (100–170 mm) Resistance (80–140 N) Pedal travel angle (0°–30°)	Five short women (1514 ± 24 mm) Five average height men (1724 ± 7 mm) Five tall men (1848 ± 44 mm)	Multi-adjustable static driving package Electro-optic 3D ELITE motion capture system
Seat position in longitudinal (X) direction	Accelerator foot plane[a] angle depending on seat height		3D pedal force sensor
Steering wheel position	Seat cushion angle depending on seat height		

[a] Acclerator foot plane is defined as the plane perpendicular to the vertical longitudinal plane (XZ) containing the line linking the ball of the foot and the heel point.

(one at each of three joints, hip, knee, and ankle) and that the pedal provides only two equations of position in horizontal and vertical directions. This implies that more than one lower limb posture is possible for a given pedal position. It is, therefore, interesting to investigate how lower limb movement is controlled during pedal operation by leg thrusting.

Lower limb movement for clutch pedal operation was studied using a multiadjustable experimental vehicle package by Wang et al. (2000). Table 8.1 summarizes the experimental conditions studied. The movements were captured using an electro-optical system with two infrared cameras operating at 50 Hz. The kinematic linkage of the lower limb is considered to be composed of the thigh, shank, and foot segments, connected by the hip, knee, and ankle joints. No translation is allowed at any joint. Joint angles were estimated from trajectories of external markers placed on the bodies of the subjects. Spatial and temporal characteristics of foot trajectories were analyzed. The most striking observations are listed as follows:

- The pedal resistance affected neither the foot spatial path nor the joint angles of the lower limb, implying that pedal location could be specified primarily by taking into account geometric constraints. Only temporal characteristics of the foot trajectory varied with the change of pedal resistance. This was confirmed by a study on heavy trucks with pedal resistance as high as 220 N by Duqueroy (2000).

- The ankle motion during the approach phase, moving the foot from its resting position to the pedal contact, seems to be independent of the variables studied. A high ankle dorsiflexion nearly reaching its motion range limit was generally observed during the approach phase. By contrast, the ankle angle at the end of travel was strongly

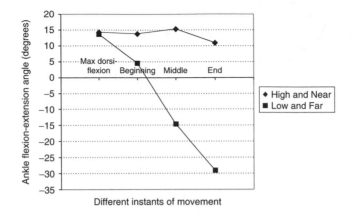

FIGURE 8.3

Influence of pedal location on the ankle flexion/extension angle during truck clutch pedal depression (data from Duqueroy 2000). The ankle flexion/extension angle α_A (group mean, $n = 15$) is shown at four different instants: maximum dorsiflexion during the approach, beginning of travel, mid-travel, and end of travel. The two curves are for two extreme pedal locations, located at the beginning of travel: high and near position ($Z = 200$ mm high, $X = 50$ mm from the accelerator heel point), and low and far position ($Z = 100$ mm high, $X = 200$ from the accelerator heel point).

dependent on pedal travel length and travel angle. During the depression phase, ankle motion can be mainly explained by collision avoidance of the heel with the floor. Duqueroy (2000) showed that the ankle angle tended to be kept constant during the whole depression phase when the pedal was positioned high and near to the body (Figure 8.3).

- When the pedal position could be adjusted in both horizontal direction and height, subjects tended to keep a knee flexion angle at midrange, close to the median of angles generally reported as comfortable (see Table 8.2). An average knee flexion angle (defined as α_k in Figure 8.2) of 62° was observed. Subjects also tended to place the pedal at a low position to avoid raising their leg, so that their heel stayed very close to the floor during the whole movement of pedal operation.

- A clear difference in the leg pivoting angle α_{PIV} (defined in Figure 8.2), which characterizes the leg opening or closing, was observed between male and female subjects. The female subjects tended to close the leg during depression, whereas the leg remained open for the male subjects (see also Table 8.3).

- The shoe–pedal contact point at the beginning of travel seemed to be invariant (around 72% of the shoe length from the heel) regardless of the pedal design parameters studied. This point is very close to the point preferred by the subjects when they were asked to locate their preferred foot contact point after having tested all trials.

TABLE 8.2

Preferred Joint Angles of the Lower Limb for Driving

			Preferred Joint Angles			Maximum Range of Motion Bullinger et al. (1991)	Comfort Range of Motion From Table 8.3
	Rébiffé (1969)	Grandjean (1980)	Ribouchon (1991) $n = 45, \pm$ SD	Porter and Gyi (1998) $n = 55$	Park et al. (2000) $n = 43$		
Trunk inclination α_S	20°–30°	—	23.4° ± 6.5°	—	—	—	—
Trunk–thigh angle α_{KHS}	95°–120°	100°–120°	131.7° ± 13.1°[c]	90°–115°	117.4° ± 7.7°	—	—
Hip angle α_H	—	—	—	—	—	—	>0°
Knee angle α_K	45°–85°[a]	50°–70°[a]	54.4° ± 10.2°[a]	42°–81°[a]	46.3° ± 8.53°[a]	5° to 130°[d]	40° to 75°
Ankle angle α_A	0°–20°[b]	0°–20°[b]	-3.7° ± 10°[b]	-23°–10°[b]	10.8° ± 8.6°[b]	20° to -40°	5° to -20°

Note: Maximum range and comfort range of motion are also summarized. The definitions of the joint angles used in this table are given in Figure 8.2. As different authors used different joint angle definitions, the data reported in Table 8.2 have been converted into angles in accordance with the definitions shown in Figure 8.2.

[a] 180° — initially published data.

[b] 90° — initially published data.

[c] Defined as the angle between the link hip–knee and the plane formed by two hip joint centers and the intervertebral joint L5/S1.

[d] The average angles of 113° and 125° are reported in Chaffin et al. (1999) for the maximum knee flexion in standing and prone position, respectively.

TABLE 8.3

Joint Angles of the Lower Limb Corresponding to the Configurations with Freely Chosen Clutch Pedal Parameters

Subject Group	At the Beginning of Travel				At the End of Travel			
	α_H	α_K	α_A	α_{PIV}	α_H	α_K	α_A	α_{PIV}
Short Women ($n = 5$)	16° ± 5.3°	70.4° ± 4.4°	-4.6° ± 5.8°	2.9° ± 8.1°	4.8° ± 4.6°	42.5° ± 4.8°	-27.2° ± 12.4°	4.9° ± 6.2°
Average men ($n = 5$)	16.5° ± 2°	73.7° ± 5.4°	6.4° ± 6.8°	-8° ± 6.9°	7.6° ± 2.3°	51.1° ± 9.1°	-16.2° ± 9.4°	-4.6° ± 5.5°
Tall men ($n = 5$)	13.7° ± 4.6°	64° ± 5.6°	2.3° ± 5.7°	-12.2° ± 5°	3.7° ± 4.3°	38.8° ± 8.8°	-22.3° ± 5.5°	-5.3° ± 5.4°
All	15.4° ± 4.1°	69.3° ± 6.3°	1.3° ± 7.4°	-5.8° ± 9.1°	5.3° ± 4°	44.1° ± 9°	-21.9° ± 10°	-1.6° ± 7.2°

Note: The average and standard deviation are shown for three groups of subjects (with different statures) at the beginning and end of pedal travel. The definitions of the joint angles used in this table are given in Figure 8.2.

Source: Wang, X. et al., INRETS-LBMC 9905, Bron, France: INRETS, 1999, 108 pp.

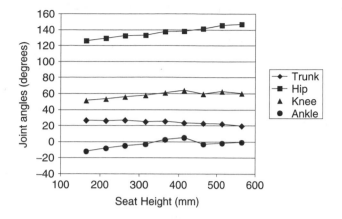

FIGURE 8.4
Influence of seat height on lower limb postures preferred for driving (group mean values from 45 subjects, 34 male and 11 female) (data from Ribouchon 1991). The angles for the trunk, knee, and ankle are defined as α_S, α_K, and α_A in Figure 8.2. The hip angle is defined as the angle between the hip–knee link and the plane formed by two hip joint centers and the intervertebral joint L5/S1. This angle is higher than the trunk–thigh angle α_{KHS} defined in Figure 8.2.

8.1.3   Preferred Lower Limb Postures and Optimal Layout of Pedals

8.1.3.1   Preferred Postures for Driving

Preferred postures for driving have been investigated by many researchers (Rebiffé 1969, Grandjean 1980, Verriest 1986, Ribouchon 1991, Porter and Gyi 1998, Park et al. 2000). In most of these experimental investigations, a multiadjustable driving rig was used, allowing subjects to choose their preferred driving postures themselves. As pedal operation mainly concerns the lower limb, only joint angles of comfort of the lower limb are summarized in Table 8.2. Interestingly, Ribouchon (1991) investigated postural comfort under the freely chosen condition (with no other constraint than looking forward for driving) as well as under the condition with imposed seat heights varying from 165 to 565 mm. Figure 8.4 shows the influence of seat height on the joint angles preferred for driving. It seems that seat height affects knee and ankle flexion–extension angles only slightly, whereas trunk inclination and trunk–thigh angle vary consistently with seat height. It should be noted that the preferred knee and ankle angles are, respectively, close to 54° and –4° (in plantarflexion), implying that the pedal should be located around these joint angles.

In the same experiment as that reported by Wang et al. (2000), each subject was also asked, after having evaluated the 30 different pedal configurations imposed by the experimental design, to freely choose all adjustable parameters including the four pedal design parameters being studied. Table 8.3 summarizes the average values of the joint angles at the beginning and end of travel for three groups of subjects (having different statures); these results are not published in Wang et al. (2000) but can be found in an internal research report (Wang et al. 1999).

The range of knee and ankle angles during depression of the pedal is within the range of preferred angles given in Table 8.2. It should be noted that these preferred angles of comfort were obtained in static situations and the dynamic effect when operating a control was not taken into account. The comfort range of motion for the knee and ankle given in Table 8.2 is therefore based on the data in Table 8.3, which resulted from the movement of depressing a fully adjustable clutch pedal. It can be noted that short-stature females had a very large plantarflexion at the end of travel. This is because the short female group had a high shoe heel (39 mm on average). The ankle joint motion data of the short female group is therefore excluded in the definition of the comfort range of ankle motion. It can also be noted that the proposed comfort range of ankle motion is within the 95th percentile values of ankle dorsiflexion and plantarflexion obtained by Nowak (1972) from an adult population of 257 women and 270 men.

8.1.3.2   *Optimal Layout of Automobile Pedals*

As the joint angles of the lower limb during pedal operation are not affected by pedal resistance, pedal location can be specified by geometric constraints, such as that the joint motions of the lower limb should be within the comfort range of motion as suggested in Table 8.2. For an ankle-operated pedal like the accelerator, the hip-to-heel distance should be determined for the knee angle to be within its comfort range. The pedal travel length should be determined according to the comfort range of motion of the ankle. To avoid relative movement between the foot and pedal, the pedal pivot should be located near the line from the heel to the ball of the foot at mid-travel.

For a leg-operated pedal like an automobile clutch pedal, the determination of its optimum layout is not straightforward as a result of kinematic redundancy of the lower limb with respect to the task of pedal operation. When the pedal position was adjustable in both horizontal and vertical directions, Wang et al. (2000) observed that the subjects preferred to position the pedal at a height such that the heel almost "slid" on the floor, thus avoiding unnecessary work to overcome gravity. If the heel–floor distance is imposed, the entire lower limb position is completely determined by pedal position if the hip pivoting movement along the line from the hip, and the pedal contact point is neglected. Using the data in Table 8.2, pedal location can be specified. For example, in the case of a seat height of 300 mm, Figure 8.5 shows maximum and comfortable reach zones of foot–pedal contact point of the two extreme subject groups (5th percentile short female and 95th percentile tall male). The heel–floor distance is fixed at 20 mm in this example. By superimposing the comfort zones of the two groups, the range of adjustment of the seat position can be determined, together with the pedal travel length as well as the travel angle, which should be close to the direction from the hip to foot–pedal contact point. Comparison of the preferred lower

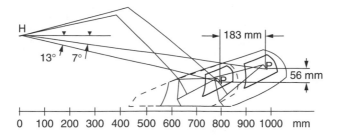

FIGURE 8.5

Maximum and comfortable foot reach zones of the two extreme subject groups (5th percentile short female and 95th percentile tall male) for a seat height of 300 mm and with the heel–floor distance fixed at 20 mm, derived from the maximum and comfortable ranges of joint motion in Table 8.2. The average preferred lower limb positions at mid-travel are also illustrated. The anthropometric dimensions are taken from the data of Rebiffé (1969). The distances from the heel to the foot-contact point (72% of shoe length) are 185 and 227 mm, respectively, for the 5th percentile short female and the 95th percentile tall male.

limb positions at the mid-travel of the two extreme groups shown in Figure 8.5 suggests that travel angle should be between 7° and 13° (or 10° for all drivers) and the seat track should have a minimum of 183 mm in the horizontal direction. It also suggests that the pedal should be adjustable in the vertical direction with a minimum range of 56 mm. This agrees with the results of the simulation study by Thompson (2001), who showed that an adjustable pedal system is required and suggested a minimum travel range of 76.2 mm rearward and 50.8 mm vertically. This allows short female drivers to be able to maintain a comfortable posture for both reaching the pedals and operating the steering wheel.

8.1.3.3 Optimal Pedal Path to Minimize Spinal Movements

Minimal body stresses for repetitive operations, such as those required during agricultural and industrial pedal use, should be a major criterion in the design of machines and the arrangement of work areas. Tractor operation often produces gross movements of the driver's trunk while depressing the clutch pedal (often at the rate of five times every minute, frequently against high resistance). Not infrequently, the tractor seat has no backrest. Similarly, operators of industrial presses often have to cope with demands for gross body movements during pedal work activity.

Highly repetitive trunk movements may lead to trunk muscle fatigue and to development of posterior ruptures in the lower two intervertebral discs (Hirsch and Nachemson 1954, Nachemson 1966, Rosseger and Rosseger 1960). The importance of providing guidelines for the design of repetitive operations, which cater for widely varying body sizes and which are based on consideration of the need to protect the body against injury, cannot be over emphasized. With this in mind, Bullock (1973, 1974, 1990, 1991) undertook a study to determine the pedal location and path in relation to the

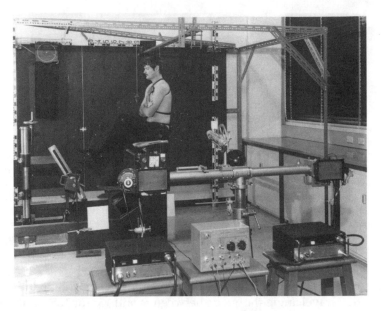

FIGURE 8.6
Experimental arrangement for a stereophotogrammetric study of spinal movements during pedal operation.

seated operator (where the seat has no backrest), which would produce minimal spinal movements during pedal depression.

Using stereophotogrammetry to measure body movements three dimensionally (Bullock and Harley 1972), Bullock (1973, 1990) studied spinal movements associated with depression of a pedal through 135 different pedal paths differing in proximity to the operator's foot on the platform, as well as in their angles with both the horizontal plane and the reference sagittal plane (the fore–aft vertical plane through the operator's hip joint). Figure 8.6 shows the experimental arrangement. For the purposes of analyzing body movements involved in the activity, "depression" was regarded as starting at the moment the subject's foot left the platform and finishing after completion of the thrust of the leg along a 10-cm path. Each travel path was related in a specific way to the seated starting position, and, by using spherical coordinates for the pedal orientations, alterations in position could be effected by making specific changes to the angulations of the hip joint when the foot was first placed on the pedal and to the extent of the operator's leg reach.

The variations of starting position fell on the segments of three spheres; the center of each was the hip joint. The radius of the first sphere was equal to the radial distance r from the hip joint to the ball of the foot when placed on the pedal in its starting position (SP). This was directly in front of the hip joint and as close as possible to the resting position of the foot on the platform while still allowing a 10 cm pedal depression movement — that is, at full

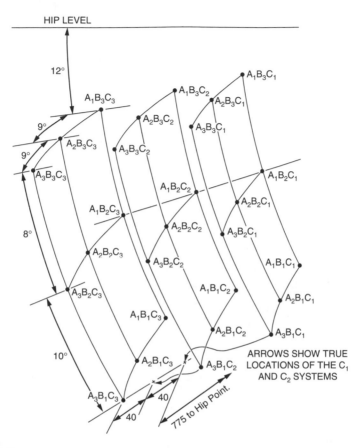

FIGURE 8.7
An "exploded" three-dimensional diagram showing the nine specified pedal locations on each of three spherical surfaces. *A* positions require varying degrees of hip abduction; *B* positions require varying degrees of hip flexion; *C* positions require varying degrees of leg reach.

depression, the pedal was within a "minimal leg reach" of the operator. This SP was defined as A_1, B_1, C_1 (Figure 8.7) and was the pedal position from which all other positions were derived. This commencing position demanded 10° of hip flexion to attain when moving from the limb position held in a correct sitting posture (that is, with hips and knees at right angles, and thighs horizontal and parallel).

Specific positions on any one of the spherical surfaces away from the SP were referenced according to whether the movement was in a lateral direction (*A*) or vertical direction (*B*). In this way, the positions A_2 and A_3 were defined lateral to the SP by abducting the leg through 9° and 18°, respectively (Figure 8.8). Attainment of commencing pedal positions on levels above the starting position (that is, at B_2 and B_3) were defined by decreasing the angle of *r* with the horizontal by 10° and 18°, respectively (that is, to 20° and 12°), as shown in Figure 8.8.

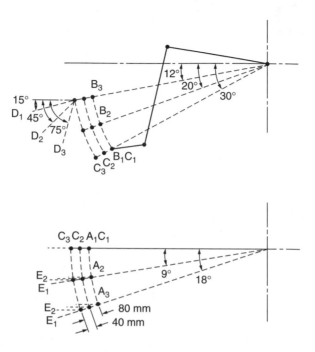

FIGURE 8.8
The specifications of pedal positions in elevation and plan views. *A* require varying degrees of hip abduction; *B* require varying degrees of hip flexion; *C* require varying degrees of leg reach; *D*, angles of pedal with horizontal; *E*, angles of pedal with reference sagittal plane.

Positions associated with leg reach were designated by the *C* nomination, shown in Figure 8.7 and Figure 8.8. The nine commencing pedal positions on each of the second and third spherical surfaces were defined by extending the leg from each position on the first spherical surface along the "hip-to-foot" line by 4 and 8 cm, respectively; the first of these positions requires a "moderate" leg reach (designated as C_2) and the latter position requires "almost maximum" leg reach after full pedal depression (designated as C_3).

Such a method of locating experimental pedal positions enabled the optimal pedal–operator relationship to be defined in terms of the magnitude of movements required to reach the pedals from the position of sitting with the feet on the platform. Also included were three angles of pedal travel path with the horizontal (D_1: 15°; D_2: 45°; D_3: 75°) and two angles with the reference sagittal plane: either that the pedal travel path be angled to that plane so that it might be continuous with the line of force from the hip to the foot on the pedal (E_1) or that it be parallel to that plane (E_2), which condition exists in present pedal arrangements. Figure 8.8 illustrates these angulations. Accordingly, from each of the possible spatial locations, the pedal could be oriented in six ways, so that its path could follow any of the combinations of pedal angles to the horizontal and to the reference sagittal planes.

In view of the absence of back support, the resistance of the pedal at the beginning of travel was set at 90 N and this was increased by 45 N per 2.5 cm of pedal movement to a maximum resistance of 270 N at the end of the pedal depression.

8.1.3.3.1 *Optimal Relationship of Pedal to Operator*

Using the criterion of minimum spinal movements, individual analyses of different combinations of the five basic variables revealed that the optimal relationship of pedal to operator is that which exists when the pedal position is as follows:

Pedal path:

- At 45° to the horizontal
- Continuous with the "hip to the foot on the pedal" line (drawn on the horizontal plane)

Spatial location:

- Anteriorly: within a minimal leg reach of the operator
- Vertically and laterally on a line between points requiring

    0° of hip abduction and 10° of hip flexion

    9° of hip abduction and 20° of hip flexion

8.1.3.3.2 *Modification of Existing Pedal–Operator Relationships*

Although recommendations regarding the optimal pedal–operator relationship are necessary when designing new work layouts, it is also important to be able to modify existing relationships. Although the angle of pedal travel path may not be changed, it should be relatively simple to alter the relationship of an individual operator to the pedal in three-dimensional space by providing adequate seat adjustments. For this reason, the locations considered to be most suitable for pedal paths fixed at specific angles were also defined by Bullock (1973, 1974, 1991):

Optimal pedal-operator relationship for a travel path parallel to the sagittal plane (E_2)

The study showed that it is preferable to align the pedal so that its travel path is continuous with the line between the hip and the foot on the pedal. However, spinal movements for such an "angled" pedal path were similar to those for the "parallel" pedal path. It could be noted that, for positions on the reference sagittal plane (designated as A_1), the pedal path is not only parallel to the sagittal plane but is also continuous with the hip to the foot on the pedal line. This particular A_1 specification is frequently included in the definition of the optimal pedal placement when satisfying various design constraints.

In a work situation where the pedal is depressed along a path parallel to the sagittal plane (E_2) it would be most suitably placed within the area bounded by the points requiring:

0° of hip abduction with 10° of hip flexion at close proximity to the operator $(A_1 \, B_1 \, C_1)$

9° of hip abduction with 10° of hip flexion at a moderate leg reach $(A_2 B_1 C_2)$

0° of hip abduction with 20° of hip flexion at a moderate leg reach from the subject $(A_1 B_2 C_2)$.

Pedal depression from a location within these boundaries should be along a 45° travel path (D_2).

Optimal locations for pedals aligned at various angles to the horizontal

A travel path at 15° to the horizontal (D_1) — Consideration of spinal movements during depression of pedals angled to the horizontal showed that the most suitable location for a pedal whose travel path must be aligned at 15° to the horizontal and parallel to the sagittal plane is in a high, medial position requiring 0° of hip abduction with 28° of hip flexion to attain it and almost maximal leg reach during depression $(A_1 B_3 C_3 D_1 E_2)$. At such a distant position, any shifting of the pedal's location should be within a short distance from this point in a downward and lateral direction (toward $A_2 B_2 C_3$).

Where pedal positions require a moderate leg reach for depression (C_2), similar A and B locations would be acceptable but not as suitable as that requiring only 0° of hip abduction and 10° of hip flexion $(A_1 B_1 C_2 D_1 E_2)$.

For pedal distances demanding only a minimal leg reach for depression (C_1), the commencing pedal position may be located on the low, medial triangular area defined by the points requiring:

0° of hip abduction with 10° of hip flexion $(A_1 B_1)$.

9° of hip abduction with 10° of hip flexion $(A_2 B_1)$.

0° of hip abduction with 20° of hip flexion $(A_1 B_2)$.

A travel path at 45° to the horizontal (D_2) — The most desirable location for a 45° pedal path at points requiring either a minimal or a moderate leg reach for depression have already been outlined in the description of the optimal pedal location for this parallel to the sagittal plane pedal alignment.

If the (45°) pedal were to be located at a position requiring a maximal leg reach on pedal depression (C_3), it could be at beginning positions demanding

0° of hip abduction with 10° or 20° of hip flexion $(A_1 B_1$ or $C_3 D_2 E_2)$

or

18° of hip abduction with 10° of hip flexion $(A_3 B_1 C_3 D_2 E_2)$.

A travel path at 75° to the horizontal (D_3) — Analyses of the effect of the pedal angle with the horizontal on the production of spinal movements showed that a steep pedal path (e.g., 75°) should be avoided as, in almost all locations, its use is associated with large range spinal movements. However, as this pedal alignment was found to exist in some tractors, the areas for its preferred location (in terms of spinal movements) should also be defined.

Pedal depression along a travel path at 75° to the horizontal and parallel to the sagittal plane should be from a starting position requiring 10° of hip flexion with 0° of hip abduction. Although a location on this low, medial line would be more desirable than higher, more lateral positions when placed at any distance from the subject, siting the pedal where it would require only a moderate leg reach for depression provides the widest area of choice ($A_1B_1C_2D_3E_2$).

8.1.3.3.3 Spatial Locations That Should Be Avoided

Results of this extensive study showed that in terms of spinal movements produced, there are certain commencing positions in which pedals should not be located no matter what pedal angles with the horizontal or with the sagittal plane are used. These include positions requiring:

18° of hip abduction with 20° or 28° of hip flexion (A_3B_2 or A_3B_3)

or

9° of hip abduction with 28° of hip flexion (A_2B_3).

In addition to these positions, other locations are disadvantageous when selected for placement of pedals at specific angles to the horizontal. For example, when an almost maximal leg reach is demanded for use of a pedal with a travel path of 15°, the pedal should not be located at a low, medial position (requiring 0° of hip abduction with 10° of hip flexion, $A_1B_1C_3D_1$). Similarly, when this extensive leg reach is required for use of a pedal with a travel path of 75° it should not be located in a high, medial position (requiring 0° of hip abduction with 28° of hip flexion, $A_1B_3C_3D_3$).

8.1.4 Static Foot Strength and Perception of Foot Force

Pedal design should take into account foot strength and also the perception of foot force. Kroemer (1971) provided a summary of maximum static forces that a sitting operator can exert on a pedal. Pedal force depends on seat type, pedal type and its position, as well as on instructions given to subjects. A strong pedal force can be exerted when a suitable backrest is provided and the pedal is nearly at seat height and at a distance that requires a knee flexion angle of 20° to 40°. A value of up to 2000 N for male maximum leg static thrust has been reported under suitable sitting conditions (Kroemer and Grandjean 1997), but in non-optimal postures the force is severely reduced.

Mortimer (1974) measured foot brake pedal force capability of drivers, and its distribution was obtained from a sample of 276 female and 323 male drivers. A thigh angle of 0° and a knee flexion angle of 160° (20° in flexion) were imposed to allow maximum force to be applied. Based on the 5th percentile female maximum brake pedal force, he suggested the force required to attain near maximum braking capability from a passenger car should not be more than 400 N. From a laboratory study with eight males and eight females, Pheasant and Harris (1982) showed that strength also depends on foot thrust direction. The forces in the direction of the hip (from the hip to the heel) were greater than those in the direction of the knee (from the knee to the heel) for all tested pedal locations. They found that the thrust force at a satisfactory driving posture, with the pedal being located at 12.5% stature below the seat reference point (SRP) and 47.5% in front (approximately leading to a hip angle of 12° to the horizontal and a knee angle of 113°) was 12% less than the maximum. The average strength of the females was 81% of the average male strength. It was also reported that left and right legs had the same strength.

Knowing the maximum strength is necessary but certainly not sufficient for pedal design. The holding duration and frequency of repetition of the pedal operation have a significant effect on fatigue biomechanical stresses. It is also required to know how people feel pedal force to determine the upper and lower limits for pedal force, especially for active pedals that make use of pedal force feedback to transmit information such as vehicle speed. In a study by Mick (1995), the relationship between the subjective sensation and the force applied to a pedal was studied to determine the upper force limit for a force feedback-based active accelerator pedal (and part of these results was published in Wang et al. 1996). The subjective perception of the pedal force was scaled using Borg's 0 to 10 category ratio scale (Borg 1982). The 0 value (rating = 0) corresponds to the resting force (F_{res}) when the foot is merely placed on the pedal without any further voluntary effort exertion. The value of 10 (rating = 10) is the maximal static force (F_{max}) that the foot can apply on the pedal without posture change. The seven intermediate levels were 0.5 (just noticeable), 1 (very weak), 2 (weak), 3 (moderate), 4 (somewhat strong), 5 (strong), 7 (very strong). In the experiment, 12 subjects aged from 24 to 53 years participated. They were grouped into four categories according to their stature and weight: "female short and light," "female average," "male average," and "male tall and heavy." Prior to experimentation, the subjects were asked to choose their preferred driving posture using a multiadjustable experimental simulation package. The meaning of the Borg CR-10 scale was then explained to them, particularly for the resting force level (0) and maximal force level (10). For each rating level of pedal force according to the Borg's CR-10 scale (presented in a random order), the subjects were asked to exert a force on a fixed pedal with the ball of their foot as if they were depressing an accelerator. The force exertion was recorded for about 5 s and the mean force was calculated during the three middle seconds. This was repeated three times. No feedback of foot force

TABLE 8.4

Main Results from the Experiment by Mick (1995) Who Studied the Perception of Static Pedal Force

	Female		Male		All
	Short and Light $n = 3$	Average $n = 3$	Average $n = 3$	Tall and Heavy $n = 3$	$n = 12$
F_{res} (rating = 0), N	10 ± 2.6	14 ± 3.7	26.3 ± 11.1	40.2 ± 5.3	20.2 ± 10.2
F_{max} (rating = 10), N	318 ± 190	276 ± 195	416 ± 116	677 ± 388	351 ± 168
Knee angle α_K	$49° \pm 10.2°$	$65° \pm 4.9°$	$48° \pm 7.5°$	$47° \pm 12.3°$	$54.2° \pm 11.6°$
Ankle angle α_A	$-5.8° \pm 15.4°$	$7.2° \pm 24.3°$	$5.6° \pm 2.9°$	$8.9° \pm 7.6°$	$3.6° \pm 14.4°$

Note: The resting (F_{res}) and maximal (F_{max}) static pedal force corresponding to ratings of 0 and 10 on the perceived exertion Borg CR-10 scale are shown, together with the knee and ankle postures adopted, for four different subject groups.

exerted was given to the subjects. The results are summarized in Table 8.4, which includes the knee and ankle angles adopted by the subjects because strength is dependent on posture. Statistical analysis of the ratings of perceived exertion of pedal force was performed to identify the least-square coefficients of a single variable regression for the power function, as shown (for the group of 12 subjects) in Equation 8.4.

$$\text{Pedal force (\%)} = 1.533 \times \text{Borg CR-10 rating}^{1.8} \text{ with } R^2 = 0.824 \quad (8.4)$$

with the pedal force normalized as $(\text{Measured force} - F_{res})/(F_{max} - F_{res}) \times 100$.

F_{res} is an objective measurement, which does not depend on the intention of the subject. It varied from 10 N for the "small and light" female subjects to 40 N for the "tall and heavy" male subjects with an average value of 20 N for all the subjects tested. A large variation of F_{max} was observed even within a same group. It varied from 276 N for the average female group to 677 N for the male tall and heavy group. From an experimental study of the discrimination of clutch pedal resistance, Southall (1985) suggested that pedal resistance should at least be required to overcome the effects of vibration and gravity. The experimental data from Table 8.4 suggest that a minimum pedal resistance of 40 N, which corresponds to the foot resting force of the "tall and heavy" male group, is required for the accelerator. As the accelerator is continuously controlled, a low pedal operation force is required. If the force is taken that corresponds to the sensation "moderate" with a rating = 2 for the "short and light" female group as the upper limit. Regression Equation 8.4 gives a force of 49 N. However, it should be noted that perception of a dynamic pedal force may be different from the perception of an isometric static force. For the clutch pedal, which is associated with a leg thrust, the resting pedal force caused by leg weight is certainly beyond the resting force with the heel supported on the floor, implying that the clutch pedal resistance should be higher than 40 N. According to Kroemer and Grandjean (1997, p. 171), pedal resistance should be higher than 60 N for pedals with a heavy tread. Further specific studies should be carried out to determine the required minimum clutch pedal resistance.

Summary

In this section, pedal operation has been treated mainly from a biomechanical point of view. In particular, the control of pedal force and lower limb movement have been analyzed during a pedal operation for both ankle- and leg-operated pedals. Preferred lower limb positions for driving have also been reviewed. Moreover, foot strength and perception of pedal force have been presented. The implications of these different aspects of pedal operation on pedal design have been discussed and practical guidelines are recommended. The main points of this section can be summarized as follows.

- For an ankle-operated pedal, its pivot, the ball of foot, and the heel point should be as closely aligned as possible if it is a pendant-type pedal. For a hinged pedal, its pivot should be close to the heel. In addition, pedal surface friction should not be too high to avoid the impression of the shoe "sticking" on the pedal.

- For a leg-operated pedal, pedal resistance has no effect on spatial characteristics of the movement of the lower limb, implying that optimum pedal location can be primarily specified by considering geometric constraints. Based on the criterion of minimization of joint moments, pedal travel angle should be in the direction of the line between the hip and the ball of the foot.

- For a leg-operated pedal, where there is no seat backrest (for example, in some agricultural tractors), the minimization of spinal movement during pedal depression is an important criterion for reducing back stresses. Based on this criterion, experimental observations have shown that it is preferable to align the travel path of the pedal so that it is continuous with the line between the hip and the contact point of the foot on the pedal, with a travel angle of 45° to the horizontal.

- For both ankle- and leg-operated pedals, their position should be specified in relation to the users' preferred postures, and the travel length and angle should be defined within the comfort range of lower limb joint motion.

- Pedal resistance should at least be high enough to overcome the effects of vibration and gravity. A minimum pedal resistance of 40 N is recommended for the accelerator, but further study is needed to determine the minimum resistance for the leg-operated clutch pedal, which will necessarily be higher than that for the accelerator pedal.

- Further studies are needed to determine the maximum acceptable pedal resistance.

Most of the studies during which these recommendations were developed used a static driving simulation package, and no real driving control task was imposed. Further studies are needed to determine optimal pedal resistance and travel with real driving tasks such as changing gear or clutch slipping. It should be noted that occupant packaging is a complex process (Roe 1993), and the whole posture, including the upper limb, should also be considered when designing foot controls. Other factors such as safety should also be taken into account.

8.2 Flooring and Standing

Rakié Cham and Mark S. Redfern

Introduction

Prolonged standing in confined spaces has often been associated with musculoskeletal problems. Those include feelings of discomfort, tiredness, and pain, and affect mostly the lower extremities and low back, but are also reported sometimes as overall body sensations. More serious health conditions such as lower extremity swelling and venous restriction have also been reported in populations required to stand for long periods of time.

One "ergonomic intervention" adopted in many occupational settings is the use of "antifatigue" mats or shoe inserts. Unfortunately, there are not enough scientific data to confirm the effectiveness of such interventions. Few laboratory or field research studies have investigated the effect of different types of shoe–floor interface on standing. Dependent measures believed to be related to standing fatigue and other medical conditions include subjective ratings of discomfort and biomechanical/physiological measures of muscle fatigue, leg volume changes, and postural movements. There is some consensus among researchers that, in general, standing on a softer floor is more "comfortable" than standing on a hard floor. However, there are significant conflicting findings regarding the quantitative effectiveness of the interventions, especially when the impact of flooring on objective biomechanical/physiological measures is considered. The reasons for the conflicting results are unknown. Furthermore, the underlying physiological mechanisms that would explain the effectiveness (or lack of) of the softer floor are unknown.

The overall goal of this section is to report on current knowledge regarding the effect of flooring on people during prolonged standing. More specifically, the first part of this section is dedicated to a summary of the epidemiological studies that have associated long-term standing with health problems. Then, the procedures (subjective and objective) and findings of laboratory and field studies are reviewed. Finally, future research directions focused on the long-term goal of optimizing ergonomic flooring interventions are suggested.

8.2.1   Epidemiological Findings: Impact of Flooring on Health Conditions

Lower extremity discomfort and pain are the most frequently reported problems associated with prolonged standing (Boussenna et al. 1982). Such complaints affect a wide range of occupations, including assembly workers and quality control inspectors (Redfern and Chaffin 1995), supermarket workers who are restricted in confined areas (Ryan 1989), health-care workers (Cook et al. 1993), and workers in the confectionery industry (Cook et al. 1993). The majority of the workers in those professions stand for long periods of time, for example, more than 4 h/day for female health-care workers in France (Estryn-Behar et al. 1990) and 70 to 80% of the work shift for Germans in the large-scale laundry industry (Hansen et al. 1998). In addition to feelings of discomfort and tiredness, standing has been found to worsen health conditions of workers diagnosed with major or minor chronic venous insufficiency including leg swelling (Krijnen et al. 1997).

Low back pain is another health problem associated with prolonged standing (Xu et al. 1997), especially among workers who spend more than 4 h a day standing (Magora 1972). Once again, their occupations include checkout supermarket personnel (Ryan 1989) and health-care workers (Cook et al. 1993, Troussier et al. 1993). In a 1-year follow-up study, MacFarlane and colleagues (1997) found that standing for more than 2 h was a predictor of low back pain among healthy workers in the female working population studied and that, in addition, standing was associated with an increased risk of recurring low back and neck pain episodes. Similar findings regarding the worsening of low back pain when standing for long periods of time were reported by Biering-Sørensen et al. (1989).

8.2.2   Findings of Relevant Published Studies

To examine the effect of prolonged standing and impact of flooring on people, investigators have selected the subjective and/or objective dependent measures shown in Table 8.5, which are believed to be associated with the health symptoms described previously. Subjective measures are discomfort or perceived tiredness ratings, while objective physiological and biomechanical measures include underfoot center of pressure (COP), movements of body segments, EMG, leg volume, and skin temperature. Review of the ten laboratory studies and one field study listed in Table 8.5 reveals sometimes similar but also disagreeing and even contradictory findings regarding the impact of flooring on standing fatigue. The similarities and differences across studies in each category of dependent measures are discussed here. Disagreements in the findings are interpreted in the light of differences in the experimental testing, data collection, and analysis procedures adopted in the various studies.

The numbers of subjects recruited in the studies in Table 8.5 ranged from 5 to 20. Subjects were tested wearing either socks only, i.e., no shoes, or a standard brand/model of footwear. Although in most laboratory studies

TABLE 8.5

Summary of Methodologies in Standing Fatigue Studies

	Rys and Konz 1988	Rys and Konz 1989	Konz et al. 1990	Zhang et al. 1991	Cook et al. 1993	Kim et al. 1994	Redfern and Chaffin 1995	Hansen et al. 1998	Krumwiede et al. 1998	Madeleine et al. 1998	Cham and Redfern 2001a
Study type and Testing duration	Laboratory study 2 h, 1 session/day	Laboratory study 1 h, 1 session/day	Laboratory study 90 min, 1 session/day	Laboratory study 2 h, 1 session/day	Laboratory study 2 h, 1 session/day	Laboratory study 2 h, 1 session/day	Field study 8 h shift/day, 5 shifts/floor	Laboratory study 2 h, 1 session/day	Laboratory study 1 h/floor, 3 sessions/day	Laboratory study 2 h, 1 session/day	Laboratory study 4 h, 1 session/day
No. of subjects	20	9	20	6	12	5	14	8	12	13	10
Independent variables	1 hard floor 1 carpet floor	1 hard floor 2 floor mats	1 hard floor 3 floor mats	1 hard floor 3 floor mats 1 hard-soled shoe 1 soft-soled shoe	1 hard floor 1 floor mat	1 hard floor 2 floor mats	1 hard floor 6 floor mats 1 shoe insert	1 hard floor 1 floor mat 1 hard-soled shoe 1 soft-soled shoe Standing v. standing and walking	1 hard floor 5 floor mats	1 hard floor 1 floor mat	1 hard floor 6 floor mats
Dependent variables related to fatigue	Subjective discomfort/ tiredness rating Foot volume Leg and foot dimensions Skin temperature Heart rate	Subjective discomfort/ tiredness rating Leg and foot dimensions Skin temperature	Subjective discomfort/ tiredness rating Leg and foot dimensions Skin temperature	Subjective discomfort/ tiredness rating COP EMG (leg) Posture Task performance	EMG (leg and back)	EMG (leg and back)	Subjective discomfort/ tiredness rating	Subjective discomfort/ tiredness rating COP Heel impact forces (walk) EMG (back) Foot volume Skin temperature Oxygen uptake Heart rate Blood pressure	Subjective discomfort/ tiredness rating	Subjective discomfort/ tiredness rating COP EMG (leg) Shank circumference Ankle movement Skin temperature	Subjective discomfort/ tiredness rating COP EMG (leg and back) Leg volume Skin temperature Task performance

Source: Redfern, M.S. and Cham, R., *American Industrial Hygiene Association Journal*, 61(5), 700–708, 2000.

each subject stood only once per day on a specific floor/shoe condition, one study tested each subject on three flooring surfaces per day (Krumwiede et al. 1998). The number of flooring conditions that were tested per study ranged from two to seven. Most studies included a hard or concrete flooring condition and at least one softer floor (carpet or mat). In addition to varying the flooring condition, three studies tested the use of shoe inserts or hard- vs. soft-soled shoes.

8.2.2.1   Subjective Measures

The subjective measures used in the studies included ratings of discomfort or tiredness collected via questionnaires and usually administered at regular time intervals ranging from 12 to 15 min (Zhang et al. 1991, Madeleine et al. 1998, Krumwiede et al. 1998) to 1 h (Cham and Redfern 2001a) during the standing period. An example of a questionnaire (used by Cham and Redfern 2001a) is shown in Figure 8.9. In the field study by Redfern and Chaffin (1995), workers rated their discomfort at the end of their 8-h shift. Subjects rated either their overall body tiredness/discomfort only (Hansen et al. 1998, Madeleine et al. 1998) or both overall body tiredness and specific body part discomfort levels for regions such as back, upper legs, knees, lower legs, ankles, and feet (Rys and Konz 1988, 1989, Konz et al. 1990, Zhang et al. 1991, Redfern and Chaffin 1995, Krumwiede et al. 1998, Cham and Redfern 2001a).

In general, researchers agree that flooring condition significantly affects workers' overall perceived tiredness and discomfort in various parts of the body including the lower back and the lower extremities (Rys and Konz 1988, 1989, Konz et al. 1990, Redfern and Chaffin 1995, Madeleine et al. 1998, Cham and Redfern 2001a). Table 8.6 and Figure 8.10 illustrate the main findings. As shown in Figure 8.10, Cham and Redfern (2001a) investigated the influence of flooring on overall body/leg tiredness and discomfort of individual body parts during standing. Ten subjects stood for 4 h on seven flooring conditions (one hard floor A and six floor mats B through G). During the third and fourth hours, the floor type had a statistically significant effect ($p < 0.05$) on a number of subjective ratings including lower leg and lower back discomfort/tiredness. On the other hand, the upper back discomfort and overall body tiredness ratings were not affected by the flooring condition. The hard floor performance was consistently rated the worst. Furthermore, *post hoc* tests indicated significant differences in some ratings not only between the hard floor and the floor mats, but also among a number of floor mats.

All of the studies in Table 8.6 that reported significant effects showed increased comfort on soft or so-called antifatigue mats compared to standing on hard concrete floors. Furthermore, significant differences in discomfort ratings were found among a number of the floor mats tested (Rys and Konz 1989, Konz et al. 1990, Redfern and Chaffin 1995, Cham and Redfern 2001a). Although the majority of the studies reported on the beneficial impact of

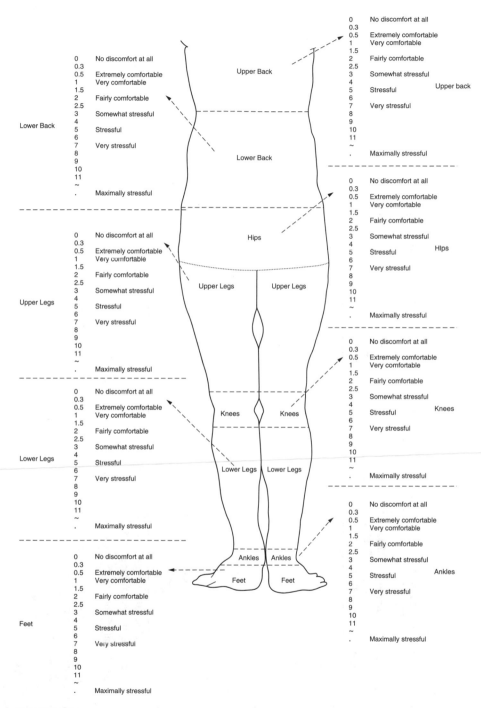

FIGURE 8.9

An example of a questionnaire (used by Cham and Redfern 2001a) for recording discomfort and tiredness ratings.

TABLE 8.6

Statistically Significant Effects ($p < 0.05$) of Floor and/or Shoes on Subjective Measures of Fatigue

Subjective Fatigue: Tiredness and Discomfort	Rys and Konz 1988	Rys and Konz 1989	Konz et al. 1990	Zhang et al. 1991	Redfern and Chaffin 1995	Krumwiede et al. 1998	Cham and Redfern 2001a
Overall fatigue	Yes, floor			No	Yes, floor/shoe insert	Inconclusive[a]	No
Overall leg fatigue				Yes,[b] only the second-order interaction term shoes × time was found statistically significant	Yes, floor/shoe insert		Yes, floor
Upper back	No	Yes, floor	No		Yes, floor/shoe insert		No
Mid back	No	Yes, floor	No			No	
Lower back	No	Yes, floor	No		Yes, floor/shoe insert	No	Yes, floor
Hips/buttocks	Yes, floor	No	Yes, floor		Yes, floor/shoe insert	No	Yes, floor
Upper legs	Yes, floor	No	Yes, floor		Yes, floor/shoe insert	Inconclusive[a]	Yes, floor
Knees					Yes, floor/shoe insert		Yes, floor
Lower legs	Yes, floor	Yes, floor	Yes, floor		Yes, floor/shoe insert	Inconclusive[a]	Yes, floor
Ankles	Yes, floor	Yes, floor	Yes, floor		Yes, floor/shoe insert	Inconclusive[a]	Yes, floor
Feet	Yes, floor	Yes, floor	Yes, floor		Yes, floor/shoe insert	Inconclusive[a]	Yes, floor

Note: Empty cells indicate that the study in the corresponding column did not investigate the impact of the flooring on the subjective fatigue variable mentioned in that specific row.

[a] A statistical analysis of variance was not performed to investigate possible significant differences in subjective responses among flooring conditions.

[b] Individual body part discomfort ratings were not considered in the analysis. Instead, all ratings were grouped in one measure that was used in the statistical analysis.

Source: Redfern, M.S. and Cham, R., *American Industrial Hygiene Association Journal*, 61(5), 700–708, 2000.

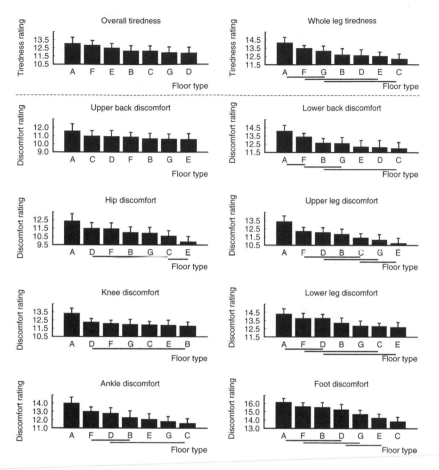

FIGURE 8.10

Impact of flooring condition on discomfort/tiredness ratings in the third and fourth hour of standing. This figure shows the average (and standard error) discomfort/tiredness ratings (higher rating indicates more discomfort/tiredness) as a function of floor type, including a hard floor A and a number of floor mats B through G. The horizontal bars show the results of *post hoc* tests. (Reprinted with permission from *Human Factors,* Vol. 43, No. 3, 2001. Copyright 2001 by the Human Factors and Ergonomics Society. All rights reserved.)

standing on softer floors when discomfort ratings are considered, others have not confirmed the effectiveness of this intervention (Zhang et al. 1991, Hansen et al. 1998). This disagreement could be due to differences in the following factors among studies:

- Duration of standing during the tests: The duration of standing varied across laboratory studies ranging from a relative short exposure of 1 h to exposure levels of 4 h. Cham and Redfern (2001a) found that testing sessions of a minimum of 3 to 4 h were needed to minimize the variability in the measures and to show statistical significance.

- Data collection methods: Various scales were used to collect discomfort ratings, ranging from 10-unit linear scales (Zhang et al. 1991, Madeleine et al. 1998, Krumwiede et al. 1998) to visual analog scales (Hansen et al. 1998) and including 5-unit linear scales (Redfern and Chaffin 1995), Corlett and Bishop scales (Rys and Konz 1988, Rys and Konz 1989, Konz et al. 1990, Zhang et al. 1991), and the Borg CR-10 scale (Cham and Redfern 2001a). The reproducibility/reliability of some of those scales (e.g., analog scales) for recording perceived tiredness and discomfort has not been fully evaluated.

- Analysis procedures: In some studies, summary measures are used instead of the direct discomfort ratings, combining the reports from all body areas into a single overall measure. This procedure could mask a significant flooring effect, which only affected one specific area of the body. Cham and Redfern (2001a), for example, found a significant impact of the flooring condition on the discomfort rating of the feet but not on that of the upper back. Thus, the relationship between the flooring condition and the combined discomfort measure (feet and upper back) may not be statistically significant.

Despite the differences in the results, the majority of the studies indicated that flooring can significantly affect standing comfort ratings. Overall, when subjective feelings of tiredness and discomfort were considered, soft shoe and soft floor interfaces were reported to outperform hard floors.

Wearing cushioned footwear also seems to alleviate discomfort and fatigue, although this type of intervention has less often been researched than floor mats. A significant relationship between the second-order interaction term of footwear × time and overall body discomfort was reported by Zhang et al. (1991), suggesting that footwear may be important for longer exposure times. Similarly, a significant effect of insoles on discomfort ratings was confirmed by Basford and Smith (1988) in a study that involved subjects standing during the majority of the workday. Redfern and Chaffin (1995) reported that viscoelastic shoe inserts improved comfort significantly compared to the performance of a hard floor and even of a number of the floor mats tested. Finally, Hansen et al. (1998) reported trends regarding the better performance of soft sole shoes compared to hard sole shoes, although the differences were not statistically significant. In summary, wearing shoes with soft soles or inserts may be beneficial; however, a rigorous comparison of the different types of interventions (mat, soft sole, insert) is needed.

8.2.2.2   *Center of Pressure and Other Biomechanical Measures*

Underfoot COP has been measured during long-term standing. The COP data indicate increased body sway, which has been assumed to be correlated (positively) with increased body stress. However, the results of investigations of the effect of flooring on COP measures are conflicting. On the one hand, some investigators have found a significantly greater amount of body sway (movement of the COP) and lateral weight shifting when standing on a hard surface or on other "uncomfortable" mats compared to standing on an "effective antifatigue mat" (Madeleine et al. 1998, Cham and Redfern 2001a). Generally, as discomfort ratings worsened, significant increases in COP shifts were recorded. It is important to note, however, that Cham and Redfern (2001a) found that significant differences in the number of weight shifts only occurred in the fourth hour of the testing session. On the other hand, studies reported by Hansen et al. (1998) and Zhang et al. (1991) have not confirmed those findings. Differences in experimental protocols may explain the differing results. In addition to the shorter testing duration in some studies (2 h vs. 4 h), other differences in the COP data collection across investigations can be identified. In particular, monitoring interval of force plate data varied from 1 min (Madeleine et al. 1998) to 1 h (Hansen et al. 1998). The time interval during which the COP positions were collected also varied, ranging from 5 s (Madeleine et al. 1998) to 10 min (Cham and Redfern 2001a).

The impact of the foot–floor interface on other biomechanical variables has been investigated, including ankle movements (Madeleine et al. 1998), the relationship between the center of gravity and foot positions (Zhang et al. 1991), and heel impact forces during a combination of standing and walking tasks (Hansen et al. 1998). All of those variables were found to be affected by the foot–floor interface conditions; however, once again, the results should be interpreted in light of the testing and processing/analysis procedures adopted in those studies.

8.2.2.3 Localized Muscle Fatigue

To assess the effect of flooring on local muscle fatigue, EMG recordings (of leg and back muscles) have been collected in some laboratory studies. Mean power frequency and amplitude of EMGs have been the dominant measures used. The findings concerning the effects of the floor–foot interface on EMG results during prolonged standing are contradictory. A number of studies have found no evidence of a foot–floor interface effect on lower back EMG (Cook et al. 1993, Cham and Redfern 2001a), while others have reported a significant impact (Kim et al. 1994). Similar mixed findings for the EMG recordings of the leg have also been reported; in some studies, no evidence of significant impact of flooring on lower extremity muscle fatigue was found (Zhang et al. 1991, Cook et al. 1993, Kim et al. 1994, Cham and Redfern 2001a), while others have found an effect (Madeleine et al. 1998). Once again, in addition to variability in exposure levels, differences in the EMG data collection and analysis procedures may be in part responsible for this apparent disagreement. Those differences include monitoring frequency of EMG data and task performed during data collection (standing or voluntary contraction), duration of EMG data collection, and choice of frequency or time domain analysis of the EMG data.

- Monitoring frequency and task performed during EMG data collections: Most studies collected muscle EMG data throughout the standing period at regular time intervals, ranging from 1 min (Cook et al. 1993, Madeleine et al. 1998) to 1 h (Hansen et al. 1998). Another way to assess leg/back EMG data was from pre- and post-standing tests using a maximal (Madeleine et al. 1998) or submaximal (Kim et al. 1994) contraction for the muscles of interest.

- Duration of EMG data collection varied from 5 s (Kim et al. 1994, Madeleine et al. 1998) to 5 min (Zhang et al. 1991).

- Frequency/time domain analyses: Both time (Zhang et al. 1991, Hansen et al. 1998, Madeleine et al. 1998) and frequency (Kim et al. 1994, Hansen et al. 1998, Madeleine et al. 1998, Cham and Redfern 2001a) domain analyses of the EMG data are found in the literature. Hansen et al. (1998) conducted both, with a time analysis indicating a significant difference in the left erector spinae EMG amplitude when pairwise comparisons among foot–floor interface conditions were performed. However, within the same study, those findings were not correlated with the results reported in the frequency domain.

8.2.2.4   Leg Swelling

Another objective variable that has been considered in flooring studies is leg swelling. It has been suggested that this variable is related to venous insufficiencies and leg tiredness. The majority of the studies have found that the type of foot–floor interface has a significant impact on leg swelling. Those include Brantingham et al. (1970), who tested subjects on two surfaces: a non-uniform resilient density flooring condition causing a slight tilt of the feet from the horizontal and a regular softened flooring surface. The experimenters alternated standing still, standing freely, and walking. Ankle movements during walking (on both flooring surfaces) were suggested to be associated with enhanced venous activity via an increase in the activity of muscles pumping fluid from the leg. During free standing, the non-uniform resilient density surface outperformed the flat surface significantly. Brantingham and colleagues explained this finding as the result of increasing activity of the lower leg muscles to maintain posture on the non-uniform surface, thus enhancing, once again, the activity of the venous pump.

Similarly, others indicated that there are significant differences in the amount of foot swelling with footwear and/or flooring conditions (Rys and Konz 1989, Madeleine et al. 1998, Hansen et al. 1998). The results of those studies were not always in agreement. For example, Madeleine and colleagues found that standing on a hard floor caused the most leg swelling, in apparent disagreement with the results of Hansen and colleagues who suggested that standing on a hard floor was better than standing on a soft mat. Hansen et al. (1998) also reported that soft-sole shoes outperformed hard-sole shoes when leg swelling was considered. In a study by Cham and Redfern (2001a) the following trends were found, although the differences were not statistically significant: the largest increase in leg volume was recorded for the hard floor and two other mat conditions that yielded the worst discomfort/tiredness ratings. Finally, Rys, Konz, and colleagues found no significant differences in foot volume when subjects stood on mats vs. a concrete floor (Rys and Konz 1988, Konz et al. 1990).

The mixed findings regarding the effect of flooring on leg swelling could be interpreted in light of the various measures of swelling used in the studies. For example, in a series of experiments, Rys, Konz, and colleagues examined the effect of flooring on foot/leg dimensions (length, width, area) and foot volume (Rys and Konz 1988, 1989, Konz et al. 1990), whereas shank circumference was recorded by Madeleine et al. (1998). Leg volume was the variable of interest in Cham and Redfern (2001a). In addition to the differences in standing exposures among studies, the monitoring frequency of foot/leg swelling was different across studies ranging from time intervals of 1 min (Madeleine et al. 1998) to only pre- and post-standing measurements (Rys and Konz 1989, Konz et al. 1990, Hansen et al. 1998, Cham and Redfern 2001a). Finally, the varied results may also be due to the different material characteristics of the flooring conditions. Thus, it is suggested that the material characteristics should be considered carefully when selecting mats or other floorings for people doing standing tasks, and that they should be reported in detail in any studies of the effects of foot–floor interfaces.

8.2.2.5   Skin Temperature

Researchers have assumed that, as postural muscles fatigue, their temperature and thus the temperature of the skin overlying them will increase. Therefore, based on this hypothesis, if the flooring condition affects "fatigue," differences in skin temperature among flooring conditions should be evident over time. This hypothesis has been confirmed by Konz et al. (1990), Rys and Konz (1989), and Cham and Redfern (2001a). However, in Cham and Redfern's (2001a) study the finding was significant for only the fourth hour of the experiment. Furthermore, the greatest increase in skin temperature was recorded when standing on floors that yielded the worst discomfort ratings, including the hard floor. Similar tendencies were reported by Madeleine et al. (1998), whereas Hansen and colleagues found no differences in the skin temperature among floor/shoe conditions (Hansen et al. 1998).

In summary, once again, results regarding the effect of foot–floor interface on skin temperature were mixed. The differences in results across studies may be attributed, again, to the differences in duration of testing and flooring material characteristics, and to the following factors:

- Different instruments were used to measure skin temperature: infrared technology (non-contact) was used by Madeleine et al. (1998) and Cham and Redfern (2001a) while a contact method using a thermistor was employed by Rys and Konz (1989) and Konz et al. (1990). The contact method may alter skin temperature recordings due to heat exchange effects between the skin and the transducer.

- The location of skin temperature recordings included the whole shank surface (Madeleine et al. 1998), specific temperature points on the calf (Rys and Konz 1989, Konz et al. 1990), foot instep (Rys and Konz, 1988, 1989, Konz et al. 1990) and foot dorsal venous arc (Hansen et al. 1998). More extensive measurements were made in the study by Cham and Redfern (2001a), in which skin temperature was collected over each of the following major muscle groups of the leg: soleus, tibialis anterior, quadriceps, and hamstrings. A control temperature was also recorded on the elbow (bony landmark with little underlying soft tissues) to remove any systematic changes in skin temperature due to possibly changing room conditions. To ensure that changes in skin temperature included in the analysis are not the result of temperature variations in the room, it is recommended either that a control temperature be recorded (as in Cham and Redfern 2001a) or that testing takes place in a temperature-controlled environment.

- The monitoring frequency ranged from 1 min (Madeleine et al. 1998) to 30 to 55 min (Rys and Konz 1989, Konz et al. 1990, Hansen et al. 1998).

8.2.2.6   Task Performance

Some investigators have hypothesized that discomfort and fatigue lead to loss of productivity and poor task performance. Two studies have tested this hypothesis. In Zhang et al. (1991), subjects performed a computer-generated visual search task during standing. Only second- and third-order interaction terms of footwear, floor and time were found to affect performance parameters significantly, indicating that the impact of foot–floor interface on task performance may become important over time. Cham and Redfern (2001a) used three tasks (a computerized version of the stroop task, target shooting, and a scanning task) and found a significant difference among flooring conditions in the performance of two of the three tasks. However, this finding was true for the third and fourth hour of standing only and no significant correlation between discomfort ratings and task performance was found. Thus, it appears that the flooring condition may have a beneficial effect on productivity for some tasks only. However, further research is needed to understand the impact of flooring on task performance more thoroughly.

8.2.2.7 Floor Material Properties

Flooring material properties thought to be related to standing comfort include thickness, compression characteristics and stiffness at specific load levels, bottom-out depth (an indication of the point at which the mat "flattens out"), elasticity and energy absorption properties (Konz et al. 1990, Kim et al. 1994, Redfern and Chaffin 1995, Hansen et al. 1998, Krumwiede et al. 1998, Cham and Redfern 2001a). Some investigators took one step farther and attempted to relate floor material properties to measures thought to be related to standing fatigue mentioned previously. For example, tiredness and discomfort were associated with (1) increased mat compression (Konz et al. 1990), (2) extremely soft or stiff materials (Redfern and Chaffin 1995), and (3) decreased elasticity, increased energy absorption, and decreased stiffness (Cham and Redfern 2001a). It is important to note that those findings hold only within the range of floor characteristics that were tested in the studies. Furthermore, some of those floor characteristics are correlated with each other and thus conclusions regarding the importance of each should be made with caution. In spite of those limitations, this type of analysis is useful in the identification of the material properties that have a significant impact on standing discomfort, tiredness, and other health problems.

Summary

Overall, current research has shown that flooring affects discomfort and fatigue. There appears to be a beneficial effect to standing on a softer floor as compared to a hard floor. Standing on a floor mat generally reduces subjective ratings of perceived tiredness and discomfort. However, floors that are "too" soft have also been shown to increase fatigue and discomfort. Thus, there appears to be some optimal range of flooring characteristics that improves standing comfort; however, current research has not yet identified this range. Experimental results concerning the impact of foot–floor interface on objective physiological or biomechanical measures are varied and sometimes contradictory across studies. Among those variables, weight shift measured by COP and normalized skin temperature were found to be positively correlated with subjective feelings of discomfort or tiredness; however, those relationships were only found to be statistically significant after a minimum of 3 h of standing. These correlations are useful as they suggest the possibility of using those objective measures to support subjective findings and to confirm that appropriate flooring may be an effective intervention for workers who stand for long periods of time. Nevertheless, it is important to note that the underlying physiological reasons for this positive correlation are not well understood and deserve to be investigated further. Finally, by affecting standing fatigue, the type of flooring may also have impact on productivity and task performance, although once again more research is needed in this area.

Differences in the methods that have been used in the various studies, including data collection procedures, testing protocols, and data processing and analysis, may be at the source of the disagreements in the findings concerning the impact of flooring on fatigue measures (both subjective and objective). The most important factor appears to be the duration of exposure. For example, it is believed that a minimum of 3 h of exposure time is required to detect significant differences in a number of measures of the effects of flooring conditions (Cham and Redfern 2001a). Laboratory tests of less than 3 h of exposure will not be sufficient to represent the impact that occurs in occupational settings. Further studies are required with long-duration exposure (8 h/day) to develop better empirical models of the relationship between flooring properties and musculoskeletal discomfort and fatigue.

Other differences in the methods that probably account for discrepancies in the findings across studies include different thermometric technologies and recording locations used for skin temperature measurements and variations in EMG data collection and analysis protocols (such as testing while standing vs. pre- and post-standing, using a maximal or submaximal reference muscle contraction, or analyzing in time vs. frequency domains). Another source of potential disagreement in the findings (e.g., in leg swelling measurements) across studies is the health status and age of the subjects participating in the studies. Finally, studies have used different flooring surfaces with different material properties, which may be another contributing factor to the varied findings.

In conclusion, standing on compliant flooring appears to reduce standing discomfort and fatigue. However, the optimal flooring characteristics have yet to be determined. A wider range of flooring material characteristics needs to be tested to identify and design the most effective floor. Finally, there is no consensus on an underlying physiological cause of discomfort and fatigue due to long-term standing. Continued research to identify the contributing factors would be of great benefit toward the development of better flooring design in the occupational setting.

8.3   Slips and Falls

Mark S. Redfern, Rakié Cham, and Brian E. Moyer

Introduction

Slips and falls are a major concern in the industrialized world. Falls are the second largest source of mortality due to unintentional injury (Fingerhut et al. 1998) and the leading cause of injury requiring medical attention (National Safety Council 1998, Warner et al. 2000). These injuries are found throughout society, including the workplace. Slips and falls are associated with the most severe occupational injuries (Courtney and Webster 1999) and are the second largest generator of workplace injuries overall (Leamon and Murphy 1995). Slipping is the dominant cause of falls. Courtney et al. (2001), in an excellent review of the epidemiology of slips and falls, reported that slips account for 32% of falls in young adults and 67% of falls sustained by elderly people. Thus, slips and falls are an important safety issue in industry and beyond.

The causes of slips and falls are complex, involving interactions of environmental and human factors. The environmental factors include properties of the floor surface (e.g., roughness, compliance, and topography) and shoe (e.g., material properties and tread), as well as other issues, such as lighting, housekeeping, and changes in elevation. Human factors include the biomechanics of gait, mental set (i.e., attention and alertness), and the health of musculoskeletal and sensory systems. These human factors result in dynamic postures that must interact with the environment to maintain balance. When there is a mismatch between the two, the result can be a slip or trip, possibly resulting in an injurious fall. An ergonomics approach to the prevention of falls is to attempt to design the environment to match the requirements of the person. This involves understanding the physical requirements of the tasks performed (such as standing, walking, or carrying) and ensuring that the environment meets the requirements for those tasks.

The purpose of this section is to describe the basic postures and dynamic postural responses of the body that occur during walking and that are relevant to slips and falls. The biomechanics of slips under varying environmental conditions are presented, including both kinematics and kinetics of the postural response. In addition, environmental factors that affect these postures are discussed. In particular, the most important environmental factor, namely, the slip resistance of the shoe–floor interface and its measurement, is discussed.

8.3.1 Biomechanics of Slips

Maintaining upright posture during locomotion is challenging and requires the successful execution of complex neuromuscular processes relying on sensory inputs (visual, vestibular, proprioceptive), information processing, and well-coordinated motor responses. The biomechanical analysis of gait on dry and contaminated floor surfaces plays a critical role in understanding numerous aspects of slips and falls to guide prevention research. Important aspects include (1) the relationship between the way humans walk and the risk of slipping, (2) the ability of the human postural control system to maintain balance during normal gait and to recover equilibrium should a slip occur, and (3) the biomechanical fidelity of slip resistance measurements (as discussed in Section 8.2.3). In particular, gait variables such as heel and foot dynamics, ground reaction forces, progression of COP, moments generated at the lower extremity joints, and kinematics of whole body postural reactions are of central interest to researchers in the field.

8.3.1.1   Classifications of Slips

The act of slipping is complex and does not always lead to a fall. In fact, a small amount of slipping at the heel typically occurs during most steps, even during normal locomotion on dry floors. Thus, a number of researchers have suggested various definitions to help classify and understand the postures and reactions involved during slips. Most definitions and classifications of slips are based on the slip distance (i.e., the sliding distance of the heel after heel contact) and the duration of movement. Slips are classified into three major categories:

1. Slipping that normally occurs during walking is usually imperceptible by the pedestrian (Leamon and Li 1990). These small slips are called "microslips" or sometimes termed "grips." They typically occur during normal locomotion on dry surfaces and are associated with slip distances below 1 cm (Perkins 1978, Strandberg and Lanshammar 1981, Leamon and Son 1989, Cham and Redfern 2002b).

2. "Macroslips with recovery" are slips that are greater than 1 cm, but do not result in a fall. These slips are usually characterized by slip distances on the order of a few centimeters (Perkins 1978, Cham and Redfern 2002b). In these cases, humans are able to perceive slipping and generate appropriate corrective reactions to regain balance. Strandberg and Lanshammar (1981) adopted a somewhat more detailed classification of macroslips with recovery that they referred to as slip-stick events. They divided these slips without falls into three groups: mini-slip (slipping motion goes unperceived by the walker), midi-slip (slip-recovery trials without "major gait disturbances") and maxi-slip (slip-recovery outcomes with large corrective responses or "near-fall" trials). In Strandberg and Lanshammar's (1981) experiments, mini-slip and midi-slip outcomes were characterized by mean slip distances of 1.2 (0.4 s.d.) cm and 5.1 (4.7 s.d.) cm, respectively, while a mean slip distance greater than 8.6 (3.7 s.d.) cm was reported for maxi-slip trials.

3. A third category is often termed the "slip-fall event," where the walker is not able to recover balance on his or her own and the heel does not come to a stop after heel contact. Slip distances greater than 10 to 15 cm are likely to result in falls (Perkins 1978, Cham and Redfern 2002b).

8.3.1.2   Heel and Foot Dynamics

The postures and dynamics of the heel and foot influence the slipping risk and are important in understanding the events that initiate and precipitate a fall. In slip events, the heel's sliding motion occurring shortly after heel contact is more pronounced than in no-slip conditions (Figure 8.11) (Strandberg and Lanshammar 1981, Cham and Redfern 2002b). The forward slip starts slightly after heel contact (about 50 to 100 ms as shown in Figure 8.11) (Perkins 1978, Strandberg 1983, Cham and Redfern 2002b). Despite the desta-bilizing nature of a slip, Cham and Redfern (2002b) have reported that subjects slipping on oily floors rotate the foot down onto the floor and reach foot-flat position at about 15% into stance regardless of the trial's outcome. As expected, recovering from slip events becomes more challenging as slip distances and peak slipping velocity of the heel increase. Strandberg (1983) suggested that a slip is likely to result in a fall if the slip distance is in excess of 10 cm or the peak sliding velocity is higher than 0.5 m/s; however, there is some disagreement in the literature regarding the threshold values beyond which a slip is considered to be unrecoverable (Brady et al. 2000). Finally, the kinematics of the heel/foot during slips on oil show a decrease of the heel's velocity, for both recovery and fall outcomes, starting approximately 190 ms after heel contact (Figure 8.11) (Cham and Redfern 2002b). This indicates an attempt by the walker to control the slipping motion of the foot. After that time, for slip-fall events, the heel accelerates again and eventually leads to a fall.

8.3.1.3 *Forces at the Shoe–Floor Interface*

Ground reaction forces recorded at the shoe–floor interface during locomo-tion on dry floors (i.e., no-slip conditions) (Figure 8.12a) have been used to assess the frictional requirements of slip-safe walking. A slip occurs when the shear forces generated by walking exceed the frictional capabilities of the foot–floor interface. Shear forces reach a local maximum shortly after heel contact and just prior to toe-off (Figure 8.12a). Thus, those are the two critical gait phases during which a slip is likely to occur. However, toe-off generates a slip in the rearward direction that is far less dangerous than the slip in the forward direction occurring shortly after heel contact. The ratio of the shear-to-normal ground reaction force (Figure 8.12b), described by Strandberg and Lanshammar (1981) as the instantaneous "friction used," may also be thought of as the "required" coefficient of friction (RCOF) on dry floors (Redfern and DiPasquale 1997, Hanson et al. 1999). The peak RCOF ($RCOF_{peak}$) occurring shortly after heel contact has been suggested to predict slip potential for various gait activities and is believed to represent the minimum frictional requirements needed (in terms of foot–floor coefficient of friction) to avoid a slip during walking (Redfern and Andres 1984, Love and Bloswick 1988, McVay and Redfern 1994, Buczek and Banks 1996). For example, studies have reported $RCOF_{peak}$ values ranging from 0.17 to 0.22 for level walking (Strandberg 1983, Perkins 1978, Redfern and DiPasquale 1997, Hanson et al. 1999, Cham and Redfern 2002a), compared to 0.45 for descending a 20° ramp (Redfern and DiPasquale 1997). In summary, a greater risk of slipping is assumed as $RCOF_{peak}$ increases beyond the measured coefficient of friction (Hanson et al. 1999).

The transfer of body weight over the base of support is not completed during slips resulting in falls (Cham 2000). This is evident in the shape of

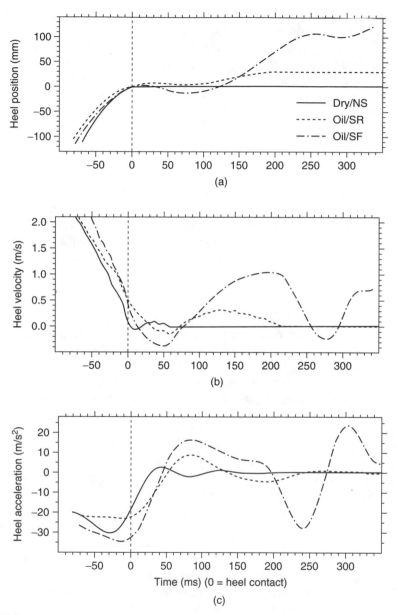

FIGURE 8.11

Typical characteristic profiles of heel dynamics in dry no-slip (NS) conditions compared to slip-recovery (SR) and slip-fall (SF) events on oily surfaces: (a) heel position along the floor, (b) heel velocity along the floor, and (c) heel acceleration along the floor. (Reprinted from *Safety Science*, 40, Cham, R. and Redfern, M.S., Heel contact dynamics during slip events on level and inclined surfaces, 559–576, Copyright (2002), with permission from Elsevier Science.)

the normal ground reaction force (Figure 8.12a and Figure 8.13a) and in the progression of the COP, which stays close to the heel in fall cases (Figure 8.13c). As expected, the lack of foot–floor friction during slip events prevents

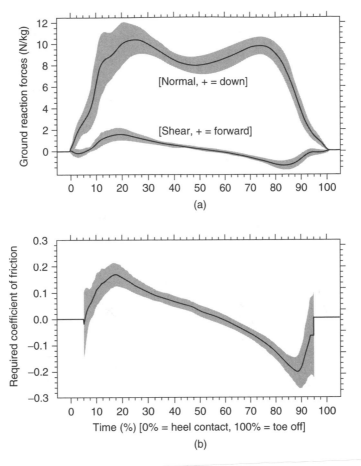

FIGURE 8.12

Mean profile (± 1 s.d.) of (a) ground reaction force normalized to body weight and (b) required coefficient of friction during the stance phase of gait on known dry vinyl floors. (Adapted from Cham 2000.)

"normal" levels of shear from being generated (Figure 8.12a and Figure 8.13b), thus significantly decreasing shear-to-normal force ratio values compared to dry frictional requirements. Finally, after a slip develops, a decrease in the shear ground reaction force occurs at between 25% and 45% of stance phase, sometimes even reversing direction (Figure 8.13b). This decrease is related to a corrective response attempting to bring the foot back toward the body (Cham and Redfern 2001b). The effects of this corrective response are also evident in the moments generated about the lower extremity joints and foot/body dynamics discussed in the next section.

8.3.1.4 Postural Reactions

Humans adopt proactive as well as reactive postural control strategies to avoid slipping and falling. Proactive strategies require knowledge of the

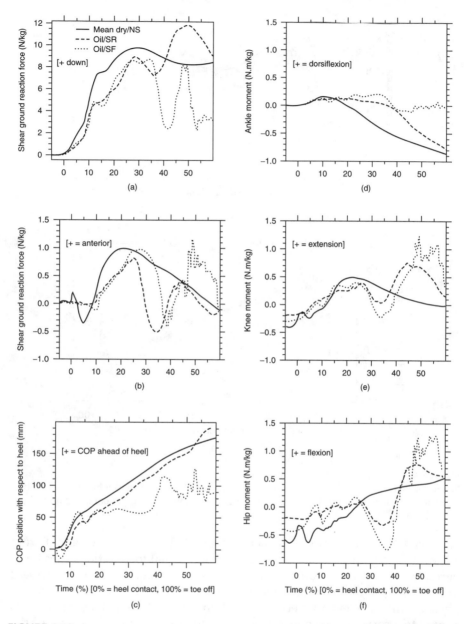

FIGURE 8.13

Mean profiles of kinetic variables on dry no-slip (NS) conditions compared to typical profile recorded during slip-recovery (SR) and slip-fall (SF) events on oily surfaces: (a) normal ground reaction force, (b) shear reaction force, (c) COP position with respect to the heel, and (d to f) lower extremity joint moments at ankle, knee, and hip (adapted from Cham 2000 and Cham and Redfern 2001b). (*Note:* Reaction forces and joint moments are normalized to body weight.) (a–c from Cham 2000, d–f reprinted from *Journal of Biomechanics*, 34, Cham, R. and Redfern, M.S., Lower extremity corrective reactions to slip events, 1439–1445, Copyright (2001), with permission from Elsevier Science.)

slipping hazard prior to physically encountering it. For example, humans adapt to "potentially" slippery surfaces; that is, anticipation of slippery surfaces and the perception of the danger affect gait biomechanics when compared to characteristics of normal gait (Cham and Redfern 2002a). Postural and temporal gait adjustments occur during the step onto a possibly slippery surface, including reductions in stance duration and loading speed to the supporting foot, shorter stride length, reduced foot–floor angle at contact and slower angular foot velocity at heel contact. The overall effect of those adaptations is a reduction in the $RCOF_{peak}$, thus reducing slip potential on potentially hazardous floors (Cham and Redfern 2002a).

Monitoring of joint moments and postural strategies during unexpected slip events is important in understanding and determining the corrective reactions that lead to a successful recovery. A slip event can be considered to be an unexpected destabilizing perturbation. To avoid a fall after a slip event, the body must generate a quick and well-coordinated corrective response of appropriate strength to reestablish balance and maintain an upright posture while continuing with the locomotion task. The joint moments generated in the lower extremity drive the postural responses seen during slips (Cham and Redfern 2001b). Plantarflexion muscle moments generated at the ankle significantly decrease with the severity of the slip (Figure 8.13d). In the most severe slips leading to falls, the ankle moment profile remains at very low levels throughout stance due to the proximity of the COP to the ankle joint. The knee and hip moments appear to be responsible for corrective reactions attempted between 25% and 45% into stance during slip events, i.e., on average between 190 and 350 ms after heel contact. In response to slipping perturbation, a significant bias toward flexor and extensor patterns (compared to nonslipping characteristics) are observed at the knee and hip, respectively, especially during slip-fall outcomes (Figure 8.13e and f). By mid-stance, the majority of slip trials are associated with compensatory reactions (Figure 8.13e and f). The corrective moments generated at the knee and hip provide enough power to result in the observed postural responses (Figure 8.14).

Characteristics of the human postural control system have been investigated using small, rapid translations of the underfoot support surface during quiet standing, usually in the anterior direction. These perturbations typically elicit postural corrective actions to maintain standing balance. These postural reactions include ankle musculature responses of the dorsiflexors to move the body's center of mass forward over the feet. There are also strong postural reactions seen at the hips during larger, faster perturbations. In general, postural reactions to small or slow perturbations result in movement about the ankle to maintain balance, termed *ankle strategies*, while reactions to larger, fast perturbations invoke hip motion or the *hip strategy* for postural correction.

Although much has been learned about the standing postural control system through perturbation protocols, it is questionable whether these

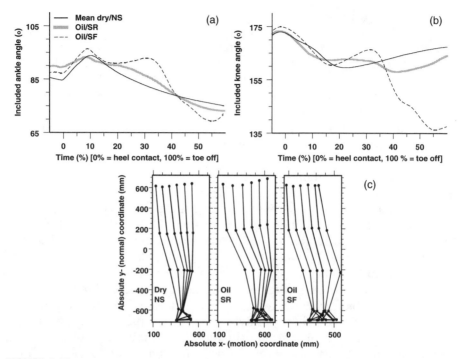

FIGURE 8.14

Mean profiles of (a) included ankle angle and (b) included knee angle on dry no-slip (NS) conditions compared to typical characteristic profiles recorded during slip-recovery (SR) and slip-fall (SF) events on oily surfaces. (c) The corresponding whole-body postural responses in sequences of postures (a and b from Cham 2000, c reprinted from *Journal of Biomechanics*, 34, Cham, R. and Redfern, M.S., Lower extremity corrective reactions to slip events, 1439–1445, Copyright (2001), with permission from Elsevier Science.)

translations of the support surface evoke corrective postural strategies similar to those recorded during real slip events (Iqbal and Pai 2000). Few of these translation-based protocols have been designed specifically to try to simulate real slip events (Hsaio and Robinovitch 1998, Tang and Woollacott 1998, 1999, Tang et al. 1998). Further, computer simulations have indicated that regions of stability generated for slips and active anterior base of support translation do not completely overlap (Pai and Iqbal 1999). In addition, knee moments, shown to be important for slip recovery, are not found in response to underfoot translational perturbations during standing. The majority of experimental standing perturbation studies describes postural strategies in terms of ankle/hip strategies, thus attributing a less dominant role to the knee joint in balance recovery following a perturbation (Horak et al. 1989, 1990, Nashner et al. 1989, Runge et al. 1999). Finally, active anterior base of support translation to specifically reproduce heel and body dynamics during naturally occurring slip events has not been proved to be viable. The differences in postural reactions between standing perturbations and actual slips during gait suggest context-specific postural strategies that are dependent on the task and conditions of the destabilizing motion.

In conclusion, findings from gait studies have become critical in slips/falls prevention research. In addition to their contribution to the field of tribology discussed in the next section, gait studies have improved our understanding of the complex relationship between gait biomechanics and slip-precipitated falls. They are essential to the identification of underlying biomechanical factors that contribute to a decreased chance of recovery after slipping. Inappropriate postural adjustments once a slip has occurred are indeed often the reason for one individual to fall on a given surface while another recovers. Furthermore, whole-body biomechanical analysis of normal gait and slipping events allows researchers to clarify the role of specific body segments during recovery reactions. These postural reactions appear to be context specific and differ from responses to perturbations during stance.

8.3.2 Environmental Factors in Slips and Falls

Postural responses during slips are highly influenced by the environment. The best method of preventing slips and falls is through appropriately adapting the environment to the postural requirements of the activities being carried out. There are numerous environmental factors that contribute to slips and falls in the workplace, and therefore contribute to the postures and postural dynamics that occur. Floor surface conditions and contaminants are believed to be the most important factors, including variations in material, roughness (Chang et al. 2001a), slope (Redfern and DiPasquale 1997), and contaminants such as water, oil, detergent, ice, or powder (Grönqvist et al. 2001). Flooring factors interact with those of the shoes, such as sole material, condition (e.g., roughness, wear, or contamination), and tread pattern. Other significant environmental factors that have been identified are inadequate lighting/visibility and obstacles. Task requirements, such as load carrying, pushing, pulling, vehicle ingress/egress, and climbing, also play a role and can be thought of as job design factors within the environment.

The risk of falling that all of these environmental factors present cannot be quantified with a single parameter. However, the slipperiness or slip resistance of a shoe–floor–contaminant interface is considered to be the most significant environmental factor related to industrial slips and falls. Therefore, measurement of slip resistance is the major method used to determine the potential for slips in industry to prevent falls. Unfortunately, there is no consensus on methodology to measure slip resistance. Currently used measurement methods include traditional triobologically based coefficient of friction measures, psychophysical measures, and biomechanical measures. The following sections describe these different methods.

8.3.2.1 *Human-Centered Approaches*

There are two approaches that incorporate a person in the slip evaluation process: psychophysical and biomechanical. The psychophysical approach relies on subjective judgments of the relative slipperiness or safety of tested shoe–floor combinations. Typically, subjects are either directly asked to rate their perceptions of the slipperiness for a task or they are asked to complete a task and some measure then quantifies their self-determined performance. An example of the latter approach require that subjects negotiate the perimeter of a triangle five times, as quickly as possible without falling down (Strandberg 1985). The time and number of falls were then used to form a measure of the slipperiness of tested shoe–floor interfaces. Skiba et al. (1986), Jung and Schenk (1989, 1990), and Jung and Rutten (1992) used a different psychophysical scheme to judge the safety of inclined surfaces — the inclination angle that subjects judged to be unsafe for walking down a ramp provided an estimate for the slip resistance of the ramp surface.

Biomechanically based methodologies are also considered to be human centered because people are still part of the slip resistance determination process; however, these approaches attempt to measure the effects of the slipperiness of a surface more directly. According to Grönqvist et al. (2001), the biomechanical approach typically employs experiments where slip risk is quantified by heel velocity or postural response (Leamon and Son 1989, Grönqvist et al. 1993, Myung et al. 1993, Cohen and Cohen 1994a,b, Hirvonen et al. 1994, Chiou et al. 2000), kinetics of slipping and falling (e.g., ground reaction forces, utilized and required friction) (Perkins 1978, Strandberg and Lanshammar 1981, Morach 1993, Hirvonen et al. 1994, McVay and Redfern 1994, Myung and Smith 1997, Redfern and DiPasquale 1997, Hanson et al. 1999, Brady et al. 2000, You et al. 2001) and slip distances and velocities (Strandberg and Lanshammar 1981, Grönqvist 1999).

Grönqvist et al. (2001) describe many other examples of both the psychophysical and biomechanics human-centered slip resistance measures and list the general strengths and weaknesses for these approaches as follows:

Advantages:

- The methods are inherently valid for the situation being examined, because human subjects are involved in the experiment, and individual behavior affects the outcome measures.

- The human factor aspect is included in the analysis and can be partly controlled (e.g., walking speed and cadence, anticipation of floor conditions vs. unexpectedly slippery surfaces).

- Biomechanical measurement data can be combined with observations of performance and/or subjective ratings (Strandberg 1985, Strandberg et al. 1985, Jung and Schenk 1989, 1990, Hanson et al. 1999).

Disadvantages:

- Human-centered experiments are time-consuming and expensive.

- They are not applicable to field applications and are most suitable for the laboratory environment.

- Inter- and intraindividual variation in gait due to anticipation of and adaptation to hazards may limit their use (Grönqvist et al. 1993).

- Subjects can adapt to certain environments, leading to changes in measured outcomes, such as the ROF$_{peak}$, kinematics, or EMG activity over repeated trials (Skiba et al. 1986, Tang et al. 1998).

- A safety harness used to protect subjects from falling can be a confounding factor in many experimental setups and may affect the measured outcomes, such as ground reaction forces, required friction, slip velocity and/or distance (Lockhart et al. 2000a,b).

8.3.2.2   Tribological Approach

The measurement of friction at the shoe–floor interface dominates the evaluation methodologies that have been used in studies, particularly in the United States. According to Strandberg (1985) there had already been more than 70 machines invented to measure slip resistance by the early 1980s. Since that time, the number of such devices has grown as technology has presented new options for data collection and as the demand for a slip meter standard has increased. However, significant controversy remains. Generally, slip resistance evaluation methods measure the coefficient of friction of the shoe–floor interface under various conditions such as dry, wet, or oily. An excellent review of these devices is presented by Chang et al. (2001b). There are two basic types of devices: those that measure the static coefficient of friction and those that measure the dynamic coefficient of friction. Examples of portable static coefficient of friction measurement devices, as described in Chang et al. (2001b), are the Drag Sled Tester, the Portable Articulated Strut Tribometer (Brungraber Mark I), and the Horizontal Pull Slipmeter. These devices apply increasing shear loads (sometimes with decreasing normal loads) to a shoe or material sample until a slip occurs. Articulated strut devices measure the strut angle at which the shoe sample mounted on the end of the articulated strut begins to slide, while others simply apply an increasing shear load to a sample attached to a sled until the sled begins to slide. The James Machine is oldest of these articulated strut testers and was used by the Underwriters' Laboratories in the 1940s to set the first minimum safety standard for a surface's available coefficient of friction. Since that time, a static coefficient of friction level of 0.5 has become commonly used as a safety criterion through consensus rather than through scientific evidence, regardless of the device used.

Dynamic coefficient of friction testing devices tend to report the ratio of shear to normal force as measured while a shoe or sample material is moving, at steady state, across a surface. Examples of portable devices are the Portable Friction Tester (Strandberg 1985), the Tortus device (Harris and Shaw 1988), and the GleitMessGerat 100 or GMG100 (Deutches Institut fur Normung 1999). Examples of laboratory dynamic testers are the French LABINRS

(Tisserand 1985, 1997), the Programmable Slip Resistance Tester–PSRT (Redfern and Bidanda 1994), the Shoe and Allied Trades Research Association (SATRA) Slip Resistance Tester (Wilson 1990), and the Slip Simulator (Grönqvist et al. 1989). Chang et al. (2001b) provide a review of these dynamic coefficient of friction testing devices. Dynamic coefficient of friction values are typically lower than static coefficient of friction values and their measured values differ from one device to the next. Justification for using a dynamic rather than a static coefficient of friction measure has been suggested by biomechanical experiments that have indicated that the human foot never comes to a complete stop during a slip (Perkins and Wilson 1983). However, typical workplace activities other than gait may more closely resemble the static coefficient of friction measurement motion profiles; therefore, which type of device to use to characterize the slip resistance of floors remains unresolved.

Other devices identified by Chang et al. (2001b) for measuring slip resistance include pendulum swing devices, such as the Siegler device and the British Portable Skid Tester (BPST), and a group of transitional friction measurement devices such as the Portable Inclinable Articulated Strut Tribometer (PIAST or Brungraber Mark II) and the Variable Incidence Tribometer (VIT). These two transitional devices use the same type of inclined strut technique employed in the static devices but impart some momentum to the tested sample.

The criteria used to evaluate the frictional capabilities of the shoe–floor interface vary, depending on the device used and the conditions tested. There is no consensus on which device is best or on a "safe" level of friction for the various activities of living. The measured coefficient of friction should always be greater than the required coefficient of friction. The required coefficient of friction, as determined by the $RCOF_{peak}$, for level surface walking is approximately 0.18, but has been reported to be within the range of 0.15 to 0.30. (Perkins 1978, Strandberg and Lanshammer 1981, Redfern et al. 2001, Grönqvist et al. 2001). The criterion for safe levels of friction has been significantly greater than the minimum required coefficients seen in these gait studies. Traditionally, a static coefficient of friction criterion for slip resistance has been a value of 0.5, although some have used values as low as 0.4. These values have been developed by historical consensus (Redfern and Rhoades 1996). Criteria for dynamic coefficient of friction measures tend to be lower, with cutoff values from 0.3 to 0.4. Examples of some of the more common criteria used were presented in Table 8.7. Although there is some variability in recommendations, most experts agree that a static coefficient of friction of 0.5 or a dynamic coefficient of friction of 0.4 is a reasonable level of friction to prevent slips and falls during walking. However, it is important to note that the value given in any criterion must take into consideration the specific tester that is being used.

TABLE 8.7

Examples of Safety Criteria for Static (SCOF) and Dynamic (DCOF) Coefficient of Friction Measurements Applied to Level Walking

ASTM (1975)		Rosen (1983)		BSI (1977)		Grönqvist et al. (1989)	
SCOF	Description	SCOF	Description	DCOF	Description	DCOF	Description
1.00	Very good	>0.60	Very safe	<0.75	Very good	>0.30	Very slip resistant
0.80	Good	0.50–0.59	Relatively safe	0.40–0.75	Good	0.20–0.29	Slip resistant
0.50	Standard for nonhazardous walkway	0.40–0.49	Dangerous	0.40	Minimum	0.15–0.19	Unsure
0.40	Poor	0.35–0.39	Very dangerous	0.40–0.20	Poor	0.05–0.14	Slippery
0.30	Hazardous	<0.34	Unusually dangerous	<0.20	Hazardous	<0.05	Very slippery

Source: Adapted from Redfern and Rhoades (1996).

Summary

Slips and falls are a major problem resulting in numerous injuries in occupational and non-occupational settings. The causation of falls is a complex interaction of environmental factors and human factors. The human factors involve postures and postural responses that are dependent on the health of the musculoskeletal and sensory systems, as well as on cognitive factors such as attention and perception of the impending environmental conditions. Small amounts of slipping normally occur during walking, but do not lead to falls. The magnitude, velocity, and acceleration of the foot relative to the floor during a slip ultimately determine the potential for a recovery or fall. However, the postural responses to foot slip during an attempted recovery can also determine the final result. These postural responses are stereotypic with rapid flexion of the knee and extension of the hip in an attempt to bring the slipping foot back toward the center of mass. This strategy is different from those reported for balance recovery during standing, suggesting that postural responses are context specific to the task and to the perceived environmental conditions. To prevent injurious falls, evaluation of the shoe–floor interface and modifications of the environmental factors are considered to be most effective. The dominant environmental factor is the frictional property of the shoe–floor–contaminant interface. Currently, there are numerous devices available to measure friction capabilities; however, there is no consensus on the best methodology. Clearly, the ergonomic approach of fitting the environment and tasks to the person is key to reducing the potential for slips and falls. Greater knowledge of the postural demands during activities and the ability to accurately and repeatably measure the slip characteristics of the environment under a variety of conditions are essential. Only then can the modification of key environmental factors be addressed to successfully reduce falling injuries.

References

ASTM (American Society for Testing and Materials), 1975. *Standard Method of Test for Static Coefficient of Friction of Polish-Coated Floor Surfaces as Measured by the James Machine*, ASTM Designation D2057-75, Philadelphia: American Society for Testing and Materials.

Basford, J.R. and Smith, M.A., 1988. Shoe insoles in the workplace. *Orthopedics*, 11, 2, 285–288.

Biering-Sørensen, F., Thomsen, C.E., and Hilden, J., 1989. Risk indicators for low back trouble. *Scandinavian Journal of Rehabilitation Medicine*, 21(3), 151–157.

Borg, G.A.V., 1982. Psychophysical bases of perceived exertion. *Medicine and Science in Sports and Exercise*, 14(5), 377–381.

Boussenna, M., Corlett, E.N., and Pheasant, S.T., 1982. The relation between discomfort and postural loading at the joints. *Ergonomics*, 25(4), 315–322.

Brady, R.A., Pavol, M.J., Owings, T.M., and Grabiner, M.D., 2000. Foot displacement but not velocity predicts the outcome of a slip induced in young subjects while walking. *Journal of Biomechanics*, 33(7), 803–808.

Brantingham, C.R., Beekman, B.E., Moss, C.N., and Gordon, R.B., 1970. Enhanced venous pump activity as a result of standing on a varied terrain floor surface. *Journal of Occupational Medicine*, 12(5), 164–169.

BSI (British Standards Institution), 1977. *British Standard Code of Practice for Stairs. BS5395*. London: British Standards Institution.

Buczek, F.L. and Banks, S.A., 1996. High-resolution force plate analysis of utilized slip resistance in human walking. *Journal of Testing and Evaluation*, 24(6), 353–358.

Bullinger, H.J., Brandera, J.E., and Muntzinger, W.F., 1991. Design, selection and location of foot controls. *International Journal of Industrial Ergonomics*, 8, 303–311.

Bullock, M.I., 1973. The Determination of Pedal Alignments for the Production of Minimal Spinal Movements Using a Stereo-Photogrammetric Method of Measurement, Ph.D. thesis, Department of Physiotherapy, University of Queensland, Australia.

Bullock, M.I., 1974. The determination of an optimal pedal–operator relationship by the use of photo-grammetry, in *Biostereometrics: Proceedings of the Symposium of Commission V,* Washington, D.C.: International Society for Photogrammetry, 290–317.

Bullock, M.I., 1990. The development of optimum worker-task relationships, in *Ergonomics: the Physiotherapist in the Workplace*, Bullock, M.I., Ed., Edinburgh: Churchill Livingston, 13–49.

Bullock, M.I., 1991. Minimising back movements during pedal depression, in *Proceedings of the 11th Congress of the International Ergonomics Association*, Paris, July 1991, 117–119.

Bullock, M.I. and Harley, I.A., 1972. The measurement of three-dimensional body movement by the use of photogrammetry. *Ergonomics*, 15(3), 309–322.

Casey, S.M. and Rogers, S.P., 1987. The case against coplanar pedals in automobiles. *Human Factors*, 29(1), 83–86.

Chaffin, D.B., Andersson, G.B.J., and Martin, B.J., 1999. *Occupational Biomechanics*, 3rd ed., New York: John Wiley & Sons.

Cham, R., 2000. Biomechanics of Slips and Falls, Ph.D. thesis, Department of Bioengineering, University of Pittsburgh, Pittsburgh, PA.

Cham, R. and Redfern, M.S., 2001a. Effect of flooring on standing comfort and fatigue. *Human Factors*, 43(3), 381–391.

Cham, R. and Redfern, M.S., 2001b. Lower extremity corrective reactions to slip events. *Journal of Biomechanics*, 34(11), 1439–1445.

Cham, R. and Redfern, M.S., 2002a. Changes in gait when anticipating slippery floors. *Gait & Posture*, 15(2), 159–171.

Cham, R. and Redfern, M.S., 2002b. Heel contact dynamics during slip events on level and inclined surfaces. *Safety Science*, 40(7–8), 559–576.

Chang, W.R., Kim, I.J., Manning, D., and Bunterngchit, Y., 2001a. The role of surface roughness in the measurement of slipperiness. *Ergonomics*, 44(13), 1200–1216.

Chang, W.R., Leclercq, S., Grönqvist, R., Brungraber, R., Mattke, U., Strandberg, L., Thorpe, S., Myung, R., Makkonen, L., and Courtney, T.K., 2001b. The role of friction in the measurement of slipperiness, Part II: Survey of friction measurement devices. *Ergonomics*, 44(13), 1233–1261.

Chiou, S., Bhattacharya, A., and Succop, P.A., 2000. Evaluation of workers' perceived sense of slip and effect of prior knowledge of slipperiness during task performance on slippery surfaces. *American Industrial Hygiene Association Journal*, 61, 492–500.

Cohen, H.H. and Cohen, M.D., 1994a. Perceptions of walking surface slipperiness under realistic conditions, utilizing a slipperiness rating scale. *Journal of Safety Research*, 25, 27–31.

Cohen, H.H. and Cohen, M.D., 1994b. Psychophysical assessment of the perceived slipperiness of floor tile surfaces in a laboratory setting. *Journal of Safety Research*, 25, 19–26.

Cook, J., Branch, T.P., Baranowski, T.J., and Hutton, W.C., 1993. The effect of surgical floor mats in prolonged standing: an EMG study of the lumbar paraspinal and anterior tibialis muscles. *Journal of Biomedical Engineering*, 15(3), 247–250.

Courtney, T.K. and Webster, B.S., 1999. Disabling occupational morbidity in the United States: an alternative way of seeing the Bureau of Labor Statistics' data. *Journal of Occupational and Environmental Medicine*, 41, 60–69.

Courtney, T.K., Sorock, G.S., Manning, D.P., Collins, J.W., and Holbien-Jenny, M.A., 2001. Occupational slip, trip and fall-related injuries — can the contribution of slipperiness be isolated? *Ergonomics*, 44(123), 1118–1137.

Deutsches Institut für Normung (DIN), 1999. Prüfung von Bodenbelagen — Bestimmung der rutschhemmenden Eigenschaft — Verfahren zur Messing des Gleitreibungskoeffizienten [Testing of Floor Coverings — Determination of the Anti-Slip Properties — Measurement of Sliding Friction Coefficient], DIN E5131, Berlin: Deutsches Institut für Normung.

Drury, C.G., 1975. Application of Fitts' law to foot pedal design. *Human Factors*, 74(1), 368–373.

Duqueroy, M., 2000. Analyse biomécanique du mouvement de débrayage sur les poids lourds [Biomechanical analysis of the movements of clutch pedal operation for trucks], DEA [Master's] thesis, DEA Physiologie du mouvement et biomécanique de la performance motrice, Université Paris XI, Paris.

Estryn-Behar, M., Kaminski, M., Peigne, E., Maillard, M.F., Pelletier, A., Berthier, C., Delaporte, M. F., Paoli, M.C., and Leroux, J.M., 1990. Strenuous working conditions and musculo-skeletal disorders among female hospital workers. *International Archives of Occupational and Environmental Health*, 62(1), 47–57.

Fingerhut, L.A., Cox, C.S., and Warner, M., 1998. International Comparative Analysis of Injury Mortality: Findings from the ICE on Injury Statistics. Advance Data from Vital and Health Statistics, Series 10, No. 303, Hyattsville, MD: National Center for Health Studies.

Glass, S.W. and Suggs, C.W., 1977. Optimization of vehicle accelerator-brake pedal foot travel time. *Applied Ergonomics*, 8(4), 215–218.

Grandjean, E., 1980. Sitting posture of car drivers from the point of view of ergonomics, in *Human Factors in Transport Research*, Part 1, Grandjean, E., Ed., London: Taylor & Francis, 205–213.

Grönqvist, R., 1999. Slips and falls, in *Biomechanics in Ergonomics*, Kumar, S., Ed., London: Taylor & Francis, 351–375.

Grönqvist, R., Roine, J., Jarvinen, E., and Korhonen, E., 1989. An apparatus and a method for determining the slip resistance of shoes and floors by simulation of human foot motions. *Ergonomics*, 32, 979–995.

Grönqvist, R., Hirvonen, M., and Tuusa, A., 1993. Slipperiness of the shoe–floor interface — comparison of objective and subjective assessments. *Applied Ergonomics*, 2(5), 258–262.

Grönqvist, R., Abeysekera, J., Gard, G., Hsiang, S.M., Leamon, T.B., Newman, D.J., Gielo-Perczak, K., Lockhart, T.E., and Pai, Y.C., 2001. Human-centered approaches in slipperiness measurement. *Ergonomics*, 44(13), 1167–1199.

Hansen, L., Winkel, J., and Jorgensen, K., 1998. Significance of mat and shoe softness during prolonged work in upright position: based on measurements of low back muscle EMG, foot volume changes, discomfort and ground force reactions. *Applied Ergonomics*, 29(3), 217–224.

Hanson, J.P., Redfern, M.S., and Mazumdar, M., 1999. Predicting slips and falls considering required and available friction. *Ergonomics*, 42(12), 1619–1633.

Harris, G.W. and Shaw, S.R., 1988. Slip resistance of floors: users' opinions, Tortus instrument readings and roughness measurement. *Journal of Occupational Accidents*, 9(4), 287–298.

Haslegrave, C.M., 1995. Factors in the driving task affecting comfort and health, in *Proceedings of the 3rd International Conference on Vehicle Comfort and Ergonomics*, Bologna, 29–31 March 1995, 223–230.

Helander, M.G. and Zhang, L. 1997. Field studies of comfort and discomfort in sitting. *Ergonomics*, 40(9), 895–915.

Hertzberg, H.T.E., and Burke, F.E., 1971. Foot forces exerted at various aircraft brake-pedal angles. *Human Factors*, 13(5), 445–456.

Hirsch, C. and Nachemson, A., 1954. New observations on the mechanical behaviour of lumbar discs. *Acta Orthopaedica Scandinavica*, 23, 224–225.

Hirvonen, M., Leskinen, T., Grönqvist, R., and Saario, J., 1994. Detection of near accidents by measurement of horizontal acceleration of the trunk. *International Journal of Industrial Ergonomics*, 14(4), 307–314.

Hoffmann, E.R., 1991. Accelerator-to-brake movement times. *Ergonomics*, 34(3), 277–287.

Horak, F.B., Diener, H.C., and Nashner, L.M., 1989. Influence of central set on human postural responses. *Journal of Neurophysiology*, 62(4), 841–853.

Horak, F.B., Nashner, L.M., and Diener, H.C., 1990. Postural strategies associated with somatosensory and vestibular loss. *Experimental Brain Research*, 82(1), 167–177.

Hsaio, E.T. and Robinovitch, S.N., 1998. Common protective movements govern unexpected falls from standing height. *Journal of Biomechanics*, 31(1), 1–9.

Iqbal, K. and Pai, Y.C., 2000. Predicted region of stability for balance recovery: motion at the knee joint can improve termination of forward movement. *Journal of Biomechanics*, 33(12), 1619–1627.

Johansson, G. and Rumar, K., 1971. Drivers' brake reaction times. *Human Factors*, 13(1), 23–27.

Jung, K. and Rutten, A., 1992. Entwicklung eines Verfahrens zur Prüfung der Rutschemmung von Bodenbelagen für Arbeitsräume, Arbeitsbereich, und Verkehrswege [The development of a process for examining slip prevention on floor surfaces in work areas, rooms and walkways]. *Zentralblatt für Arbeitsmedizin, Arbeitsschutz, Prophylaxe und Ergonomie*, 42(6), 227–235 [in German with English summary].

Jung, K. and Shenk, H., 1989. Objektivierbarkeit und Genauigkeit des Begehungsverfahrens zur Ermittlung der Rutschemmung von Bodenbelagen [Objectification and precision of locomotion for determining the slip prevention of floor surfaces]. *Zentralblatt für Arbeitsmedizin, Arbeitsschutz, Prophylaxe und Ergonomie*, 39(8), 221–228 [in German with English summary].

Jung, K. and Shenk, H., 1990. Objektivierbarkeit und Genauigkeit des Begehungsverfahrens zur Ermittlung der Rutschemmung von Schuhen [Objectification and precision of locomotion for determining the slip prevention of shoes]. *Zentralblatt für Arbeitsmedizin, Arbeitsschutz, Prophylaxe und Ergonomie*, 40(3), 70–78 [in German with English summary].

Kim, J.Y., Stuart-Buttle, C., and Marras, W.S., 1994. The effects of mats on back and leg fatigue. *Applied Ergonomics*, 25(1), 29–34.

Konz, S., Wadhera, N., Sathaye, S., and Chawla, S., 1971. Human factors considerations for a combined brake-accelerator pedal. *Ergonomics*, 14(2), 279–292.

Konz, S., Bandla, V., Rys, M., and Sambasivan, J., 1990. Standing on concrete vs. floor mats, in *Advances in Industrial Ergonomics and Safety II*, Das, B., Ed., London: Taylor & Francis, 991–998.

Krijnen, R.M., De Boer, E.M., Ader, H.J., and Bruynzeel, D.P., 1997. Venous insufficiency in male workers with a standing profession. Part 2: diurnal volume changes of the lower legs. *Dermatology*, 194(2), 121–126.

Kroemer, K.H.E., 1971. Foot operation of controls. *Ergonomics*, 14(3), 333–361.

Kroemer, K.H.E. and Grandjean, E., 1997. *Fitting the Task to the Human*, 5th ed., London: Taylor & Francis.

Krumwiede, D., Konz, S., and Hinnen, S., 1998. Floor mat comfort, in *Advances in Occupational Ergonomics and Safety*, Kumar, S., Ed., Amsterdam: IOS Press, 159–162.

Leamon, T.B. and Li, K.W., 1990. Microslip length and the perception of slipping, presented at Twenty Third International Congress on Occupational Health, 22–28 September, Montreal, Canada.

Leamon, T.B. and Murphy, P.L., 1995. Occupational slips and falls: more than a trivial problem. *Ergonomics*, 38, 487–498.

Leamon, T.B. and Son, D.H., 1989. The natural history of a microslip, in *Advances in Industrial Ergonomics and Safety I*, Mital, A., Ed., London: Taylor & Francis, 633–638.

Lockhart, T.E., Smith, J.L., Woldstad, J.C., and Lee, P.S., 2000a. Effects of musculoskeletal and sensory degradation due to aging on the biomechanics of slips and falls, in *Proceedings of the XIVth International Ergonomics Association/44th Human Factors and Ergonomics Society Congress*, Vol. 5, Santa Monica, CA: Human Factors and Ergonomics Society, 83–86.

Lockhart, T.E., Woldstad, J.C., Smith, J.L., and Hsiang, S.M., 2000b. Prediction of falls using a robust definition of slip distance and adjusted required coefficient of friction, in *Proceedings of the XIVth International Ergonomics Association/44th Human Factors and Ergonomics Society Congress*, Vol. 4, Santa Monica, CA: Human Factors and Ergonomics Society, 506–509.

Love, A. and Bloswick, D., 1988. Slips and falls during manual handling activities, in *Proceedings of the Annual Conference of the Human Factors Association of Canada*, Edmonton, Alberta, Canada, September, 133–135.

MacFarlane, G.J., Thomas, E., Papageorgiou, A.C., Croft, P.R., Jayson, M.I., and Silman, A.J., 1997. Employment and physical work activities as predictors of future low back pain. *Spine*, 22(10), 1143–1149.

Madeleine, P., Voigt, M., and Arendt-Nielsen, L., 1998. Subjective, physiological and biomechanical responses to prolonged manual work performed standing on hard and soft surfaces. *European Journal of Applied Physiology and Occupational Physiology*, 77(1–2), 1–9.

Magora, A., 1972. Investigation of the relation between low back pain and occupation. III. Physical requirements: sitting, standing and weight lifting. *Industrial Medicine and Surgery*, 41(12), 5–9.

McVay, E.J. and Redfern, M.S., 1994. Rampway safety — foot forces as a function of rampway angle. *American Industrial Hygiene Association Journal*, 55(7), 626–634.

Mick, F., 1995. Eine experimentelle Untersuchung der Korrelation von Komfortempfinden und Handhabungskräften an Padalen, Diplomarbeit, Technische Universität, Munich.

Morach, B., 1993. Quantifierung des Ausgleitvorganges beim menschlichen Gang unter besonderer Berücksichtigung der Aufsetzphases des Fusses [Quantification of the anterior progression in human stride with a specific consideration of heel strike of the foot], Fachbereich Sicherheitstechnik der Bergischen Universität — Gesamthochschule Wuppertal, Wuppertal [in German].

Morrison, R.W., Swope, J.G., and Halcomb, C.G., 1986. Movement time and brake pedal placement. *Human Factors*, 28(2), 241–246.

Mortimer, R.G., 1974. Foot brake pedal force capability of drivers. *Ergonomics*, 17(4), 509–513.

Myung, R. and Smith, J.L., 1997. The effect of load carrying and floor contaminants on slip and fall parameters. *Ergonomics*, 40(2), 235–246.

Myung, R., Smith, J.L., and Leamon, T.B., 1993. Subjective assessment of floor slipperiness. *International Journal of Industrial Ergonomics*, 11, 313–319.

Nachemson, A., 1966. The load on lumbar discs in different positions of the body. *Clinical Orthopaedics and Related Research*, 45(March–April), 107–122.

Nashner, L.M., Shupert, C.L., Horak, F.B., and Black, F., 1989. Organization of posture controls: an analysis of sensory and mechanical constraints. *Progress in Brain Research*, 80, 411–418.

National Safety Council (NSC), 1998. Accident Facts, Itasca, IL: NSC.

Nowak, E., 1972. Angular measurements of foot motion for application to the design of foot pedals. *Ergonomics*, 15(4), 407–415.

Pai, Y.C. and Iqbal, K., 1999. Simulated movement termination for balance recovery: can movement strategies be sought to maintain stability in the presence of slipping or forced sliding? *Journal of Biomechanics*, 32(8), 779–786.

Park, S.J., Kim, C.B, Kim, C.J., and Lee, J.W., 2000. Comfortable driving postures for Koreans. *International Journal of Industrial Ergonomics*, 26, 489–497.

Perkins, P.J., 1978. Measurement of slip between the shoe and ground during walking, In *ASTM Special Technical Publication, Walkway Surfaces: Measurements of Slip Resistance*, Philadelphia, PA: American Society for Testing and Materials, 71–87.

Perkins, P.J. and Wilson, M.P., 1983. Slip resistance of shoes — new developments. *Ergonomics*, 26(1), 73–82.

Pheasant, S.T. and Harris, C.M., 1982. Human strength in the operation of tractor pedal. *Ergonomics*, 25(1), 53–63.

Poock, G.K., West, A.E., Toben, T.J., and Sullivan, J.P.T., 1973. A combined accelerator-brake pedal. *Ergonomics*, 16(6), 845–848.

Porter, J.M. and Gyi, D.E., 1998. Exploring the optimum posture for driver comfort. *International Journal of Vehicle Design*, 19(3), 255–266.

Rebiffé, R., 1969. The driving seat: its adaptation to functional and anthropometric requirements, in *Proceedings of a Symposium on Sitting Posture*, Grandjean, E., Ed., London: Taylor & Francis, 132–147.

Redfern, M.S. and Andres, R.O., 1984. The analysis of dynamic pushing and pulling: required coefficient of friction, in *Proceedings of the 1984 International Conference on Occupational Ergonomics*, May, Toronto, Ontario, Canada, 569–571.

Redfern, M.S. and Bidanda, B., 1994. Slip resistance of the shoe floor interface under biomechanically relevant conditions. *Ergonomics*, 37(3), 511–524.

Redfern, M.S. and Chaffin, D.B., 1995. Influence of flooring on standing fatigue. *Human Factors*, 37(3), 570–581.

Redfern, M.S. and Cham, R., 2000. The influence of flooring on standing comfort and fatigue. *American Industrial Hygiene Association Journal*, 61(5), 700–708.

Redfern, M.S. and DiPasquale, J., 1997. Biomechanics of descending ramps. *Gait & Posture*, 6(2), 119–125.

Redfern, M.S. and Rhoades, T.P., 1996. Fall prevention in industry using slip resistance testing, in *Occupational Ergonomics Theory and Applications*, Bhattacharya, A. and McGlothlin, J.D., Eds., New York: Marcel Dekker, 463–476.

Redfern, M.S., Cham, R., Gielo-Perczak, K., Grönqvist, R., Hirvonen, M., Lanshammar, H., Marpet, M., Pai, C.Y., and Powers, C., 2001. Biomechanics of slips. *Ergonomics*, 44(13), 1138–1166.

Ribouchon, S., 1991. Confort postural du conducteur assis: étude des situations de confort optimal et des situations observées sous constrainte de hauteur d'assise, Ph.D. thesis, Université de Paris-Sud, Paris.

Roe, R.W., 1993. Occupant packaging, in *Automotive Ergonomics*, Peacock, B. and Karwowski, W., Eds., London: Taylor & Francis, 11–42.

Rosen, S.I., 1983. *The Slip and Fall Handbook*, Columbia, MD: Hanrow Press.

Rosseger, R. and Rosseger, S., 1960. Health effects of tractor driving. *Journal of Agricultural Engineering Research*, 5(3), 241–275.

Runge, C.F., Shupert, C.L., Horak, F.B., and Zajac, F.E., 1999. Ankle and hip postural strategies defined by joint torques. *Gait & Posture*, 10(2), 161–170.

Ryan, G.A., 1989. The prevalence of musculo-skeletal symptoms in supermarket workers. *Ergonomics*, 32(4), 359–371.

Rys, M. and Konz, S., 1988. Standing work: carpet vs. concrete, in *Proceedings of the Human Factors Society 32nd Annual Meeting*, Anaheim, CA, 24–28 October, 522–525.

Rys, M. and Konz, S., 1989. An evaluation on floor surfaces, in *Proceedings of the Human Factors Society 33rd Annual Meeting*, Denver, 16–20 October, 517–520.

Skiba, R., Wieder, R., and Cziuk, N., 1986. Evaluation of results by measuring the coefficient of friction with the aid of an inclined plane. *Kautschuk und Gummi Kunststoffe*, 39(10), 907–911.

Snyder, H.L., 1976. Braking movement time and acceleration–brake separation. *Human Factors,* 18(2), 201–204.

Southall, D., 1985. The discrimination of clutch-pedal resistances. *Ergonomics,* 28(9), 1311–1317.

Strandberg, L., 1983. On accident analysis and slip-resistance measurement. *Ergonomics,* 26(1), 11–32.

Strandberg, L., 1985. The effect of conditions underfoot on falling and overexertion accidents. *Ergonomics,* 28, 131–162.

Strandberg, L. and Lanshammar, H., 1981. The dynamics of slipping accidents. *Journal of Occupational Accidents,* 3, 153–162.

Tang, P.-F. and Woollacott, M.H., 1998. Inefficient postural responses to unexpected slips during walking in older adults. *Journal of Gerontology. Series A, Biological Sciences and Medical Sciences,* 53(6), M471–M480.

Tang, P.-F. and Woollacott, M.H., 1999. Phase-dependent modulation of proximal and distal postural responses to slips in young and older adults. *Journal of Gerontology. Series A, Biological Sciences and Medical Sciences,* 54(2), M89–M102.

Tang, P.-F., Woollacott, M.H., and Chong, R.K., 1998. Control of reactive balance adjustments in perturbed human walking: roles of proximal and distal postural muscle activity. *Experimental Brain Research,* 119(2), 141–152.

Thompson, D.D., 2001. The determination of the human factors/occupant packaging requirements for adjustable pedal systems, in *Digital Human Modeling for Vehicle and Workplace Design,* Chaffin, D.B., Ed., Warrendale, PA: Society of Automotive Engineers, 101–111.

Tisserand, M., 1985. Progress in the prevention of falls caused by slipping. *Ergonomics,* 28(7), 1027–1042.

Tisserand, M., Sauliner, H., and Leclercq, S., 1997. Comparison of seven test methods for the slip resistance of floors: contributions to developments of standards, in *Proceedings of the 13th Triennial Congress of the International Ergonomics Association,* Vol. 3, Seppala, P., Luopajarvi, T., Nygard, C.H., and Mattila, M., Eds., Helsinki: Finnish Institute of Occupational Health, 406–408.

Troussier, B., Lamalle, Y., Charruel, C., Rachidi, Y., Jiguet, M., Vidal, F., Kern, A., De Gaudemaris, R., and Phelip, X., 1993. Incidence socio-économiques et facteurs pronostiques des lombalgies par accident du travail dans le personnel hospitalier du CHU de Grenoble [Socioeconomic incidences and prognostic factors of low back pain caused by occupational injuries among the hospital personnel of Grenoble University Hospital Center]. *Revue du Rhumatisme (Edition Française),* 60(2), 144–151.

Verriest, J.P., 1986. A tool for the assessment of inter segmental angular relationships defining the postural comfort of a seated operator, in *Passenger Comfort Convenience and Safety,* SAE Paper 860057, Warrendale, PA: Society of Automotive Engineers, 71–83.

Wang, X., Mick, F., Vernet, M., and Freigneu, F., 1996. Experimental investigation of the relationship between the sensation and force applied to a pedal, in *Advances in Applied Ergonomics,* Özok, A.F. and Salvendy, G., Eds., West Lafayette: USA Publishing Istanbul, 572–575.

Wang, X., Le Breton-Gadegbeku, B., Verriest, J.P., Deleurence, P., Piechnick, B., and Chanut, O., 1999. Etude expérimentale et analytique de l'influence de différents paramètres sur le mouvement de la jambe lors d'une action de débrayage et sur le confort perçu. Report INRETS-LBMC 9905, Bron, France: INRETS, 108 pp.

Wang, X., Verriest, J.P., Le Breton-Gadegbeku, B., Tessier, Y., and Trasbot, J., 2000. Experimental investigation and biomechanical analysis of lower limb movements for clutch pedal operation. *Ergonomics,* 43(9), 1405–1429.

Warner, M., Barnes, P.M., and Fingerhut, L.A., 2000. Injury and poisoning episodes and conditions: National Health Interview Survey, 1997. Advance Data from Vital and Health Statistics, Series 10, No. 303, Hyattsville, MD: National Center for Health Statistics.

Wilson, M.P., 1990. Development of SATRA slip test and tread pattern design guidelines, in *Slips, Stumbles and Falls: Pedestrian Footwear and Surfaces,* Gray, B.E., Ed., ASTM STP 1103, Philadelphia, PA: American Society for Testing and Materials, 113–123.

Xu, Y., Bach, E., and Orhede, E., 1997. Work environment and low back pain: the influence of occupational activities. *Occupational and Environmental Medicine,* 54(10), 741–745.

You, J.Y., Chour, Y.L., Lin, C.J., and Su, F.C., 2001. Effect of slip on movement of body center of mass relative to base of support. *Clinical Biomechanics,* 16, 167–173.

Zhang, L., Drury, C., and Woolley, S., 1991. Constrained standing: evaluating the foot/floor interface. *Ergonomics,* 34, 175–192.

9

Shoulder Girdle and Upper Arm

CONTENTS

Introduction

This chapter contains two sections on the shoulder girdle and upper arm. The first section describes the anatomy of this region, muscle functions and kinematics (in particular the so-called scapulohumeral rhythm), as well as some of the biomechanical models developed for analysis of this musculo-skeletal system.

0-415-27908-9/04/$0.00+$1.50
© 2004 by CRC Press LLC

The second section reports on risk factors for musculoskeletal discomfort and discusses how these result from upper arm elevation, posture of the upper arm with respect to the trunk, and forearm posture, including the effects of forearm supports. It also pays special attention to the influence of psychosocial factors on the risk of musculoskeletal injury.

9.1   Shoulder Girdle

DirkJan Veeger and Frans C.T. Van der Helm

Introduction

This section focuses on the biomechanics of the shoulder and arm, related to the identification of stresses and strains, and discusses the special anatomy of the arm and its consequences. Special attention is paid to the measurement of shoulder–arm kinematics. Last, a few biomechanical models developed for the analysis of the shoulder and arm are described.

9.1.1   Anatomy

9.1.1.1   Bony Structures

The human arm, in contrast to the human leg, is not specialized. It has not evolved into an extremity that is especially suitable for a specific category of tasks. The arm can be used for a large diversity of tasks, varying from manipulation of small objects to handling of heavy materials. In addition, the human arm has a large range of motion. From an anatomical view, the difference in function between arms and legs is easily visible in the difference in structure between the shoulder girdle and the pelvis. Figure 9.1 illustrates the elements of the arm and shoulder girdle.

The large range of motion of the arm results from its connection to the scapula, which has a loose connection to the trunk. Because the scapula is able to slide and rotate over the surface of the rib cage, it is possible to move the base of the arm, the glenohumeral joint. This leads to a great increase in the reach of the arm (Figure 9.2). Of course, the intermuscular coordination of the muscles that connect the scapula to the trunk is extremely important in this regard.

A second reason for the large range of motion of the arm is that the glenohumeral joint is shaped like a small and shallow saucer (the glenoid) and a large cup (the humeral head). Cup and saucer are connected by strong, but loose, ligamentous tissue. The joint structure allows rotations in three directions, as well as some translation. As a result, the range of motion of the arm would be considerable even if the scapula were fixed (Figure 9.2).

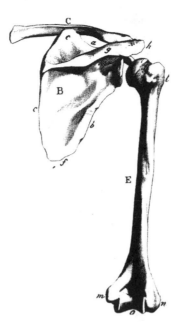

FIGURE 9.1

Frontal view of the bony parts of the human shoulder. B = shoulder blade or scapula, C = clavicle or collarbone, E = upper arm or humerus, f = inferior angle of the scapula, h = acromion. (From Del Medico, G., 1811. *Anatomia per Uso dei Pittori e Scultori.* Rome, Presso Vincenzo Poggioli.)

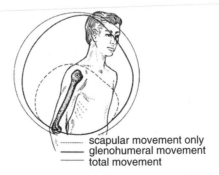

scapular movement only
glenohumeral movement
total movement

FIGURE 9.2

Range of motion curves for the distal end of the humerus, projected as a sphere with the shoulder as center. The dotted line illustrates the contribution to motion by movement of the scapula relative to the trunk, the thick line the contribution from rotations in the glenohumeral joint. The thinner line illustrates the movement range that is possible when all joints are involved. (From Benninghoff-Goertler, 1964. *Lehrbuch der Anatomie des Menschen I*, 9th ed., Munich: Urban & Fischer. With permission.)

Despite the fact that the glenohumeral joint is loose, spontaneous (sub)luxation seldom occurs. It is assumed that this is the result of muscular control by the rotator cuff muscles. It is evident that good coordination between these muscles is highly important.

9.1.1.2   Muscle Function

Intermuscular coordination is essential for providing a stable base for the hand (Figure 9.3), and for producing the necessary external forces. For the upper arm to perform a movement, the scapula must be moved to the optimum position and stabilized. Then, the upper arm must be stabilized in the glenohumeral joint by muscular effort. The requirement for these stabilizing effects increases with the need for accuracy in the movement. To position the scapula, both outward- and inward-rotating muscles must be activated. The outward-rotating movement of the scapula is brought about by combined action of the trapezius (upper part) and serratus anterior (lower part) (as shown in Figure 9.4), whereas the rhomboids and levator scapulae, aided by the serratus anterior (upper part) and the clavicular connection, can act as inward rotators (as shown in Figure 9.5). It is very taxing to position the scapula in a outwardly rotated position, because the outwardly rotating muscles must lift the scapula first and in addition must exert force acting in opposition to the inwardly rotating muscles that stabilize the scapula.

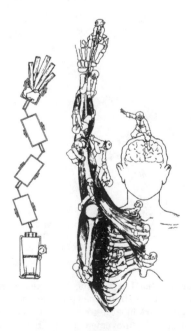

FIGURE 9.3
Illustration of the stability problem related to arm elevation. (From Codman, E.A., 1934. *The Shoulder.* Boston: Thomas Todd. With permission from Krieger Publishing Company.)

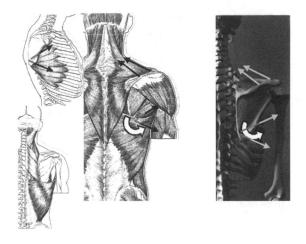

FIGURE 9.4

Illustration of the outward rotating muscle couple for the scapula. The responsible muscles are the descending part of trapezius (top arrows) and the lower parts of serratus anterior (bottom arrows). (Drawings courtesy of Dr. A. Aarås.)

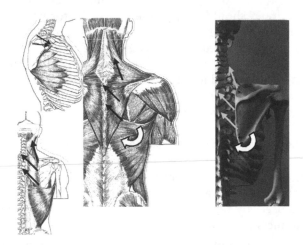

FIGURE 9.5

Illustration of the inward rotating muscle couple for the scapula. The responsible muscles are the levator scapulae (top arrow), both rhomboids (which are hidden under the trapezius muscle), and the upper parts of serratus anterior (indicated on left insert). (Drawings courtesy of Dr. A. Aarås.)

The rotator cuff muscles at the glenohumeral joint are formed by the supraspinatus, infraspinatus, subscapularis, and teres minor. These muscles have their origin on the scapula and their insertion on or near the humeral head. This muscle group is especially suited for stabilization of the upper arm in the glenohumeral joint. The supraspinatus draws the head of the upper arm into the joint so that the head is pressed toward and slides a little downward on the glenoid socket of scapula when the upper arm lifts. Both the infraspinatus and subscapularis draw the upper arm toward the trunk and press the head of upper arm toward the glenoid.

The high demand for stability of the glenohumeral joint increases when the effect of gravity increases, as when the arm is raised above shoulder height. In such a position, the need for balance, illustrated in Figure 9.3, becomes most pronounced. As a consequence, it can be expected that muscle stresses will increase when arm postures above shoulder height are adopted (although only when the trunk is upright).

9.1.2 Shoulder–Arm Kinematics

9.1.2.1 Measurement of Kinematics

Because of the anatomy and function of the upper extremity, the movements of the arm and shoulder can only rarely be simplified to a planar motion. As a consequence, either three-dimensional interpretation of planar data or three-dimensional recording of posture is necessary. Of course, the latter is the better option. A complicating factor in the measurement of shoulder function is that, for a thorough understanding, the position and displacements of the scapula should also be measured. Unfortunately, the scapula moves beneath the skin during arm elevation, as can easily be checked by palpation. As a consequence, the movements of skin markers identifying anatomical landmarks will not show the actual scapular movements. This complicates the use of standard *in vivo* measurement techniques with video, opto-electronic, or electromagnetic recording systems. There are three options to overcome this problem:

1. Quasi-static measurements, where a particular task is divided into different steps, or snapshots, in which the orientations of trunk, scapula, and arm are measured. These measurements can be done by palpation (Johnson and Barnett 1996) or, when an opto-electronic system is used, by repositioning of markers to the new locations of the anatomical landmarks.

2. Measurement of trunk and arm motion and estimation of scapula position from regression equations. These regression equations (Veeger et al. 1993, Pascoal et al. 2000, De Groot and Brand 2001) are based on the existence of a so-called scapulohumeral rhythm (see below) and describe the orientation and position of the scapula on the basis of trunk and arm orientations and, if necessary, of the external load on the hand.

3. Ignoring the problem by assuming that the shoulder is a thoraco-humeral joint. For some cases this can be acceptable, for example, when an indication of the net torque around the glenohumeral joint is needed. Doorenbosch et al. (2001) estimated that the functional rotation axis for arm movements in the sagittal plane was at the

humeral head, while for movements in the frontal plane this axis was located about 13 cm medially relative to the acromion (Figure 9.1). However, when shoulder motions are described on the basis of functional rotation axes, interpretation of the resulting values concerning muscle function becomes highly unreliable, as these motions are not the actual joint rotations. As a consequence, the relative contributions of muscles to the resulting movement cannot be calculated.

9.1.2.2 *Three-Dimensional Kinematics*

One of the major difficulties in measuring shoulder posture and kinematics is the three-dimensional description of motions. Positions of the shoulder joint are commonly described in terms of degrees of humeral elevation relative to the principal anatomical planes: flexion–extension in the sagittal plane and abduction–adduction in the frontal plane, combined with axial rotation in the transversal plane. Although this terminology is generally accepted, it does not allow for unambiguous description of all arm positions and movements and additional terms such as "horizontal abduction" are usually introduced. In addition, description of many arm positions in three dimensions requires a strict order of rotations. A striking example, known as Codman's paradox (Codman 1934), can illustrate this. A position with the hand placed on the head with the hand palm downward (shown in Figure 9.6) cannot be unambiguously described as either endo- or exo-rotated. Starting the description from the standard anatomical position (with the arm vertical and the palm of the hand facing forward), the newly adopted position of the arm can be described as a combination of 90° internal rotation and 180° forward flexion of the arm (with 90° elbow flexion), but also as a combination of 180° abduction and 90° external rotation (and 90° elbow flexion). For a clear and unambiguous description of the posture of the shoulder joint, it is better to adopt a strict order, preferably following the order suggested by Pearl et al. (1992), where shoulder posture is described

FIGURE 9.6
Illustration of Codman's paradox. In the position shown, the arm can be seen as exo-rotated as well as endo-rotated, depending on the rotations that were made to reach this position.

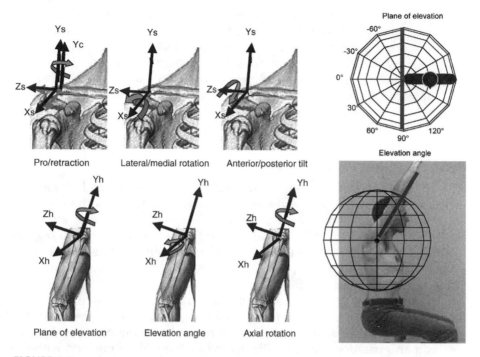

FIGURE 9.7
Definitions of rotations for the scapular motions (top left) and thoracohumeral motions (bottom left). The rotations are in fact in the Euler decomposition order. Humerus orientations can be described as the movement of the elbow over the surface of a sphere, with the shoulder in the center of that sphere (right).

in terms of three angles relative to a sphere where the elbow travels over the surface of that sphere and the center is formed by the shoulder (Figure 9.7): first, the angle related to the plane of elevation, subsequently the angle of elevation within that plane, and finally the angle of axial rotation. This description order is in fact comparable to a sequence of Euler rotations in the vertical–outward–local vertical order (see Karduna et al. 2000).

A second difficulty in the description of kinematics is the definition of the local coordinate systems of the separate segments. The exact angles and movements depend on the way in which the local coordinate system of a segment is defined, which depends on the anatomical landmarks used for this process. A list of definitions of anatomical landmarks is given in Table 9.1. One of the major difficulties is the definition of the coordinate system for the humerus, for which two distal landmarks can easily be defined but for which it is virtually impossible to define a well-identifiable proximal landmark because the humeral head is completely covered by muscles (the muscles of the rotator cuff, as well as the deltoid muscle). The best solution for this problem is to define the local landmark as the glenohumeral rotation

TABLE 9.1

Definitions of Anatomical Landmarks

Segment	Abbreviation	Definition
Thorax	IJ	Deepest point of incisura jugularis (suprasternal notch)
	PX	Processus xiphoideus, most caudal point on sternum
	C7	Processus spinosus of 7th cervical vertebra
	T8	Processus spinosus of 8th thoracic vertebra
Clavicle	SC	Most ventral point on sternoclavicular joint
	AC	Most dorsal point on acromioclavicular joint (shared with scapula)
Scapula	AC	Most dorsal point on acromioclavicular joint (shared with clavicula)
	TS	Trigonum spinae scapulae, midpoint of triangular surface on medial border of the scapula in line with the scapular spine
	AI	Angulus inferior, most caudal point of scapula
	AA	Angulus acromialis, most laterodorsal point of scapula
	PC	Most ventral point of processus coracoideus
Humerus	GH	Glenohumeral rotation center, estimated by regression or motion recordings
	EM	Most caudal point on medial epicondyle
	EL	Most caudal point on lateral epicondyle

center (more or less equal to the center of the humeral head). Methods for determining this landmark have been described by Meskers et al. (1998), Veeger (2000), and Stokdijk et al. (2000).

For more detailed descriptions of three-dimensional kinematic recording procedures of the arm and shoulder, the reader is referred to Van der Helm (1997).

9.1.2.3   Scapulohumeral Rhythm

As mentioned above, the scapula shows large displacements with respect to the skin during arm movements and is therefore difficult to measure. There exists, however, a scapulohumeral rhythm (Inman et al. 1944; Figure 9.8), which is defined as the rate of contribution of thoracoscapular rotation and glenohumeral rotation to the elevation angle of the arm. This rhythm appears to be stable both between and within subjects, and to be relatively insensitive to the simultaneous exertion of, or resistance to, an external load. The rhythm sometimes even appears to be stable under conditions deviating from normal conditions, such as fatigue or even impingement (Karduna et al. 2002). The scapulohumeral rhythm can therefore be used to estimate the scapula orientation when the thoracohumeral angle is known. In an extensive study comprising different movements, velocities, and external loads, Pascoal et al. (2000) concluded that the scapula position can be reliably reconstructed and that, when arm orientation and external load are known, the scapula orientation can be predicted from arm posture and external load (Table 9.2).

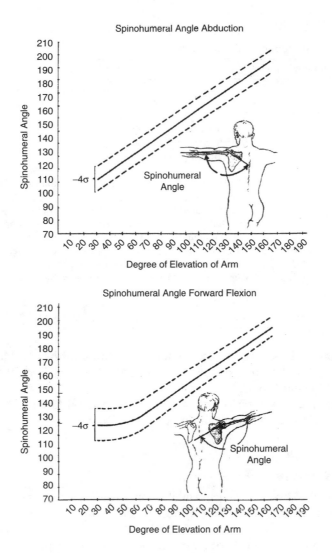

FIGURE 9.8
The scapulohumeral rhythm. The spinohumeral angle is the angle between the spine of the scapula and the upper arm. (Top) Rhythm during abduction. (Bottom) Rhythm during arm flexion. (From Inman, V.T. et al., *Journal of Bone and Joint Surgery,* 26, 1–30, 1994. Copyright *The Journal of Bone and Joint Surgery.* With permission.)

The maximal range of motion (ROM) for the arm is thus the consequence of reorientation of the scapula, as well as of rotations of the humerus relative to the scapula. Table 9.3 provides an overview of the joint rotations related to the ranges of motion for different types of movements. From this table it follows that, although in theory the maximal elevation for the thoraco-humeral angle is around 180°, which occurs in anteflexion, the actual maximum movement within the glenohumeral joint is only 131°. The rotation of the scapula is responsible for the other 50°.

TABLE 9.2

Coefficients of the Linear Regression Model of the Scapulohumeral Rhythm

	Constant	Plane of Arm Elevation (°)	Arm Elevation Angle (°)	Initial Orientation of Scapula (°)	Load in the Hand (kg)	Explained Variance
S_y	14.759	0.1060	−0.0736	0.677 (Si_y)	−0.183	0.78
S_z	−11.008	−0.0811	0.4130	0.730 (Si_z)	0.811	0.87
S_x	0.4330		0.0400	0.973 (Si_x)		0.79

Note: S_y, S_z, and S_x are the scapula rotations: protraction, laterorotation, or outward rotation and anterior spinal tilt, respectively (see also Figure 9.7). Data and angle definitions are based on the conventions proposed by the International Shoulder Group (Van der Helm 1997).

Source: Pascoal, A.G. et al., *Clinical Biomechanics*, 15(S1), S21–S24, 2000.

TABLE 9.3

Average Maximal Joint Angles of 24 Subjects in Eight Different Range of Motion Movements

Range of Motion Movements	Elevation Angle of Humerus, Relative to the Scapula (scapulohumeral angle) (°) Mean	SD	Axial Rotation of Humerus (°) Mean	SD	Elbow Angle (0° is full extension) (°) Mean	SD
Anteflexion	131	10				
Retroflexion	50	8				
Abduction	132	10				
Adduction	54	19				
Endo-rotation	66	14	51	26		
Exo-rotation	55	18	82	17		
Elbow flexion					148	14
Elbow extension					13	34

Definitions of movements:

Ante-/retroflexion and abduction:	Maximal elevation of the extended arm in the sagittal plane for ante/retroflexion (ideally −90° or +90° plane of elevation), or in frontal plane for abduction (ideally +0° plane of elevation).
Adduction:	Maximal elevation of the extended arm in the contralateral frontal plane.
Endo-/exo-rotation	Maximal internal/external rotation of the flexed arm in front of the body at a humerus elevation angle of 65°.

SD = standard deviation.

Source: Magermans, D.J. et al., 2001, in *Proceedings of the Third Congress of the International Shoulder Group*, Chadwick, E.K.J. et al., Eds., Newcastle-upon-Tyne, 4–6 September, 2000, Delft, the Netherlands: Delft University Press, 101–104.

9.1.3 Biomechanical Modeling

Biomechanical modeling is the translation of a biological system to a mechanical analogy. Models can vary from simple models ("inverted pendulum model" for the study of human balance) to extremely complex models. Each level of complexity has its merits and drawbacks. The simplest models that are available for the evaluation of load on the shoulder and arm are linked-segment models. Linked-segment models describe the upper extremity as a series of bars and linkages, representing segments and joints, but do not include muscles or their mechanical representatives, so-called actuators. These models can be designed as "generic" models, with, for example, a three degrees of freedom elbow joint, but can also include more specific information on joint properties, for example, in the form of two specific axes for the elbow (Veeger et al. 1997). Because of the difficulty related to the measurement of the position of the scapula, the scapula is usually ignored, in which case the shoulder is modeled as a functional thoracohumeral joint.

A linked-segment model can only be used for the estimation of net moments and net forces, but obviously does not allow for the calculation of muscle or ligament forces, or for the quantification of joint contact forces. The linked-segment models require kinematic input on the trunk and arm and the input of any external forces. Many of these models have been developed. One of the best-known examples is the three-dimensional static strength prediction (3-D SSPP) model (Chaffin and Erig 1991).

Musculoskeletal models include actuators, which are mechanical representations of muscles. These models comprise specific segment, joint, and muscle characteristics. In addition to the output that can be obtained from linked-segment models, as described above, musculoskeletal models can be used for calculation of joint contact forces and individual muscle forces. In these calculations they can account for the forces needed to position the scapula, and the forces needed to prevent glenohumeral dislocation. They do not, however, allow for the more detailed estimation of muscle contributions in terms of reaction to fatigue and metabolic behavior. Very few musculoskeletal models of the shoulder girdle (and arm) are available. Examples of those models can be found in Van der Helm (1994), Högfors et al. (1995), or Charlton and Johnson (2001). Figure 9.9 illustrates another model, in this case for wheelchair propulsion.

In principle, metabolic behavior can be modeled with so-called muscle models, because they contain mechanically and physiologically valid descriptions of muscle behavior, and it should be possible to use those models for evaluation of the effects of fatigue, for example, repetition. To date, these models are still in the developmental phase.

In most cases in ergonomics, data collection for the application of musculoskeletal models is complicated and quite often not really necessary. Recent research (Praagman et al. 2000) has shown that a strong correlation exists between net moment around the glenohumeral joint and contact force in that joint. When it is assumed that joint contact force can be used as an indicator for mechanical load on the shoulder (as it directly represents a possibly damaging process, but also indirectly represents the load on the shoulder muscles), it appears the net shoulder moments can be used that

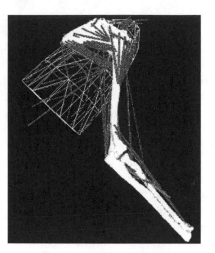

FIGURE 9.9
Illustration of a shoulder–elbow model, as used for the analysis of wheelchair propulsion.

for an indication of mechanical workload. In most ergonomic studies the net shoulder moments will mainly be determined by the effect of gravity of the arm (and thus arm position) and, when applicable, also the external weight carried, and less by arm accelerations. Only in tasks such as drilling will other external forces and torques be important. As a consequence, from a mechanical viewpoint, the highest external moments and thus also the highest workload at the shoulder will be found in conditions in which the arm is extended and 90° elevated. A warning is warranted here: the fact that the mechanical load is highest at 90° arm elevation does not imply that the "physiological" load is also the highest. 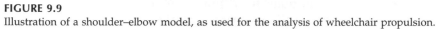 This load is dependent on muscle characteristics. Muscles will be less capable of producing force at lengths shorter than their optimum length. This implies that, although the mechanical load for arm positions higher than 90° elevation is lower, the physiological load might be higher! This is then due to the fact that the muscles responsible for the external moment must produce the necessary force under less favorable conditions. Muscle models that account for the physiological characteristics of muscles will be able to produce more accurate predictions.

If more specific information is needed, a combination of linked-segment models and (surface) electromyography (EMG) is also an option. The estimated net moments can then be interpreted with the help of EMG data. It should, however, be kept in mind that EMG in itself is highly unreliable for the estimation of muscle forces, as EMG is intrinsically dependent on muscle length and shortening velocity, and also on the choice of electrode, electrode position, and skin preparation. EMG is, on the other hand, a reasonable indicator of strain on a muscle, especially when the EMG amplitude can be expressed as "average rectified value," or "root-mean-square" value relative to a maximal voluntary contraction.

In everyday ergonomic practice it is quite common to use linked-segment models in an "implicit" manner: the work place is evaluated through a simple kinematic analysis, without taking the effect of gravity and external forces into account. Although this is not impossible, it should be kept in mind that this method is very sensitive to interpretation errors. Ignoring gravity effects and external forces can easily lead to misinterpretation of the actual load on the musculoskeletal system.

To date, biomechanical modeling has mainly been focused on so-called high-force conditions, where peak stresses are thought to be the major risk factor. Models can be used to calculate peak external torques, or to estimate peak forces on muscles. None of the models is, however, capable of calculation of maximal strain related to overuse injury. The maximum force that a muscle can produce is usually limited to a theoretical level, which is based on average muscle size and theoretical maximum force per unit muscle mass. How this relates to strain is yet unknown, but might become known in the future.

"Long-exposure" conditions have not been the subject of modeling research, because, to date, valid mechanical models for the effects of fatigue have not been developed. As a consequence, much effort has been invested in the development of methods that quantify fatigue based on EMG analysis. Examples of these are the estimation of the "mean frequency" and "median frequency" of a Fourier analysis based on a (statically obtained) EMG signal. Both mean frequency and median frequency decrease when the muscle becomes fatigued. Although frequency changes reflect the occurrence of muscle fatigue, they have not been useful in determining when complaints of fatigue are likely to occur in conditions in which there is a low exposure to external forces. It appears that the analysis of the occurrence of "gaps" in the EMG signal (i.e., periods when the measured part of a muscle does not show activity) of a statically loaded muscle (usually the trapezius muscle) might predict the development of long-term overuse injury (Veierstedt et al. 1993). These relationships are, however, not very strong.

Musculoskeletal models are more useful in laboratory studies than in field studies, as the measurement procedures are quite complex. It is expected, now that models are becoming more easily available and more easily applicable, that more data will be published on the mechanical load on the upper extremity in different tasks. A recent example of this is the study by De Vries et al. (2001), who investigated the effect of handle position on the load on the shoulder when pushing four-wheeled containers. The results are shown in Figure 9.10. This study led to the conclusion that, for the shoulder, a handle at shoulder height would be most beneficial and would lead to far lower strain on the rotator cuff muscles. This high handle position, however, led to a far larger strain on the elbow extensors. In fact, this study proved what has often been ignored until recently: that load reduction at one level of the human musculoskeletal system might be quite unfavorable for other parts of the body (see also Kuijer 2002, for a comparison of load on the lower back and load on the shoulder).

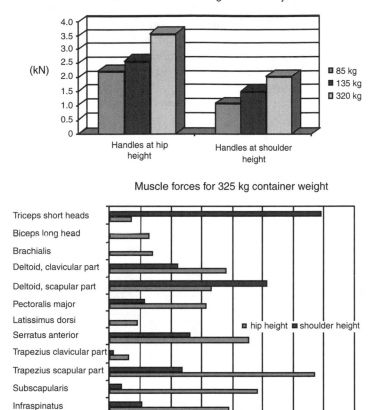

FIGURE 9.10

Results for the mechanical load on the shoulder during two-handed pushing of four-wheeled containers, with weights from 85 to 320 kg. The glenohumeral joint contact force was higher when pushing at hip height. This was related to higher forces in the rotator cuff muscles (as shown in the lower figure). When pushing at shoulder height it appears that the elbow extensors are more involved. (From De Vries, W.H.K. et al., 2001, in *Proceedings of the Third Congress of the International Shoulder Group*, Chadwick, E.K.J. et al., Newcastle-upon-Tyne, 4–6 September, 2000, Delft, the Netherlands: Delft University Press, 92–95. With permission.)

Summary

The human shoulder and arm have a large range of motion as a result of the special anatomy of these structures and the special role of the scapula and clavicle in the shoulder girdle. The consequences of this are that the intermuscular coordination needed for stabilization of the arm is of great importance and that working with the arms above shoulder height (when working in an upright posture) will lead to relatively high muscle stresses.

The measurement of arm–shoulder motion is difficult and requires three-dimensional methods. The orientation of the scapula, in particular, is difficult to measure, but it can be directly quantified by palpation or reconstructed from the orientations of the arm and trunk, based on the existence of a scapulohumeral rhythm.

Estimation of the mechanical load on the arm and shoulder is also still difficult. Biomechanical models of the shoulder and arm are scarce and highly complex. State-of-the-art models are now capable of estimating stresses on muscles and some ligaments, but the translation of these values into measures of (risk of) damage is still not possible. In addition, the application of these models in ergonomic practice is quite complicated. Until models are developed that can account for muscle characteristics and fatigue, the relatively simple procedure of estimating net moments around joints appears to be the method with the best combination of validity and feasibility.

9.2   Upper Arm

Arne Aarås

Introduction

Static muscle load in the neck and shoulder, the high frequency of repetitive movements of the upper arm, and the high force requirements of these movements seem to be predictors for onset of musculoskeletal discomfort in the upper part of the body (Kaergaard et al. 2001). Frost et al. (2001) found that the occurrence of shoulder disorders among workers in repetitive work increased with frequency of movements of the upper arm. They found a rate ratio for shoulder tendinitis of 4.5 (95% CI: 1.44 to 14.14) when comparing a group of workers with the highest exposure with the group of lowest exposure (referents). The incidence of neck and shoulder syndrome has been reported to increase with repetition rate as well as with high shoulder force and with percentage of time with static workload in the shoulder (Mikkelsen et al. 2001). Duration of repetitive movements of the upper arm (as in computer use and particularly with computer mouse use) was found to be associated with neck and shoulder symptoms. Jensen and Christensen (2001) found an odds ratio for duration of work to be 2.14 (95% CI: 1.4 to 3.3) when comparing full-time computer use vs. computer use of less than one quarter of the work time of women operators. This is supported by a study by Brandt et al. (2001). They found that reports of moderate to severe pain in the neck and right shoulder were associated with work with input devices such as a mouse for VDU workers. The study was based on a self-report questionnaire regarding neck pain and a clinical examination. The results were adjusted for age, gender, and work stress but not for other physical aspects of the work such as work posture. Christensen et al. (2001) studied single motor unit activity patterns in the trapezius muscle. They found that, in order to reduce high activity in the trapezius, repetitive finger movements must be limited. Furthermore, high repetitive finger movements activated co-contraction in neck and upper limb muscles. Another mechanism for work-related myalgia is proposed by Larsson et al. (2001). They found a significantly lower capillarization in the trapezius muscle (the blood supply by small vessels) in cleaners suffering from trapezius myalgia than in cleaners without myalgia. Reduced blood flow may influence the oxidative metabolism in the muscle fibers. Kilbom (1994), in a review paper, assessed the risk for musculoskeletal discomfort in repetitive work. She concluded that a high risk exists when repetitive work has a cycle time of less than 30 s (i.e., a frequency of two movements per minute or more), especially in combination with exertion of high force. She further concluded that a high risk also exists when the fundamental cycle consists of more than 50% total cycle time and the repetitive work is performed for more than 1 h per day. Greater risk of musculoskeletal disorder is also expected when the upper arm has a frequency of ten movements per minute or more. The risk increases considerably when the external force exerted is high or when there is a high static load or extreme posture (Kilbom 1994). More details about risk factors for musculoskeletal discomfort are given in the following subsections. Section 9.2.1 deals with the effects of gravity on the musculoskeletal system, due to the posture of the upper arm, Section 9.2.2 deals with the effects of the arm posture in relation to the trunk, and Sections 9.2.3 and 9.2.4 describe studies indicating reduction in muscle load by supporting the forearm in a neutral position. The last section deals with the importance of psychosocial factors in the development of musculoskeletal discomfort.

9.2.1 Upper Arm Elevation

Working posture and movements of the upper arm as well as external load in the hands influence the physical load on the musculoskeletal system of the arm. EMG can be used to illustrate how the different shoulder muscles are loaded to balance the shoulder moment when moving the upper arm. Sigholm et al. (1984) studied the muscle load in several shoulder muscles regarding flexion and abduction of the upper arm (where flexion of the upper arm is elevation in the sagittal plane while abduction of the upper arm is elevation in the coronal plane). They calculated the percentage increase in the integrated EMG amplitude when the load in the hand was increased by 1 kg during flexion and abduction (as shown in Table 9.4). In particular the stabilizing rotator cuff muscles (infraspinatus and supraspinatus) were found to be influenced to a great extent by increasing the hand load.

TABLE 9.4

Percentage Increase of EMG Level
in Various Shoulder Muscles When
Hand Load Is Increased by 1 kg

Muscle	Flexion	Abduction
Deltoid		
Anterior	21	16
Medial	17	9
Posterior	14	17
Infraspinatus	35	41
Supraspinatus	22	20
Trapezius upper	15	17

Note: The table indicates the mean values
(from nine subjects) over 21 different
arm positions with the hand above waist
level. In each position the hand was
loaded with 0, 1, or 2 kg.

Source: Sigholm, G. et al., *Journal of Orthopaedic
Research*, 1, 379–386, 1984.

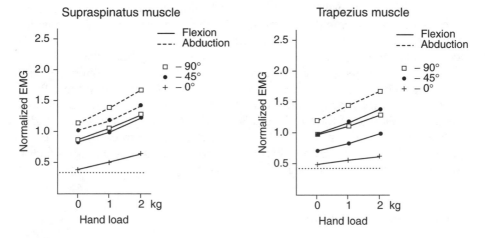

FIGURE 9.11
The muscle load for the supraspinatus muscle (left graph) and the trapezius muscle (right graph) depending on the flexion and abduction angles in the glenohumeral joint and the external load in the hand. (Reprinted from *Journal of Orthopaedic Research*, 1, Sigholm, G. et al., Eletromyographic analysis of shoulder muscle load, 379–386, Copyright (1984), with permission from The Orthopaedic Research Society.)

The activity in the shoulder muscles also depends on how high the upper arm is lifted. Figure 9.11 illustrates the integrated EMG signal level for the trapezius and supraspinatus muscles when the load in hand is increased from 0 to 1 kg and from 1 to 2 kg. On the *y*-axis are the individual EMG recordings normalized for each muscle, where each integrated (rms) value is divided by the mean of all values recorded from the particular muscle.

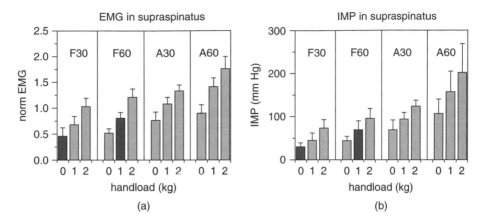

FIGURE 9.12

(a) Normalized rms-values of the EMG; (b) intramuscular pressure from the same isometric test positions. Muscle activity (EMG) and intramuscular pressure (IMP) in the supraspinatus muscle at shoulder flexion 30° and 60° with a flexed elbow (F), and at shoulder abduction 30° and 60° with a straight elbow (A). In each position the hand was loaded with 0, 1, or 2 kg. The two separately shaded columns indicate the approximate work situation in the assembly type of work (flexion 30° without hand load) and welding work (flexion 60°, 1 kg hand load). The bars indicate ± 1 standard deviation about the mean value. (From Järvholm, U. et al., *Ergonomics*, 34(1), 57–66, 1991. With permission from Taylor & Francis (http://www.tandf.co.uk).)

Figure 9.11 shows that the shoulder muscles are loaded differently during flexion. The supraspinatus muscle is heavily loaded at an angle of 45° flexion but is only lightly loaded when the upper arm is lifted to 90°, while the activity of trapezius muscle increases more in this angle interval. Sigholm et al. (1984) concluded that the degree of lifting of the upper arm is the most important parameter for deciding the load on the shoulder muscles. The stabilizing shoulder muscles such as the rotator cuff muscles showed greater dependence on load in the hand compared with the muscles moving the upper arm in the glenohumeral joint (deltoid, pectoralis major, biceps brachii, coracobrachialis, latissimus dorsi, and triceps brachii). This means that the upper extremity is not constructed to carry a burden, but is very mobile in moving the hand within extensive reach envelopes. These results are supported by studies of Järvholm et al. (1991). They found that the intramuscular pressure as well as the EMG muscle activity increased in the supraspinatus muscle both with increasing flexion and increasing abduction, as shown in Figure 9.12. Järvholm et al. (1991) have documented that an intramuscular pressure of 40 mmHg significantly reduced the muscle blood flow. They also showed that in simulated welding work (flexion 60°, 1 kg hand load) this threshold is exceeded. Welding in this arm position has been shown to induce EMG signs of localized muscle fatigue in the supraspinatus for both experienced and inexperienced workers (Kadefors et al. 1976).

Aarås (1994) showed that the shoulder moment seems to influence the incidence of musculoskeletal sick leave. Table 9.5 shows the recorded postural angles of the upper arm relative to vertical (as group median values)

TABLE 9.5

Postural Angles of the Upper Arm (flexion/extension and abduction/adduction), Flexion Angles in the Elbow Joint, and Weight of the Hand Tool for Assembly Workers Working under Different Work Systems

Work System	No. of Subjects	Flexion (+ve) and Extension (–ve)		Abduction (+ve) and Adduction (–ve)		Flexion in Elbow	Weight of Hand Tool
		Median	Range	Median	Range	Median	(N)
8B sitting	6	19°	7°–44°	8°	5°–14°	80°	8.5
8B standing	6	17°	8°–43°	5°	–3°–12°	75°	
10C sitting	3	8°	0°–28°	7°	0°–8°	70°	3.5
10C standing	3	–11°	–14°–8°	7°	2°–19°	90°	
11B sitting	2	11°	8°–14°	–1°	–4°–2°	70°	3.5
11B standing	4	13°	–11°–15°	–4°	–8°–2°	70°	
DF sitting	6	39°	11°–58°	60°	23°–77°	90°	0.2

Source: Aarås, A., *Ergonomics,* 37, 1679–1696, 1994.

for assembly workers working under different work systems (coded as 8B, 10C, 11B, and DF). The work tasks for all systems were production of parts for telephone exchanges. Mostly, the work consisted of connecting wires with bare ends to the needle-shaped terminals on a frame by using a wrapping gun. The wire was placed in a wrapping gun and then the gun was positioned on the terminal. Thereafter, the wrapping gun spun the wire around the terminal. The size of the frame was different for the four work systems and the weights of the tools used differed, but otherwise the work was almost identical. The group median value of the flexion angle in the elbow joint was estimated from video recordings (Table 9.5). The weight of the hand tool for the different work systems is presented in Table 9.5. The static shoulder moments (Nm) were calculated for the 50th percentile of the joint angle values (i.e., that angle has a higher value for 50% of the recording time). These were the angles of the upper arm in the glenohumeral joint and the flexion of the forearm in the elbow joint. In addition, the weights of the tool and the arm were considered. The static shoulder moments are shown in Figure 9.13. The assembly workers at the 8B and DF work systems had to perform work with higher shoulder moments (6.2 and 5 Nm, respectively), compared with workers at the 10C and 11B work systems (1.4 and 3.7 Nm, respectively). The incidence of musculoskeletal sick leave was higher for the 8B and DF work tasks compared with the 10C and 11B work tasks (Westgaard et al. 1986). Furthermore, the incidence of musculoskeletal sick leave was very similar for workers at 8B and DF work system, corresponding to a similar shoulder moment.

Kadefors (1994) similarly reviewed the prevalence of musculoskeletal illness among welders and concluded that there is an excessive risk for development of shoulder pain due to inflammatory processes in the rotator cuff for welding at or above shoulder level. In Sweden, welders had an excess risk of 90% of early retirement compared with the male workforce in general. Diseases of the musculoskeletal system constituted the most prevalent diagnosis

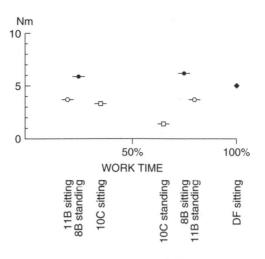

FIGURE 9.13

The static shoulder moment (Nm) calculated on the basis of the group median value of postural angles of the upper arm and flexion in the elbow joint for different work systems. The time in sitting and standing position as a percentage of total work time is indicated on the x-axis. (From Aarås, A., *Ergonomics*, 37, 1679–1696, 1994. With permission from Taylor & Francis (http://www.tandf.co.uk).)

among welders who retired early: the percentage was as high as 49%. Herberts and Kadefors (1976) found the prevalence rate of supraspinatus tendinitis to be significantly higher among welders, who often had arm work above shoulder height, than among white-collar workers. They found that the supraspinatus muscle is particularly strained in work at or above shoulder level.

Other studies support the above results (Chaffin 1973, 1987, Hagberg 1981); i.e., sustained, elevated arm work, especially when supporting a load, repeated extended upper arm reaches, both in the horizontal and vertical planes, create load moments at the glenohumeral joint that are related to an increase in shoulder muscle fatigue and to the incidence of shoulder tendinitis and/or bursitis. Bjelle et al. (1981) found that workers who suffered from soft tissue rheumatism had a significantly longer duration and higher frequency of abduction or flexion of the upper arms compared with their control groups. The elevation of the arm above 60° was found to be significantly more frequent, as well as sustained for a longer duration, in workers with acute shoulder neck pain than in matched controls. Hagberg and Wegmann (1987) concluded that welders had an increased risk of acquiring shoulder tendinitis, with the rate ratio found to be 10 in welders, plate workers, and pooled groups who worked with elevated arms. Furthermore, the same researchers found that, when working at shoulder level, the odds ratio for rotator cuff tendinitis was 11. Kilbom and Persson (1987) also demonstrated that shoulder abduction was related to the onset of symptoms of neck/shoulder pain.

There is a consensus that static muscle load should be kept at a minimum while the dynamic work pattern should be increased (Kilbom 1988). Static muscle load can be reduced by improving the working postures and by reducing the holding time of the postures. Rose et al. (1992) found that when subjects were allowed to decide on the duration of static work themselves, they stopped for a pause at approximately 20% of maximum holding time (MHT). In ISO standard 11226 (ISO 2000), 20% MHT is recommended as the maximum acceptable holding time. Because of a linear relationship between % MHT and discomfort, this implies a score of 2 on the Borg 10-point pain scale (Borg 1982). Kilbom et al. (1986) found that the number of flexing or abducting movements of the upper arm per hour were negatively related to neck and shoulder symptoms. A longitudinal epidemiological study has indicated that the introduction of more dynamic work in terms of more flexion of the upper arm reduced the incidence of musculoskeletal illness in the neck and shoulder (Aarås et al. 1990).

9.2.2   Upper Arm Position with Respect to the Trunk

Several studies have shown that the direction of the elevated upper arm (flexion or abduction) in relation to the upright trunk may play a role with respect to shoulder load. Kilbom et al. (1986) suggested a stronger relationship between upper arm abduction and the severity of musculoskeletal disorder compared with that for upper arm flexion. Figure 9.11 shows that muscle load was higher in upper arm abduction than in upper arm flexion (Sigholm et al. 1984).

Several researchers have reported that visual display unit (VDU) operators prefer a backward tilting posture of the back, as shown in Figure 9.14. This trunk posture is accompanied by an elevated upper arm position, suggesting that the operators do not want to have the upper arm retroflexed (i.e., that they do not work with their elbow behind the line of the trunk when viewed from the side). In a study of preferences for layout of the VDU workplace, Taiwanese VDU users of both PC and computer-aided design (CAD) workstations preferred to sit in a pronounced backward leaning posture with the upper arm slightly elevated and an elbow angle greater than 90° (Hsu and Wang 1999). This is also in line with study by Grandjean et al. (1983) who found that the majority of their subjects preferred trunk inclinations between 100° and 110°. The angles they adopted for upper arm flexion were within a 95% confidence interval of 13° to 33° from vertical. The inclination of the chair backrest also significantly affects upper arm flexion. On average, at a backrest inclination of 15°, the upper arms are flexed by 5° more compared with posture with an upright backrest (Delleman 1999).

When performing deskwork, a posture with an upper arm abduction angle of 15° to 20° or less and a flexion angle of 25° or less is highly desirable (Chaffin and Andersson 1984) (Figure 9.15). These recommendations are based on a study by Chaffin (1973). He used EMG frequency spectrum shifts and subjective pain ratings to measure endurance times for various elevation positions of the arm.

FIGURE 9.14

The posture with the back tilted backward 10° to 15° and the cervical spine vertical gives the lowest static muscle load in the shoulder. (Reprinted from *Applied Ergonomics*, 32, Aarås, A., Horgen, G., Bjørset, H.-H., Ro, O., and Walsøe, H., Musculoskeletal, visual and psychosocial stress in VDU operators before and after multidisciplinary ergonomic interventions. A 6 years prospective study — Part II, 559–571, Copyright (2001b), with permission from Elsevier.)

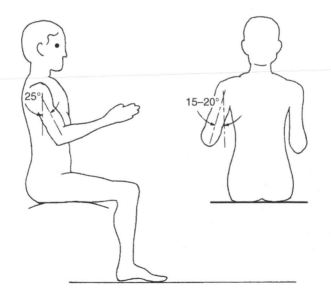

FIGURE 9.15

Flexion angles of 25° and abduction angles of 15° to 20° are acceptable for the upper arm in the glenohumeral joint. (From *Occupational Biomechanics*, Chaffin, D.B., and Andersson, G.B.J., Copyright (1984). This material is used by permission of John Wiley & Sons, Inc.)

Aarås et al. (1988) recommended that arm flexion should be less than 15° and arm abduction less than 10°. This recommendation is based on a study of sick leave due to shoulder injuries of female assembly workers who performed the wrapping operation with only small flexion angles (work systems coded as 11B and 10C in Table 9.5). Their incidence of musculo-skeletal sick leave was low and approximated the incidence of such illness for a group of female workers without continuous workload. These results indicate that the above values of arm position are beginning to approximate an acceptable arm position of continuous work tasks, when the external load is low (Aarås 1994). However, to suggest a threshold level for acceptable mechanical load on the shoulder is very complicated. Both intensity and duration of the load as well as the duration of necessary pauses between periods with prolonged mechanical load should be assessed. Further research is needed for establishing safe limits.

9.2.3   Forearm Support

Forearm support is documented to reduce muscle load in the upper part of the body as well as in the low back. Andersson and Ørtengren (1974) reported more than 25 years ago that forearm support reduced the intradiscal pressure of the lumbar spine. Aarås et al. (1997) measured EMG from the upper part of the trapezius muscle during a laboratory study of data-entry work. The trapezius muscle load was significantly lower when the subjects were sitting with forearm supported than when sitting or standing without forearm support (as shown in Figure 9.16). The static load on the right trapezius muscle was 0.8% maximum voluntary contraction (MVC) when sitting with support of the forearm and 3.6% MVC when sitting without support. There was also a tendency to lower trapezius load for standing versus sitting without support. This may be due to lower flexion of the upper arm in the glenohumeral joint in standing compared with sitting (Aarås et al. 1988).

These results indicate clearly the importance of supporting the forearm on the tabletop when operating the keyboard. Similar results were also found when using the mouse as an input device. The results from this laboratory study (Aarås et al. 1997) were confirmed in a real work situation in a field study (Aarås et al. 1998). This study had a parallel group design with two intervention groups, in technical (T) and software (S) departments, and one control group (C) of VDU male operators. Approximately 50 software engineers participated in each group. The work task was very similar for all the three groups. One of the aims of the study was to investigate how support of the forearms on the tabletop would affect the development of pain level of the VDU operators. The results of the study after 2 years are shown in Figure 9.17. The average intensity of shoulder pain during the previous 6 months was compared before and after modifying the workplace to allow the operators to support their forearms. The pain level was assessed on

Right Trapezius
Static Load (P = 0.1)

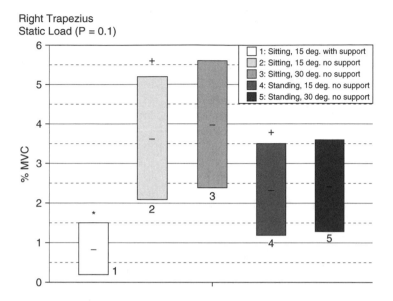

FIGURE 9.16

Static trapezius load (mean and 95% confidence interval, $n = 20$) for five postures for data entry work (with and without forearm support); 15° and 30° indicate the gaze angle to the center of the screen. Note: An * indicates a significant difference when compared with the column containing a +. (From Aarås, A. et al., *Ergonomics*, 40, 1255–1268, 1997. With permission from Taylor & Francis (http://www.tandf.co.uk).)

100 mm visual analog scale (VAS). No significant differences were found between the three groups regarding shoulder pain during the 6 months before intervention. At 6 months after the intervention the S group reported a significant reduction in shoulder pain ($p = 0.02$), and there was a tendency to reduction in shoulder pain in the group T although the difference was not statistically significant ($p = 0.08$). No significant changes were found in the control group C ($p = 0.92$). The intervention groups reported a significantly lower intensity of shoulder pain compared with the control group ($p = 0.02$).

Therefore, supporting the forearms on the tabletop in front of the VDU seems of fundamental importance for reducing muscle load and musculoskeletal pain. This statement is supported by a study of Milerad and Ericson (1994). They showed that arm support, but not hand support, was of significant importance for the load on three shoulder muscles (trapezius, supraspinatus, and deltoid) as shown in Table 9.6. Kylmäaho et al. (1999) showed that narrow keyboard tables, chairs without armrests or with non-adjustable armrests, and non-adjustable desks resulted in less frequent supporting of the forearm and a smaller area of support of the forearm in both mouse and keyboard work. Supporting only the wrist during VDU work was found to increase the trapezius load compared with no support (Bendix and Jessen 1986). This is confirmed by Fernstrøm et al. (1994) who found

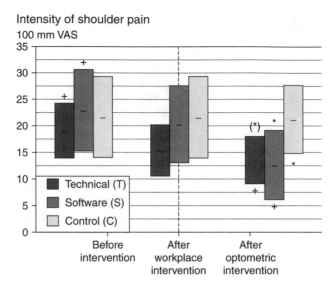

FIGURE 9.17

Intensity of shoulder pain during the previous 6 months for two intervention groups, technical (T) and software (S), and a control (C) group. The values are given (as mean, with 95% confidence interval) before intervention, after a workplace intervention, and after an optometric intervention. On the VAS scale a rating of 0 represented no pain and 100 represented very severe pain. Note: An * indicates a significant difference when comparing the column containing a plus sign +. (*) indicates a difference very close to significant difference. (Reprinted from *Applied Ergonomics*, 29(5), Aarås, A., Horgen, G., Bjørset, H.-H., Ro, O., and Thoresen, M., Musculoskeletal, visual and psychosocial stress in VDU operators before and after multidisciplinary ergonomic interventions, 335–354. Copyright (1998), with permission from Elsevier.)

TABLE 9.6

Significance Level (statistically significant, $p < 0.05$, or trend, $p < 0.10$) in Mean EMG in Three Shoulder Muscles for Comparisons between Arm Support, Hand Support, and No Support in a Low Precision Task with Force Level 7.5 N

Shoulder Muscle	Arm Support vs. No Support	Hand Support vs. No Support	Arm Support vs. Hand Support
Deltoid	0.01	0.07	0.06
Trapezius dominant	0.02	NS	NS
Trapezius/supraspinatus	0.003	NS	0.05

Note: NS = not significant ($p \geq 0.10$).

Source: Milerad, E. and Ericson, M.O., *Ergonomics*, 37(2), 255–264, 1994.

that the use of a palm rest in keyboard-only work did not decrease shoulder or forearm muscle strain. Järvholm et al. (1991) showed that support of the arm by a suspension device during assembly work reduced supraspinatus muscle load (normalized EMG) by 20% and intramuscular pressure by 34%.

9.2.4 Forearm Posture

A pronated position of the forearm seems clearly to influence the postural load and the discomfort in the shoulder. This statement is supported by several studies showing that neutral position of the forearm and wrist/hand is important to reduce musculoskeletal discomfort. Aarås and Ro (1997) found that there were significantly more micropauses in trapezius muscle activity when working with a neutral position of the forearm compared with a pronated position ($p = 0.005$). The number of periods per minute when the trapezius load was below 1% MVC was 37 (27.8 to 70.9) vs. 27 (11.6 to 41.5), given as group median with 95% confidence interval of median. A field study was performed to evaluate whether the participants with pain in the shoulder would experience a change in the development of musculoskeletal pain in the upper part of the body. The intervention consisted of permitting a more neutral position of the forearm instead of the pronated one when using a traditional computer mouse (Aarås et al. 1999). The group of 67 subjects was randomly divided into intervention and control groups. Their work tasks were software engineering, bookkeeping, and secretarial work. The results are shown in Figure 9.18. After working with a neutral position of the forearm for 6 months, a significant reduction of shoulder pain was reported by the intervention group, while the control group, who continued with a work situation in which their forearm posture was pronated, reported only small changes in the level of pain. Figure 9.18 shows the intensity of

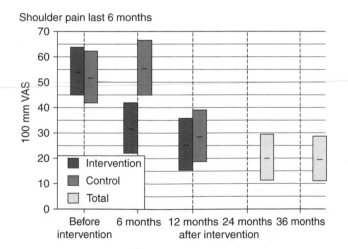

FIGURE 9.18

Intensity of average shoulder pain during the previous 6 months before intervention, at 6 months after the first group had received the intervention, and at 12 months when both intervention groups worked with a neutral position of the hand using the computer mouse. After 12 months intervention and control groups were combined. The values are given as group mean with 95% confidence interval. (Reprinted from *International Journal of Industrial Ergonomics*, 30(4–5), Aarås, A., Dainoff, M., Ro, O., and Thoresen, M., Can a more neutral position of the forearm when operating a computer mouse reduce the pain level for visual display unit operators? 307–324, Copyright (2002), with permission from Elsevier.)

shoulder pain reported during a 3-year period from the time of the initial intervention. At 6 months after the initial intervention, the control group was also given the intervention mouse, allowing each operator to use a more neutral position of the forearm during the remaining 2.5 years of the study. After 6 months of use, a significant reduction was reported of the control group's average of intensity of pain of the shoulder. The intervention group that received the initial intervention did not report any significant changes in pain in any of the body areas from 6 to 12 months of the study period; i.e., the initial reduction in pain level obtained was maintained (Aarås et al. 2001a). In fact, the reduction in pain level was still maintained by both groups after a 3-year follow-up (Aarås et al. 2002).

9.2.5 Psychosocial Factors

To reduce musculoskeletal discomfort, it is very important to evaluate psychosocial factors in addition to posture and movements. Psychological stress is reported to increase the static muscle load of the neck and shoulder girdle muscles (Wærsted et al. 1987). Van den Heuvel et al. (2001) found in a cohort study of 1738 workers in the Netherlands that a 3-year cumulative incidence of pain in neck and upper extremity was 31.9%. Incidence of pain in the neck and shoulder was 23.5%. After adjustment for physical and individual factors, an increased risk was observed for workers with high job demands (odds ratio: 2.2; C.I. 1.1 to 4.5) and for workers with low co-worker support (odds ratio: 2.8; C.I. 1.2 to 6.3). Hagberg et al. (2001) studied constrained work in a study population of 9798 workers representative of the Swedish work force (in the year 1999). Approximately one third of these spend at least half of their working time at the computer display. Constrained work was defined repetitive VDU work for at least 25% of the working time, during which the workers could not decide the work pace for more than half the time and could not take short breaks any time for more than half the time. The results showed strong associations of musculoskeletal symptoms in the neck and arms with constrained work, psychosocial stress, and poor management. Psychosocially stressful work was defined in terms of having to skip lunches at least once a week and having no time for conversation or to think of things other than work for at least half of the time. Poor management meant that the employees mostly did not have, or never had, any possibility of support and encouragement from supervisors when there were problems at work and that the supervisor had never, or seldom, shown any appreciation in the previous 3 months. The prevalence rate ratios for the three risk factors (constrained work, psychosocial stress, and poor management), when adjusted for each other, were 1.4, 1.3, and 1.2, respectively, for women and 1.7, 1.5, and 1.2, respectively, for men. Hagberg et al. concluded that prevention against musculoskeletal symptoms should target both physical and psychosocial and management style factors.

A subjective feeling of tenseness can be related to musculoskeletal pain, according to a study by Vasseljen et al. (1995a). They found high correlation between pain and perceived general tension ($r = 0.66$, $p < 0.01$). The same researchers have studied various work factors among subjects with pain and controls without pain (Vasseljen et al. 1995b). Linear discrimination analysis was used to identify the combination of variables that best differentiated cases and controls. Perceived general tension was found to be the strongest association with shoulder and neck pain in both manual and office workers. Vasseljen et al. (1995b) concluded that in most sedentary work, primarily involving mental demands, the risk of neck and shoulder pain was almost exclusively related to psychosocial and individual factors. In work with mainly biomechanical or physical demands, risk was almost exclusively related to muscle activity.

Van Galen (2001) studied physical stress, psychological stress, and muscular co-contraction. They found that both auditory stress and mental load led to increased overall muscle co-contraction and to increased overall muscle activation. These findings are supported by the fact that enhanced muscle activation and co-contraction are especially prevalent in individuals who suffer from upper extremity complaints (Bloemsaat et al. 2001). Hyldtoft et al. (2001) studied the relations between the ability of neuromuscular coordination in the arm and shoulder muscles during a low graduated isometric handgrip test, using visual feedback from a PC monitor. They found that women with poor neuromuscular coordination had a higher level of muscle activity (in the trapezius muscle as well as in the extensors and flexors of the forearm) than women with good neuromuscular coordination (median EMG amplitude: 1.6% vs. 0.4% maximum EMG).

Crenshaw et al. (2001) studied the effect of different attitude states on proprioception, i.e., movement and position perception. Suggestions were administered under hypnosis to enhance the effect. Both positive and negative suggestions were used. Typical suggestive statements were "you are having a good day," "you feel calm and relaxed," "you are having a bad day," and "you feel stressed and irritated." They examined the upper arm position for different attitude states and found that negative suggestions caused a significant reduction in proprioception acuity. They considered that workers with negative attitudes were more at risk of developing work-related problems.

Summary

High static muscle load in the neck and shoulder as well as high frequency of repetitive movements of the upper arm and high force requirements of these movements must be minimized to reduce musculoskeletal illness in the upper part of the body. The degree of elevation of the upper arm (especially abduction) is the most important parameter for deciding the load on the shoulder muscles. In particular, the rotator cuff muscles (supraspinatus and infraspinatus) are influenced to a great extent by increasing any load in the hand. The shoulder moment has been shown to influence the incidence of musculoskeletal sick leave. Increased frequency of movements of the upper arm influences the level of pain in the neck and shoulder. Forearm support reduces the muscle load in the neck and shoulder as well as discomfort in these body areas. Furthermore, working with a more neutral position of the forearm and wrist (avoiding pronation) is also important to reduce musculoskeletal pain in the shoulder region. Psychological stress, mental load, and lack of support from management and co-workers are important psychosocial factors for the risk of development of discomfort in the musculoskeletal system.

References

Aarås, A., 1994. The impact of ergonomic intervention on individual health and corporate prosperity in a telecommunication environment. *Ergonomics*, 37, 1679–1696.

Aarås, A. and Ro, O., 1997. Workload when using a mouse as an input device. *International Journal of Human–Computer Interaction*, 9(2), 105–118.

Aarås, A., Westgaard, R.H., and Stranden, E., 1988. Postural angles as an indicator of postural load and muscular injury in occupational work situations. *Ergonomics*, 31, 915–933.

Aarås, A., Westgaard, R.H., Stranden, E., and Larsen, S., 1990. Postural load and the incidence of musculoskeletal illness, in *Proceedings of the NIOSH Conference Promoting Health and Productivity in the Computerized Office*, Sauter, S., Dainoff, M., and Smith, M., Eds., London: Taylor & Francis, 68–93.

Aarås, A., Fostervold, K. I., Ro, O., and Thoresen, M., 1997. Postural load during VDU work: a comparison between various work postures. *Ergonomics*, 40, 1255–1268.

Aarås, A., Horgen, G., Bjørset, H.-H., Ro, O., and Thoresen, M., 1998. Musculoskeletal, visual and psychosocial stress in VDU operators before and after multidisciplinary ergonomic interventions. *Applied Ergonomics*, 29(5), 335–354.

Aarås, A., Ro, O., and Thoresen, M., 1999. Can a more neutral position of the forearm when operating a computer mouse reduce the pain level for visual display unit operators? A prospective epidemiological intervention study. *International Journal of Human–Computer Interaction*, 11(2), 79–94.

Aarås, A., Dainoff, M., Ro, O., and Thoresen, M., 2001a. Can a more neutral position of the forearm when operating a computer mouse reduce the pain level for visual display unit operators? A prospective epidemiological intervention study. Part II. *International Journal of Human–Computer Interaction*, 13(13), 13–40.

Aarås, A., Horgen, G., Bjørset, H.-H., Ro, O., and Walsøe, H., 2001b. Musculoskeletal, visual and psychosocial stress in VDU operators before and after multidisciplinary ergonomic interventions. A 6 years prospective study – Part II. *Applied Ergonomics*, 32, 559–571.

Aarås, A., Dainoff, M., Ro, O., and Thoresen, M., 2002. Can a more neutral position of the forearm when operating a computer mouse reduce the pain level for visual display unit operators? *International Journal Industrial Ergonomics*, 30(4–5), 307–324.

Andersson, B.J.G. and Ørtengren, R., 1974. Lumbar disc pressure and myoelectric back muscle activity during sitting: II. Studies on an office chair. *Scandinavian Journal of Rehabilitation Medicine*, 3, 115–121.

Bendix, T. and Jessen, F., 1986. Wrist support during typing — a controlled electromyographic study. *Applied Ergonomics*, 17, 162–168.

Benninghoff-Goertler, 1964. *Lehrbuch der Anatomie des Menschen I*, 9th ed., Munich: Urban & Fischer.

Bjelle, A., Hagberg, M., and Michaelson, G., 1981. Occupational and individual factors in acute shoulder-neck disorders among industrial workers. *British Journal of Industrial Medicine*, 38, 356–363.

Bloemsaat, J.G., Van Galen, G.P., Ruijgrok, J.M., Van Eijsden-Besseling, M.D., and Timmers, R.M., 2001. Effects of cognitive load on forearm EMG response activity in individuals suffering from work related upper extremity disorder (WRUED), in *Programme and Abstract Book of the Fourth International Scientific Conference on Prevention of Work-Related Musculoskeletal Disorders*, September 30–October 4, Amsterdam, 160.

Borg, G., 1982. A category scale with ratio properties for intermodal and interindividual comparisons, in *Psychophysical Judgement and the Process of Perception*, Geissler, H.-G., Ed., Berlin: VEB Deutscher Verlag der Wissenschaften, 25–34.

Brandt, L.P.A., Lassen, C.F., Mikelsen, S., Andersen, J.H., Kryger, A.I., Overgaard, E., Butcher, I., and Thomsen, J.F., 2001. Neck and shoulder disorders among technical assistants and machine technicians in Denmark — the Nudata study, in *Programme and Abstract Book of the Fourth International Scientific Conference on Prevention of Work-Related Musculoskeletal Disorders*, September 30–October 4, Amsterdam, 130.

Chaffin, D.B., 1973. Localized muscle fatigue-definition and measurement. *Journal of Occupational Medicine*, 15, 346–354.

Chaffin, D.B., 1987. Occupational biomechanics: a basis for workplace design to prevent musculoskeletal injuries. *Ergonomics*, 30, 321–329.

Chaffin, D.B. and Andersson, G.B.J., 1984. *Occupational Biomechanics*, New York: John Wiley & Sons.

Chaffin, D.B. and Erig, M., 1991. 3-Dimensional biomechanical static strength prediction model sensitivity to postural and anthropometric inaccuracies. *IEEE Transactions*, 23, 215–227.

Charlton, I.W. and Johnson, G.R., 2001. Application of spherical and cylindrical wrapping algorithms in a musculoskeletal model of the upper limb. *Journal of Biomechanics*, 34, 1209–1216.

Christensen, H., Søgaard, K., Søgaard, G., and the Procid Group, 2001. Recommendations for healthier computer work based on scientific evidence provided under "PROCID," in *Programme and Abstract Book of the Fourth International Scientific Conference on Prevention of Work-Related Musculoskeletal Disorders*, September 30–October 4, Amsterdam, 193.

Codman, E.A., 1934. *The Shoulder*, Boston: Thomas Todd.

Crenshaw, A., Sandberg, M., Hagner, I.-M., Finer, B., and Johansson, H., 2001. Negative suggestions (enhanced via hypnosis) cause a significant reduction in proprioceptive acuity, in *Programme and Abstract Book of the Fourth International Scientific Conference on Prevention of Work-Related Musculoskeletal Disorders*, September 30–October 4, Amsterdam, 95.

De Groot, J.H. and Brand, R., 2001. A three-dimensional regression model of the shoulder rhythm. *Clinical Biomechanics*, 16(9), 735–743.

De Vries, W.H.K., Veeger, H.E.J., Hoozemans, M.J.M., and Kuijer, P.P.F.M., 2001. Load on the shoulder in pushing four-wheeled containers, in Chadwick, E.K.J., Veeger, H.E.J., Van der Helm, F.C.T., and Nagels, J., Eds., *Proceedings of the Third Congress of the International Shoulder Group,* Newcastle-upon-Tyne, 4–6 September 2000, Delft, the Netherlands: Delft University Press, 92–95.

Delleman, N.J., 1999. Working Postures. Prediction and Evaluation, Thesis, Soesterberg, the Netherlands: TNO Human Factors.

Del Medico, G., 1811. *Anatomia per Uso dei Pittori e Scultori,* Rome: Presso Vincenzo Poggioli.

Doorenbosch, C.A.M., Mourits, A., Veeger, H.E.J., Harlaar, J., and Van der Helm, F.C.T., 2001. Determination of functional rotation axes during elevation of the shoulder complex. *Journal of Orthopaedic and Sports Physical Therapy,* 31, 133–137.

Fernstrøm, E., Ericson, M.O., and Malker, H., 1994. Electromyographic activity during type-writer and keyboard use. *Ergonomics,* 37, 477–484.

Frost, P., Andersen, J.H., and the PRIM Study Group, 2001. Occurrence of shoulder disorders among workers in repetitive work, in *Programme and Abstract Book of the Fourth International Scientific Conference on Prevention of Work-Related Musculoskeletal Disorders,* September 30–October 4, Amsterdam, 189.

Grandjean, E., Hünting, W., and Pidermann, M., 1983. VDT workstation design: preferred settings and their effects. *Human Factors,* 25, 161–175.

Hagberg, M., 1981, Electromyographic signs of shoulder muscular fatigue in two elevated arm positions. *American Journal of Physical Medicine,* 60, 111–121.

Hagberg, M. and Wegman, D.H., 1987. Prevalence rates and odds ratios of shoulder-neck diseases in different occupational groups. *British Journal of Industrial Medicine,* 44, 602–610.

Hagberg, M., Ekman, A., Andersson, A., and Wigaeus Tornqvist, E., 2001. Neck and arm symptoms in relation to management style, psychosocial stress and constrained work among computer users in the Swedish workforce, in *Programme and Abstract Book of the Fourth International Scientific Conference on Prevention of Work-Related Musculoskeletal Disorders,* September 30–October 4, Amsterdam, 151.

Herberts, P. and Kadefors, R., 1976. A study of painful shoulder in welders. *Acta Orthopaedica Scandinavica,* 44, 381–387.

Hogfors, C., Karlsson, D., and Peterson, B., 1995. Structure and internal consistency of a shoulder model. *Journal of Biomechanics,* 28, 767–777.

Hsu, W.-H. and Wang, M.J., 1999. The comparison of preferred settings between PC and CAD workstation, in Bullinger, H.-J. and Ziegler, J., Eds., *Conference Book: Human–Computer Interaction,* Vol. 1, Mahwah, NJ: Lawrence Erlbaum Associates, 36–40.

Hyldtoft, C.A., Jensen, C.R., and Jørgensen, K., 2001. Motor control in low graduated isometric handgrip strength mobilisation using feedback from a PC monitor, in *Programme and Abstract Book of the Fourth International Scientific Conference on Prevention of Work-Related Musculoskeletal Disorders,* September 30–October 4, Amsterdam, 146.

Inman, V.T., Saunders, J.B.D.M., and Abbot, L.C., 1944. Observations on the function of the shoulder joint. *Journal of Bone and Joint Surgery,* 26, 1–30.

ISO, 2000. *ISO Standard 11226, Ergonomics — Evaluation of Static Working Postures,* Geneva: International Organization for Standardization.

Järvholm, U., Palmerud, G., Kadefors, R., and Herberts, P., 1991. The effect of arm support on supraspinatus muscle load during simulated assembly work and welding. *Ergonomics,* 34(1), 57–66.

Jensen, C. and Christensen, H., 2001. Duration of computer use and musculoskeletal symptoms, in *Programme and Abstract Book of the Fourth International Scientific Conference on Prevention of Work-Related Musculoskeletal Disorders*, September 30–October 4, Amsterdam, 148.

Johnson, G.R. and Barnett, N.C., 1996. The measurement of three-dimensional movements of the shoulder complex. *Clinical Biomechanics*, 11, 240–241.

Kadefors, R., 1994. Welding and musculoskeletal disease: a review, in *Proceedings of the 12th Triennial Congress of the International Ergonomics Association*, Vol. 2, Toronto, Canada: Human Factors Association of Canada, 72–74.

Kadefors, R., Petersen, I., and Herberts, P., 1976. Muscular reaction to welding work: an electromyographic investigation. *Ergonomics*, 19, 543–558.

Kaergaard, A., Andersen, J.H., and the PRIM Health Study Group, 2001. Follow-up study of the role of individual, physical and psychosocial factors in the onset of neck and/or shoulder pain, in *Programme and Abstract Book of the Fourth International Scientific Conference on Prevention of Work-Related Musculoskeletal Disorders*, September 30–October 4, Amsterdam, 132.

Karduna, A.R., McClure, P.W., and Michener, L.A., 2000. Scapular kinematics: effects of altering the Euler angle sequence of rotations. *Journal of Biomechanics*, 33(9), 1063–1068.

Karduna, A.R., Ebaugh, D., Michener, L.A., and McClure, P.W., 2002. The pattern of scapulohumeral motion — how stable is it? in *Proceedings of the World Congress of Biomechanics*, Calgary, 4–9 August.

Kilbom, Å., 1988. Intervention programmes for work-related neck and upper limb disorders: strategies and evaluation. *Ergonomics*, 31, 735–747.

Kilbom, Å., 1994. Repetitive work of the upper extremity: Part I. Guidelines for the practitioner. *International Journal of Industrial Ergonomics*, 14, 51–57.

Kilbom, Å. and Persson, J., 1987, Work technique and its consequences for musculoskeletal disorders. *Ergonomics*, 30, 273–279.

Kilbom, Å., Persson, J., and Jonsson, B., 1986. Disorders of cervicobrachial region among female workers in the electronic industry. *International Journal of Industrial Ergonomics*, 1, 37–47.

Kuijer, P.P.F.M., 2002. Effectiveness of Interventions to Reduce Workload in Refuse Collectors, Ph.D. thesis, University of Amsterdam, Amsterdam.

Kylmäaho, E., Rauas, S., Ketola, R., and Viikari-Juntura, E., 1999. Supporting the forearm and wrist during mouse and keyboard work: A field study, in Bullinger, H.-J. and Ziegler, J., Eds., *Conference Book: Human–Computer Interaction*, Vol. 1, Mahwah, NJ: Lawrence Erlbaum Associates, 23–26.

Larsson, B., Kadi, F., Bjørk, J., and Gerdle, B., 2001. Capillary supply and moth-eaten fibers in biopsies from female cleaners with and without trapezius myalgia and healthy controls, in *Programme and Abstract Book of the Fourth International Scientific Conference on Prevention of Work-Related Musculoskeletal Disorders*, September 30–October 4, Amsterdam, 163.

Magermans, D.J., Smits, N.C.M.A, Chadwick, E.K.J., Veeger, H.E.J., and Van der Helm, F.C.T., 2001. Functional evaluation of the shoulder, in *Proceedings of the Third Congress of the International Shoulder Group*, Newcastle-upon-Tyne, 4–6 September 2000, 101–104.

Meskers, C.G.M., Van der Helm, F.C.T., Rozendaal, L.A., and Rozing, P.M., 1998. *In vivo* estimation of the glenohumeral joint rotation center from scapular bony landmarks by linear regression. *Journal of Biomechanics*, 31, 93–96.

Mikkelsen, S., Andersen, J.H., Bonde, J.P., and Fallentin, N., 2001. A longitudinal study on repetition force and posture and the development of neck, shoulder, elbow and wrist disorders: the Prim study, in *Programme and Abstract Book of the Fourth International Scientific Conference on Prevention of Work-Related Musculoskeletal Disorders*, September 30–October 4, Amsterdam, 84.

Milerad, E. and Ericson, M.O., 1994. Effects of precision and force demands, grip diameter, and arm support during manual work: an electromyographic study. *Ergonomics*, 37(2), 255–264.

Pascoal, A.G., Van der Helm, F.C.T., Correia, P.P., and Carita, I., 2000. Effects of different arm external loads on the scapulo-humeral rhythm. *Clinical Biomechanics*, 15(S1), S21–S24.

Pearl, M.L., Harris, S.L., Lippitt, S.B., Sidles, J.A., Harryman, D.T., and Matsen, F.A., 1992. A system for describing positions of the humerus relative to the thorax and its use in the presentation of several functionally important arm positions. *Journal of Elbow and Shoulder Surgery*, 1, 113–118.

Praagman, M., Stokdijk, M., Veeger, H.E.J., and Visser, B., 2000. Predicting mechanical load of the glenohumeral joint, using net joint moments. *Clinical Biomechanics*, 15, 315–321.

Rose, L., Ericson, M., Glimskär, B., Nordgren, B., and Ørtengren, R., 1992. Ergo-index-development of a model to determine pause needs after fatigue and pain reactions during work, in *Computer Applications in Ergonomics, Occupational Safety and Health*, Mattila, M. and Karwowski, W., Eds., Amsterdam: Elsevier, 461–468.

Sigholm, G., Herberts, P., Almstrøm, C., and Kadefors, R., 1984. Electromyographic analysis of shoulder muscle load. *Journal of Orthopaedic Research*, 1, 379–386.

Stokdijk, M., Nagels, J., and Rozing, P.M., 2000. The glenohumeral joint rotation centre *in vivo*. *Journal of Biomechanics*, 33, 1629–1636.

Van den Heuvel, S.G., Bongers, P.M., Blatter, B.M., and Hoogendoorn, W.E., 2001. Work-related psychosocial risk factors in relation to pain in neck and upper extremities, in *Programme and Abstract Book of the Fourth International Scientific Conference on Prevention of Work-Related Musculoskeletal Disorders*, September 30–October 4, Amsterdam, 220.

Van der Helm, F.C.T., 1994. A finite element musculoskeletal model of the shoulder mechanism. *Journal of Biomechanics*, 27, 551–569.

Van der Helm, F.C.T., 1997. A standardized protocol for recordings of the shoulder, in *Proceedings of the First Conference of the International Shoulder Group*, Delft, 26–27 August, 7–12.

Van Galen, G.P., 2001. Physical stress, psychological stress and muscular co-contraction as ingredients for an etiological model of work-related upper extremity disorders (WRUEDs), in *Programme and Abstract Book of the Fourth International Scientific Conference on Prevention of Work-Related Musculoskeletal Disorders*, September 30–October 4, Amsterdam, 244.

Vasseljen, O., Mørk Johansen, B., and Westgaard, R.H., 1995a. The effect of pain reduction on perceived general tension and EMG recorded trapezius muscle activity in workers with shoulder and neck pain. *Scandinavian Journal Rehabilitation Medicine*, 27, 243–252.

Vasseljen, O., Westgaard, R.H., and Larsen, S., 1995b. A case-control study of psychological and psychosocial risk factors for shoulder and neck pain at the workplace. *International Archives Occupational Environmental Health*, 66, 375–382.

Veeger, H.E.J., 2000. The position of the rotation center of the glenohumeral joint. *Journal of Biomechanics*, 33, 1711–1715.

Veeger, H.E.J., Van der Helm, F.C.T., and Rozendal, R.H., 1993. Orientation of the scapula in a simulated wheelchair push. *Clinical Biomechanics*, 8, 81–90.

Veeger, H.E.J., Yu, B., An, K.-N., and Rozendal, R.H., 1997. Parameters for modeling the upper extremity. *Journal of Biomechanics*, 30, 647–652.

Veeger, H.E.J., Rozendaal, L.A., and Van der Woude, L.H.V., 2002. Load on the shoulder in low intensity wheelchair propulsion. *Clinical Biomechanics*, 17, 211–218.

Veiersted, K.B., Westgaard, R.H., and Andersen, P., 1993. Electromyographic evaluation of muscular work pattern as a predictor of trapezius myalgia. *Scandinavian Journal of Work, Environment & Health*, 19, 284–290.

Wærsted, M., Bjørklund, R., and Westgaard, R.H., 1987. Generation of muscle tension related to a demand of continuing attention, in *Work with Display Units '86*, Knave, B and Wideback, P.G., Eds., Amsterdam: North Holland, 288–293.

Westgaard, R.H., Wærsted, M., Jansen, T., and Aarås, A., 1986. Muscle load and illness associated with constrained body posture, in *The Ergonomics of Working Postures*, Corlett, N., Wilson, J., and Manenica, I., Eds., London: Taylor & Francis, 5–18.

10

Forearm and Hand

CONTENTS

Introduction

This chapter contains two sections on the forearm and the hand. The first section outlines the coupling of the hand with tools and/or control actuators for work. It begins with the functional anatomy of this distal part of the upper extremity, describes classifications for hand posture (particularly ways of grasping), discusses hand–handle interface characteristics (grip span, geometry, surface conditions, and use of gloves), and measures of strength, and ends with methods for implementation of various relevant demands into specifications for selecting or designing hand tools and machine controls.

The second section describes effective and healthy postures, movements, and forces of the elbow, forearm, and wrist for the performance of manual tasks. Evaluation criteria for posture are established on the basis of data on joint range of motion, comfort, strength, and pressure in the carpal canal. There is also discussion of the effects of local contact stresses and of the interactions between elbow, wrist, and finger postures, as well as of the effects of movement velocities and accelerations.

10.1 Hand–Handle Coupling

M. Susan Hallbeck and Roland Kadefors

Introduction

The aim of this section is to outline the coupling of the hand with tool handles and/or control actuators for work. This work of the hand may be static or dynamic and may use the fingers, hand, and/or wrist. For example, if a worker must use the thumb on top of a pipette to draw a fluid, then move the pipette to another test tube, and then use the thumb again to release the fluid, he or she will have a particular set of motions that will define the job. The thumb may be placed in an awkward position; the grip may be too large for comfort for a small female hand or too small for comfort for a large male hand, depending on the size of the pipette. With gloves, the coupling between the hand and the pipette may be slick or sticky. The movement path may induce accuracy or actuation problems. Repetition and force may be high when performing the task. The job may have problems that involve both static and dynamic coupling of the fingers, hand, and/or forearm. This section discusses the factors associated with using a hand tool for work.

The hand is an actuator that is used both for gross work and precise movement, as well as an artistic and communication device. This is especially true for the workplace where many tasks are actuated or performed by the fingers, in conjunction with the hand and/or wrist. Grasp and pinch are the main hand–handle couplings in industrial tasks. Therefore, they will be the main hand–handle couplings referred to in this section.

Although these tasks are commonplace, the study of the effect of the tasks on the motion or posture has not been common. Delleman (1999) presented an overview of studies showing the effect of grip type and orientation on posture. The tasks have typically been studied from the other direction — looking at or constraining the posture and motion and assessing strength.

10.1.1 Functional Anatomy

Wrist flexion and extension forces measured externally at the hand are a combination of forces acting between the forearm and hand, generated by exertions of agonistic and antagonistic muscle groups within both the hand (intrinsic) and the forearm (extrinsic) and by wrist-dedicated (carpi) muscles. The majority of forces exerted by the hand come from the digital (digiti) muscles residing in the forearm and crossing the wrist via sheathed tendons to power flexion and extension of the digits.

The muscles that control the wrist serve two functions in the hand. They provide the fine adjustment of the hand into its functioning position and, once this position is achieved, they stabilize the wrist to provide a stable working platform for the hand and digits (Hazelton et al. 1975). If wrist movement is considered in relation to digit movement, two independent actions are found to be possible. When the wrist-dedicated muscles stabilize the joint, the extrinsic and intrinsic digit muscles can alter the position of the digit(s). Conversely, when digit posture is stabilized, the wrist can be positioned over a large range of motion (Flatt 1961).

In the normal hand, strong flexion of the extrinsic digital flexors produces a synergistic contraction of the wrist extensors (Kaplan and Smith 1984). The digital flexors and extensors contribute to the wrist action, particularly under loads. In such cases, the digital muscles develop reaction forces against the object held (or within the hand itself if the fist is clenched) and add their contractile forces to the wrist action (Taylor and Schwarz 1955). This stabilizing antagonism is demonstrated in maximal exertions where nearly all forearm muscles contract to stabilize the wrist (Amis et al. 1979). Participation of the muscle sets is proportional to the loading applied (Amis et al. 1979). Muscles whose tendons merely cross the wrist (secondary wrist movers) en route to the digits cannot be relied upon for their wrist moving and stabilizing characteristics, as wrist function may at times conflict with their primary digit functions (Brand 1985). This dual innervation may be bypassed when subjects are instructed to extend the wrist with maximal effort while their fingers are loose and flaccid (Ketchum et al. 1978).

The extrinsic digital muscles are used for finger and thumb force generation and movement, but also for wrist torque generation and movement. This dual function can cause difficulties when a worker is exerting force in the hand or torque about the wrist while moving the hand or wrist (Brand 1985) because of the multiple tasks that the secondary wrist movers are required to perform.

10.1.2   Description of Hand Posture

Napier (1956) divided hand posture into two categories: prehension, in which an object is held partly or entirely within the compass of the hand, and nonprehensile posture, in which objects are pushed or lifted. Napier (1956) further divided the prehension category into power and precision grasp. Power grasp involves both digital and palmar opposition (shown in the top three grips of Figure 10.1) while precision grasp involves pinching opposition between thumb (digit I) and the tip or pad of one or more digits (shown in the bottom four grips of Figure 10.1). Landsmeer (1962) redefined precision grip, stating that because only the digits were involved in the

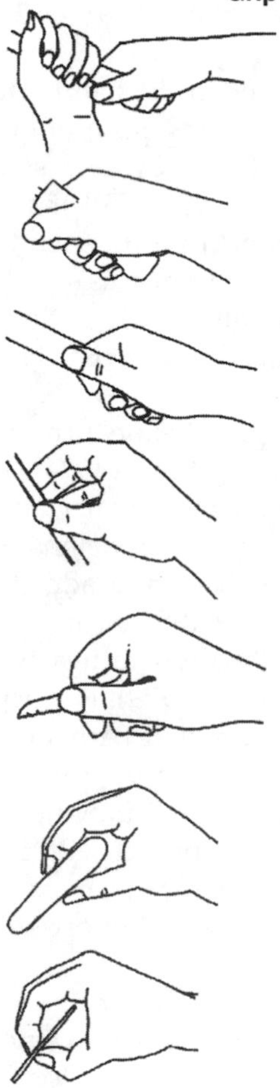

Grip Type	Women, N	Men, N	Force
Two-handed grip	540	900	
Transverse power grip	300	500	
Diagonal power grip	150	250	
Multi-finger pinch	100	150	
Key pinch	50	80	
Three-jaw chuck pinch	50	80	
Finger tip pinch	35	50	Precision

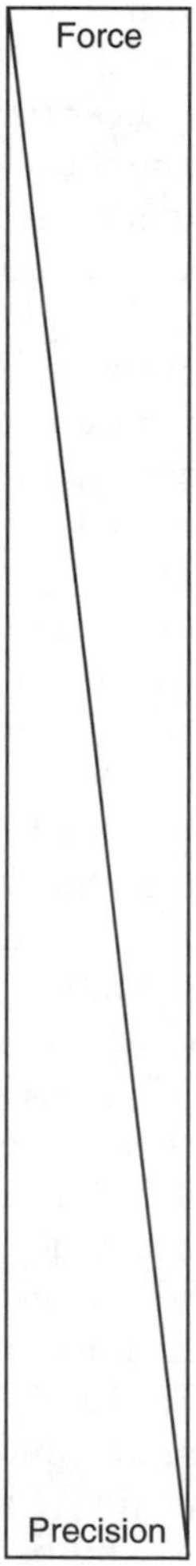

FIGURE 10.1
Grip type with strength values for women and men, showing the trade-off between force and precision. (Adapted from Wikström et al. 1991.)

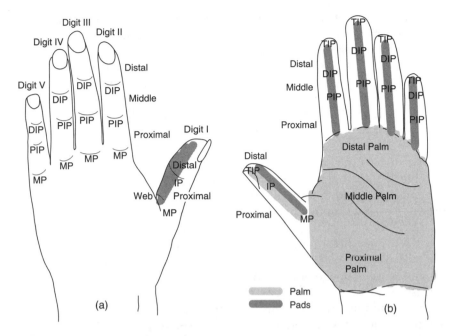

FIGURE 10.2
Classification by contact surfaces: (a) dorsal surface, (b) palmar (volar) surface. (Adapted from Casanova and Grunert 1989.)

prehension (without the palm) it should be termed precision handling or manipulation. These concepts were useful at the time they were presented, but they hardly describe the full range of prehensile postures of the hand in a reproducible fashion. Kroemer (1986) classified hand–handle couplings by the involvement of the digits and palmar surfaces of the hand. This was further refined by Casanova and Grunert (1989) who classified prehension patterns solely by contact surfaces, such as the digits in contact and the areas on the digital pads or palm. Thus, holding a key (often called a key or lateral pinch as shown in the fifth grip in Figure 10.1) would be termed a digit I (thumb) pad distal–digit II pad distal and radial proximal prehension. The areas of contact are shown in Figure 10.2 as adapted from Casanova and Grunert (1989).

10.1.3 Hand–Handle Characteristics

Hand–handle coupling is not solely the function of the contact surface of the digits and palm. The handle geometry (size, shape, etc.), handle surface conditions (slickness, temperature, texture, padding, etc.), and interference from personal protective equipment (such as gloves, discussed in Section 10.1.5) also affects the quality of the hand–handle coupling. The effect of handle geometry has been evaluated by many researchers (Ayoub and Lo Presti 1971, Terrell and Purswell 1976, Bobjer 1989). The handle geometry can be the size of the handle with respect to the hand as discussed in Section 10.1.6 or the shape of the handle. The cross-sectional shape of a handle affects how the compressive forces on the hand are distributed. For example, with a contoured handle with ridges between the fingers, there should be a greater contact area for the palmar aspect of the fingers. However, this is true only if the handle is sized perfectly for the user; otherwise the ridges may be directly under the fingers causing pressure points. These compression or pressure points can cause calluses, irritation, or fatigue over time as discussed in Section 10.2. This may also be the case with a handle that is too narrow, such as the handles on a pair of scissors. The small surface contact area gives a high compressive force over a small area, resulting in short-term pain and perhaps long-term numbness. Thus, the handle surface should not introduce pressure points to the hand or fingers.

Handle surface conditions have shown that higher frictional surfaces allow higher force exertions especially when pushing or pulling in line with the long axis of the handle (Jorgensen et al. 1989, Hallbeck et al. 1990). Slick surfaces cause slippage of the hand or fingers with respect to the handle. In contrast, a textured surface may allow the worker to relax the hand from the handle and still retain a sense of control of the tool. The surface may be hard or cushioned. Cushioning may cause loss of perceived control, which may, in turn, cause workers to grasp the tool harder to feel that they have control, as shown in Section 10.1.5.

Cold may interfere with hand–handle coupling because the surface area of the hand results in more rapid chilling of the tissues than for body parts with lower ratios of surface area to volume. The cold is a specific problem for the hand in that many of the muscles' tendons crossing the wrist on the way to the fingers are sheathed with synovial fluid lubricating their movement. The synovial fluid in the tendon sheaths increases in viscosity as it cools; therefore, the internal forces may be greater when the tendons must be moved through a thicker fluid. This may translate into higher internal carpal tunnel forces and eventual disability due to illness and injury, such as carpal tunnel syndrome.

10.1.4   Grasp, Pinch, and Wrist Strength

The number and strength of individual digits involved in force generation affect the strength magnitude (Berg et al. 1988). The digital, wrist, arm, and whole-body posture also all affect the magnitude of the force. The digital posture helps define the type of prehension, such as the three-jaw chuck (Figure 10.3) or, in the Casanova and Grunert (1989) definition, digit I pad distal–digit II pad distal–digit III pad distal, as shown in Figure 10.2. Keenan (1998) also found that the number of digits affected the output force when using a trigger actuator such as a drill or nut-runner. The index finger trigger force was approximately 40% of the four-finger trigger force (grip strength) and the two-fingered trigger force was approximately 70% of the four-finger

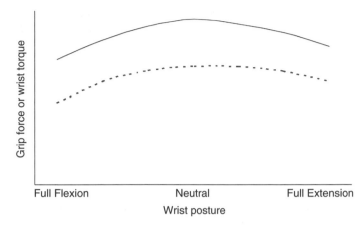

FIGURE 10.3
Representation of grip force and wrist torque over the ROM of the wrist. The solid line represents static strength in static wrist postures and the dashed line represents dynamic strength with angular wrist movement at greater than 30°/s. (Adapted from Wilhelm and Hallbeck 1997.)

trigger force. The subjects in his experiment felt most comfortable and secure with the index finger trigger for aiming the tool but, once actuated, preferred the four-finger trigger overall. Similarly, Lee and Cheng (1995) found in their study that, while the trigger force increased as the number of digits employed increased, the force required to hold the tool decreased with a multifinger trigger.

The effect of forearm, wrist, and finger posture on the strength of the exertion is due to the angle at which muscles act during maximal voluntary exertion; as this changes, the capacity to generate force varies (Hazelton et al. 1975, Pryce 1980, Eastman Kodak Company 1983, Brand 1985, Miller and Wells 1988, O'Driscoll et al. 1992). The angles at which the extrinsic digital flexors approach the digits change with wrist position (Brand 1985). The forearm complex has been optimized for strength in flexion, but not in extension. The hand contains more flexors than extensors and they are more powerful. When the physical capacity of the wrist muscles is assessed, the wrist flexors have more than twice the capacity of the extensors (Åstrand and Rodahl 1986). In forceful grasp exertion, the extensor muscles fatigue faster than the flexors (Kilbom et al. 1993). Thus, the highest grasp, pinch, or wrist strength would occur where the length–tension relationships of the muscles used in an exertion are maximized, for example, with the wrist in the neutral to slightly extended position. Additionally, the discomfort from exerting a wrist torque as low as 10% to 20% of maximum is worse during flexion than extension with repeated wrist exertions (Carey and Gallwey 2002). Practically, in an industrial task requiring great wrist torque exertion, the wrist should be in a neutral position.

Putz-Anderson (1988) reports that the grasp strength generated while in a neutral wrist angle is reduced to 75% at 45° wrist extension, to 60% at 45°

wrist flexion, and to 45% at 65° wrist flexion. Similarly, Hallbeck and McMullin (1993) found the strength at 45° wrist extension to be 82% of the force exerted in a neutral wrist position and that at 45° wrist flexion to be 72%.

Three-jaw chuck pinch strength in wrist extension (up to 30°) has not been found to differ significantly from that in the neutral wrist position (Anderson 1965, Kraft and Detels 1972); however, a wrist posture with 15° flexion was found to significantly reduce three-jaw chuck pinch strength (Kraft and Detels 1972). This has led to some saying that slight extension is a functionally neutral posture.

Terrell and Purswell (1976) found that pronation of the forearm significantly reduced the maximal grasp force in a neutral wrist position (88% of that in a neutral forearm position), while even a fully supinated forearm posture did not reduce the grasp force. More recent studies on pinch grip, such as those of Woody and Mathiowetz (1988), found key (lateral) pinch strength to be significantly higher in a neutral forearm position than in full pronation (tested in an arm posture with a 90° elbow angle and between 0° and 30° hand extension). Woody and Mathiowetz (1988) also found that, for palmar-pad pinch grip, forearm positions with full pronation did not significantly change maximal pinch forces from those in a neutral forearm position; however, they did not test any supinated postures. Power grip strength and forearm position were studied by Marley and Wehrman (1992). When compared with strength at a 90° included elbow angle, their results indicated a significant decrease in power grip strength with the forearm fully pronated (82% of that in a neutral posture), but no practically significant difference with the forearm fully supinated. A study by Zellers et al. (1992) demonstrated that arm posture at the shoulder significantly affected force generation with more neutral or relaxed shoulder postures allowing the highest force exertions and decrement with the abduction of the arm in any direction. The shape of the relationship between the grip force or wrist torque that can be exerted and wrist posture is basically an inverted U shape, as shown in Figure 10.3, with lower forces and torques at the extreme extensions than at the full flexions. The dynamic strength is reduced with movement of the wrist and, as the speed increases, the strength decreases further. This is shown in the simplified diagram (Figure 10.3), which represents the relationship for both static and dynamic hand strength or wrist torque (Wilhelm and Hallbeck 1997).

Force exertion direction also affects grip force and wrist torque under both isometric and isokinetic conditions and angular velocity is one of important factors in the isokinetic condition because grip force and wrist torque dramatically reduced even in slow velocities (Hallbeck et al. 1990, Lehman et al. 1993, Wilhelm and Hallbeck 1997, Jung and Hallbeck 2002). It has been shown that angular wrist velocities as low as 5°/s reduce maximal grasp force by as much as 40% (Lehman et al. 1993). An angular velocity of 30°/s yields another significant decrease in strength (Wilhelm and Hallbeck 1997, Jung and Hallbeck 2002). Lehman et al. (1993) also suggest that as angular velocity increases, maximal grasp force decreases; however, beyond 50°/s, grasp force does not decrease significantly.

There also seems to be a trade-off between force and precision. The precision of the exertion is due to the involvement of finer muscle units and an example is writing with a pencil or pen in a multifinger pinch grip rather than trying to write with the pencil or pen in a power grip; however, hammering with high force is easier using a power grasp than using a multifinger pinch grip. The force vs. precision trade-off is shown in Figure 10.1, which presents strength data collected for different grip types by Wikström et al. (1991).

10.1.5 Gloves

Gloves are used as personal protection equipment by a wide variety of people in many industries. However, the use of gloves often results in a reduction of capabilities. Grasp capabilities may be reduced due to interference of the glove in closing the hand around an object, the possible decreased friction between the glove and a handle, and the interference of the glove in tactile feedback (Cochran et al. 1986). The interference of the glove in closing the hand may be due to bunching of glove material at the joints or interphalangeal spacing by the material.

Armstrong (1986) suggested that more grip force is required to exert a given amount of force with gloves than without gloves, so that the muscles must exert greater force when wearing gloves than they would with a bare hand (Cochran et al. 1986, Sudhakar et al. 1988). The net result of using gloves is a trade-off between hand protection and task performance.

Grasp force exertion was found to decrease with increasing layers of gloves. Hertzberg (1955) reported a 20% to 30% reduction from bare hand grasp force in pilots wearing Air Force gloves. Using five commercially available gloves, Cochran et al. (1986) determined that the forces generated with the bare hand were significantly higher than those with any of the gloves. Hallbeck and McMullin (1993) found that grasp force differed significantly between conditions barehanded (0 mm) or wearing thermal knit gloves (2 mm), mesh gloves (3 mm) or gloves with a layered combination of thermal knit and mesh (5 mm). The decrement in force when wearing gloves compared to barehanded ranged from 9% with the thermal glove to 18% with the layered combination.

Gloves also add bulk between the fingers causing interphalangeal spacing or spreading of the fingers. Results from a study employing five interphalangeal spacings (0 to 10 mm) indicated that the loss of grasp force due to increased finger spacing is significant (Hallbeck et al. 1994). In a related set of studies on interphalangeal spacing, Ramakrishnan et al. (1994) found the effect of interphalangeal spacing on maximal grasp force to be significant, with the loss of grasp generally increasing as spacing increased over the nine interphalangeal spacing conditions from 0 (barehanded) to 10 mm. From 0% to 26% less force could be generated, on average, as measured on a hand dynamometer. In studies by Hallbeck and McMullin (1993) and Kamal et al. (1992), pinch force (three-jaw chuck and lateral) was not found to be affected by gloves.

The pinches shown in Figure 10.1 demonstrate that the glove would not affect the joint closure or separate the digits used, as gloves would when bunching during a grasp.

The tactile feedback may be important for handle grip force and control (Fellows and Freivalds 1991) or accuracy (Hägg and Hallbeck 2001). A study was performed to evaluate decreased feedback, simulated by padding a tool handle, which showed that subjects grasped the tool tighter than without the padding (Fellows and Freivalds 1991). With the foam grip, the tool grip force within the hand was greater than necessary for the task due to the deformation of the foam and a "loss of control" feeling in the subjects. However, most subjects strongly preferred the foam grips (Fellows and Freivalds 1991). This stronger-than-required grip can reduce the range of motion for the tool, as discussed in Section 10.1.3, and increase forces in the hand due to the muscular forces in the hand that must control the tool conflicting with the forces that move the hand or wrist. Care must be taken, as shown in Section 10.1.2, that the hand–handle coupling does not place high pressures on certain portions of the fingers, hand, or wrist.

The hand–handle coupling location affects the decrement in strength due to gloves. If the hand–handle coupling occurs at the distal phalanges of digits 1, 2, and 3, strength is expected to be minimally affected by the glove interference (Hallbeck and McMullin 1993); however, if the coupling is in the palm and pads of whole digits, the decrement will be large. The thicker and less pliable the glove material, the greater the strength decrement due to gloves.

10.1.6   Grip Span

Fransson and Winkel (1991) summarized 12 grip span studies published between 1954 and 1990. Optimal grip spans in the majority of studies range from 4.5 cm on the lower end to 8.9 cm on the upper end (one study disclosed a 3.8 cm optimum grip span for females). During this period Wang (1982) also studied the effect of grip span on strength. The dynamometer employed in his study was set at spans of 3.5, 4.7, and 6.0 cm. He found that the highest force was in the 3.5 to 4.7 cm range. After the period covered by Fransson and Winkel (1991), Crosby and Wehbe (1994) reported an inverted U-shaped graph with the maximal force at the 4.7 cm span for females and 6.0 cm span for males. Härkönen et al. (1993) found that the average (including both genders) impulse force was maximal at 6.0 cm. Similar results have been reported by Ramakrishnan et al. (1994) who examined the effect of grip spans of 3.5, 4.7, and 6.0 cm on static grasp force and found that the values, averaged over males and females, did not differ between 4.7 and 6.0 cm grip spans, but that a grip span of 3.5 cm yielded a significantly (38%) lower grasp force. Ramakrishnan et al. (1994) found a grip span of 6.1 cm to yield a significantly higher force than a 7.4 cm grip span, averaged over both males and females. The gender differences in terms of maximal grip force may be due in part to the differences in hand size, as shown in Figure 10.4. The figure also demonstrates

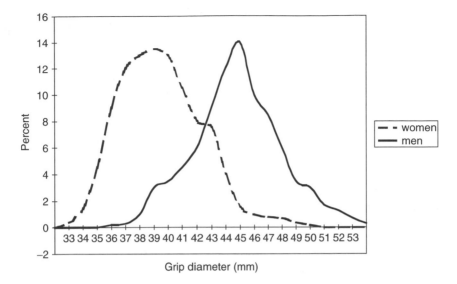

FIGURE 10.4

Distributions of grip diameters (in thumb and index finger pinch) for males and females. (Adapted from Sperling 1989.)

that there is a large overlap between males and females with respect to grip span, which is related to handle diameter. Many males with small hands are therefore likely to benefit from development of tools for females. The practical implication is that the size of the handle diameter should be based on the optimized grip span for better force generation, and conversely lower fatigue at a given level of holding exertion.

Several studies have focused on the strength of the so-called trigger finger(s). Radwin and Oh (1991) studied handle and trigger size on power tools and found that the highest average grip strength was produced at a trigger distance of between 5 and 6 cm. In another study, Oh and Radwin (1993) further defined that for small and medium-sized hands the 5 cm trigger distance was optimal, and that for large hands the 6 cm trigger distance was optimal. Lee and Cheng (1995) studied trigger types and distances on a power tool. They found that the subjects could produce the highest average resultant force at the 5.0 cm trigger distance for both a one- and two-finger trigger. Björing and Hägg (2000) and Keenan (1998) found that the effect of trigger distance on force exertion was much like the grip span results. Björing and Hägg (2000) defined the optimal distance from the web of the hand (thumb crotch) to the trigger finger(s) to be 3.9 cm (range 3.6 to 4.3 cm) for males and 3.3 cm (range 2.9 to 3.7 cm) for females.

10.1.7 Implementation

Tools and control actuators have to be designed with due consideration to the hand–object coupling. This has been subject to due attention within the standardization process. For example, ISO (1984; for machine tools) specifies the direction of movement of manual control actuators relative to the controlled component. The European standard on "Safety of Machinery — Ergonomic Requirements for the Design of Displays and Control Actuators" (CEN 2000) contains a classification of force/torque requirements where control actuators are used frequently or for long durations. Different grip types are defined as (1) contact grip (typically operation of pushbuttons, etc.), (2) pinch grip, and (3) clench or power grip. For each type of grip, recommended minimum dimensions of manual control actuators are given. For contact grip the maximum recommended linear force is 10 N (finger and thumb) and 20 N (hand). The maximum recommended torque for the contact grip is 0.5 Nm. Maximum recommended linear actuation force in pinch grip is 10 to 20 N depending on direction of force. For actuating torque in pinch grip the corresponding figures are 1 to 2 Nm. In power grip, the maximum recommended force is 35 to 55 N, depending on direction. The pulling direction is generally the one allowing the highest force.

The Swedish Hand Tool Project (Kardborn 1998) developed specifications that reflect a full range of aspects that need to be taken into account for hand–handle coupling. This project had a participative design, and involved end users as well as designers, manufacturers, dealers, and ergonomists. Demand specifications were developed and were organized by (1) demands in user terms, (2) ergonomic demands or factors, and (3) technical demands. The project also included a checklist relevant to all kinds of handheld tools and machines, and references to existing standards and authoritative publications (Kadefors and Sperling 1993). It is appropriate in the development of the hand tool specifications to take into account this checklist, shown in Figure 10.5.

An example can be given of the specification for plate shears, as follows. Plate shears are, in the present context, a tool for cutting thin plate (iron, stainless steel, copper, or brass). The shears may be right or left turning. Tools analyzed in this example are shears for cutting up to 1.0 mm stainless steel plate or, alternatively, 1.5 mm iron plate or 2.0 mm copper plate.

10.1.7.1   Demands in User Terms

Examples of demands by users include:

- The tool shall be designed for work not only in the optimal working zone but also in awkward postures of the fingers, wrist, forearm, and shoulder.
- The tool shall be able to be used for a working day without excessive pain or discomfort.
- The tool shall be able to be used by all users having normal hand function, irrespective of hand size, dexterity, gender, or age.
- The tool shall be suitable for use outdoors.
- The tool shall be suitable to carry in the pocket.

FIGURE 10.5

Hand Tool Checklist of Factors Relevant to the Development
and Selection of All Tools

1	Be safe to handle for the user
2	Be easy to understand
3	Fit the work task
4	Fit the working object and working space
5	Avoid damage to the working object
6	Be easy to pick up and put down
7	Be easy to store
8	Be easy to carry
9	Provide an adequate grip
10	Fit right and left hand
11	Fit the hand size and force capacity of the user group
12	Provide a possibility for grip variation
13	Provide a possibility of two-handed operation whenever necessary
14	Provide the required force output
15	Provide the required precision
16	Be effective
17	Require low force input
18	Be as light as the function requires
19	Be balanced
20	Be stable
21	Counteract static force load
22	Distribute the pressure on the grip surface
23	Provide adequate grip friction
24	Provide adequate visual access for carrying out the work
25	Provide grip comfort
26	Provide climatic comfort
27	Provide feedback of completed work
28	Be easy to recognize
29	Be hygienic
30	Counteract allergy
31	Be attractive

Source: Adapted from Kadefors and Sperling (1993).

10.1.7.2 Ergonomic Factors

The ergonomic factors, both mandatory and desirable, are summarized for
the plate shears in Table 10.1. These are based on the condition that the tool
is used more than half the workday on and off, or for 30 min continuously.

10.1.7.3 Technical Demands

For technical demands of the tool, consultation of the appropriate technical
standards is necessary. Of particular importance are the material properties
of the cutting edges with respect to resistance to wear and possibilities for
sharpening, and also that the edge design allows optimization of angular
relationships between the cutting edges during cutting. The edges should
be lightly serrated for ease of cutting.

TABLE 10.1

Ergonomic Factors for Plate Shears

Mandatory Demands	
Force	The tool shall be designed so as to require less than 30% of maximal grip force in cutting work. The expected grip force capacity of a female operator is approximately 300 N, of a male operator 500 N. In two-handed operation the expected grip forces are approximately 540 and 900 N, respectively.
Local pressure	The handle of the tool shall be designed so as to allow use for 30 min continuously without perceived pain or discomfort at the hand surface. The radius of the handle shall not be smaller than 10 mm. The local pressure in the palm surface shall not exceed 450 kPa for female operators or 650 kPa for male operators.
Hand posture	The grip span in maximal force exertion shall be in the range 5–6.5 cm. Work in the optimal working zone shall be possible without extreme wrist deviation or flexion/extension.
Desirable Demands	
Handle size	The handle should be designed so as to allow room for the entire hand in the grip applied.
Surface friction	The surface of the handle should have friction properties eliminating risk of slipping due to work with greasy hands.
Weight	The weight of the tool should be as low as possible, taking into account the function of the tool.

Source: Adapted from Kilbom et al. (1993).

Summary

For the use of the hand in work situations, it is important to know the effective output of the hand or wrist with movement. For movement with hand actuation, the wrist posture should be kept as near neutral as possible (consideration should be given to forearm, elbow, and upper arm posture as well). The grip span or diameter of the tool should be based on the hand size of the worker. The hand should not exert excessive force, especially with awkward postures. The speed of hand movement reduces the amount of torque that can be exerted and the magnitude of the grip force that can be exerted. The main type of tool used in industrial settings is the hand tool. These are typically held with a power grip or pinch. The actuation type helps define the position of the fingers and thumb. The actuation mechanism (button, switch, trigger, etc.) defines the type of movement required. Hand tools and the hand–handle coupling may cause discomfort and performance decrements, and if the movements are awkward or require high force, high repetition rate, or long duration to perform the job, they may induce illness or injury. In addition to these factors, the size of the grip (length and diameter) on the tool, the number of digits involved in the grip, the movement path and the tool handle characteristics may be exacerbators for illness and injury.

In the design of manual tasks to be carried out in work situations, it is imperative that due consideration be given to the basic characteristics of the hand–object coupling. The type of grip that is required in the task determines the force and precision that can be achieved by the operator. There is a trade-off between these modalities: a transverse power grip can be exerted with a maximum force of 500 N in men (300 N in women) at low precision, whereas in a high precision pinch grip the average maximum force that can be produced is 50 N in men, 35 N in women. These characteristics result from the basic biomechanics and inherent sensory feedback of the wrist–hand system. The grasp force output is, however, further mediated by modifiers, for example, the wrist posture (forearm, elbow, and upper arm posture may also have an effect). A neutral (slightly extended) wrist position should be chosen in situations requiring high grip or pinch force. In operation of hand tools (e.g., pliers) consideration should be given to the effect on force output capability allowed by the grip diameter and grip span. There is a marked average gender difference in hand size that affects grip dimensions and, consequently, the design requirements for tool handles or controls. There are international standards outlining, for example, the maximum force and torque outputs that should be allowed in control of actuators, using different types of grips.

10.2   Elbow, Forearm, and Wrist

Richard Wells

Introduction

The aim of this section is to outline effective and healthy postures, movements, and forces of the elbow, forearm, and wrist for the performance of manual activities. Evaluation criteria include performance, comfort, and musculoskeletal injury risk.

The epidemiological research addressing the relationship between work exposure and work-related musculoskeletal disorders (WMSD) has been extensively reviewed in Hagberg et al. (1995), Bernard (1997), and NRC (2001). These reviews found strong and consistent relationships between many types of WMSD and workplace exposures including force exerted, heavy lifting, awkward postures, repetitive work, vibration, and local contact stresses and their combinations, as well as potentiating factors such as cold. High contact stresses may lead to occlusion of blood flow and temporary numbness or local pain (Fransson-Hall and Kilbom 1993). In addition to these physical factors, psychosocial factors have also been identified; however, these are not considered further in this section (for a discussion of this particular topic, refer to Section 9.2.5).

The forces and movements present in the body tissues during work may be thought of as the final common pathway by which work may lead to musculoskeletal disorders (Wells et al. 1997). These have been termed internal exposure. This is conceptualized as a time-varying force applied to particular tissues for a period of time. Mechanical exposure is thus conceptualized as having three dimensions: amplitude (level), frequency (time variation pattern), and duration of exposure (Winkel and Mathiassen 1994, Hagberg et al. 1995).

10.2.1 Posture Definition

Figure 10.6 illustrates the bony structures involved and definitions of the various joint angles (American Academy of Orthopaedic Surgeons 1965, ISO 2000). The definitions are in agreement except for the term *wrist deviation* compared to *wrist abduction*.

10.2.2 Importance of Posture of the Elbow, Forearm, and Wrist for Comfort, Performance, and Injury

Posture and its time-varying pattern (movement) give important information to help design and evaluate work. Posture itself may give rise to forces on body tissues, alter the tissue response to the tissue loads, and affect performance. External forces enter into this picture by acting simultaneously with posturally induced loads. Figure 10.7 illustrates the combined effect of external loads and postures. Figure 10.8 shows in more detail the areas where contact stresses are to be avoided. In general, posture may have four types of effect on comfort, injury, and performance, as shown in Table 10.2. The first three affect loading on body tissues while the last affects their capacity. These factors apply to all body regions but specific examples are given for the distal upper limb where available.

10.2.3 Evaluation of Working Postures, Movements, and Forces

Considering the risk factors for musculoskeletal injury identified in the literature, it is difficult to imagine any work without hazards present; one can quickly fall into the mind-set that work is inherently dangerous and that the observation of a bent wrist during work implies substantial risk of developing musculoskeletal disorder. The distinction between hazard and risk is important: hazard is a condition that may cause injury or illness; the risk of injury is given by the hazard and the conditions (time, etc.) under which the person is exposed to the hazard. Even for wrist flexion close to an individual's range of motion (ROM), the risk is low if the motion is infrequent. In fact, the adoption of "extreme postures" for short periods of time is probably beneficial; they are called stretch breaks! This reinforces the importance of the time variation pattern and duration in the definition of mechanical exposure.

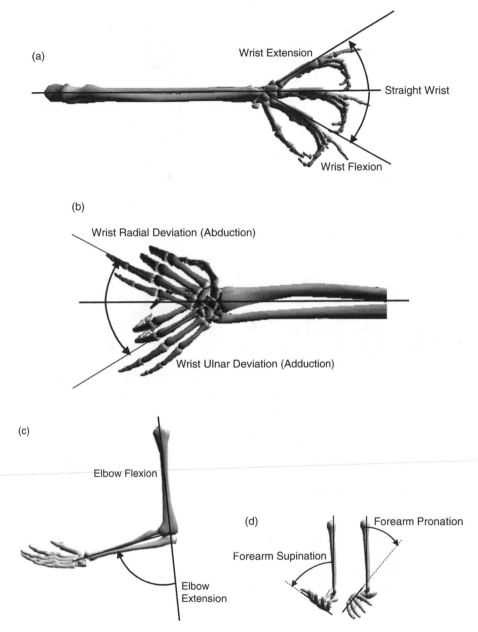

FIGURE 10.6
Angle definitions for the wrist (a, b), the elbow (c), and the forearm (d).

10.2.4 Evaluation Criteria for Posture

The changes in comfort and performance with various postures have been summarized above. A healthy and effective posture is one in which perfor-

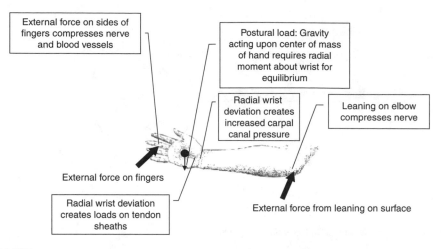

FIGURE 10.7
Example of the combined effects of external forces and postures on comfort and performance.

mance is not impaired appreciably (compared to the posture with maximal performance) and in which risk factors for discomfort and injury do not exceed benchmark values. Factors considered include carpal tunnel pressure, discomfort and reduction in grip force, and specific deleterious combinations of postures and movements identified in the literature. Aspects of performance that are affected by arm posture include maximum force exertion, endurance, and task execution time. For example, it has been seen that non-neutral wrist postures, especially flexed postures, are associated with reduced power grip (Pryce 1980).

Figure 10.9 through Figure 10.12 illustrate the ranges of posture resulting from the application of the criteria developed. The clear central area gives a sense of healthy and effective posture. The shading is the range of maximum limits in the literature and the start of the dark shading is the upper bound over all sources. Each joint motion is presented separately, as in general there are insufficient data available to rate combinations of postures.

ROM can be taken from the normative data of the American Academy of Orthopaedic Surgeons (1965), Boone and Azen (1979), and Kee and Karwowski (2001). As Boone and Azen included children as well as adults as participants, only those participants over 19 years old have been included here.

A range of posture in which comfort was rated as good or better was estimated from published studies. Many studies reported statistically significant changes in strength with posture; however, biologically important grip force decrement of greater than 20% was arbitrarily chosen.

Carpal canal pressure is increased with postural deviations from a straight wrist position. A maximum pressure threshold of 30 mmHg was taken (Lundberg et al. 1982), as chronic pressures above this level lead to nerve dysfunction and injury. Carpal canal pressures were estimated from Keir et al. (1997), Werner et al. (1997), and Rempel et al. (1998) in the condition

TABLE 10.2

Effects of Posture on Upper Limb Injury, Comfort, and Performance

Posture	Effects of Posture	Example
1. Extreme joint ROM	Extremes of joint ROM causing the joint to be supported by passive tissues (ligaments and elastic tissue in muscles): a. Passive tissue may be at risk of injury under unexpected loads and muscular response may be delayed. b. Active contraction of other muscles may be required to maintain this posture against the elasticity of stretched passive tissues. c. Other effects below are maximized in extreme posture.	a. Holding a load with the elbow fully extended leads to pain in flexor muscles and the elbow joint (Rose et al. 2000). Full flexion of the lumbar spine leads to creep of ligaments (McGill and Brown 1992). b. Holding the fingers straight with a straight wrist requires tiring contraction of wrist and finger extensor muscles (Keir et al. 1996, Keir and Wells 2002). c. See also examples in other categories.
2. Non-extreme: Postural load change	A non-extreme posture causing a moment of force due to gravitational loading of associated segments. This moment is supplied by mainly muscular forces but creates loads in many body tissues. It is commonly termed postural load.	Holding the wrist extended while typing requires an extensor moment (Keir and Wells, 2002).
3. Non-extreme: Geometry change and resulting effects	A non-extreme posture causing changes in geometry and function of the tissues of the joint or associated limb including extra loading on tissues: a. Compression of nerve and blood vessels b. Change in compartment pressures c. Change in angles of pull of tendons d. Changes in joint loads e. Changes in blood supply f. Changes in muscle performance g. Changes in comfort	a. Compression of median nerve occurs via flexor tendons with wrist flexion (Keir et al. 1997). b. Increases in carpal canal pressures are seen with deviations of the wrist from a straight position, especially extension (Keir et al. 1997). c. The geometry of the tendons passing over the wrist changes as the wrist deviates from straight leading to higher loads on tendon sheaths (Tichauer 1966, Moore et al. 1991, Keir and Wells 1999). d. Forcible supination (driving screws) with a straight elbow increases loads on head of radius with resulting pain (Tichauer 1966). e. Holding arms raised affects muscle blood perfusion independent of muscle activity (Holling and Verel 1957). f. Grip force is much reduced with a flexed wrist (Pryce 1980, Miller and Wells 1988). As the posture changes active (and passive) muscle force changes. Most joints have a range under which moment production is maximal (Kulig et al. 1985).

TABLE 10.2 (continued)

Effects of Posture on Upper Limb Injury, Comfort, and Performance

Posture	Effects of Posture	Example
		g. Uncomfortable with more than 20° wrist ulnar abduction (Hünting et al. 1981, Kee and Karwowski 2001).
4. Non-extreme: Tolerance change	A non-extreme posture may cause a decrease in the ability of the joint or tissue to withstand loading.	In flexion and/or twist the lumbar disc is able to withstand 30–40% less load before failure (Gunning et al. 2001).

of not exerting a force with the hand. The first study used cadaver preparations and the latter two measured pressures *in vivo* using an indwelling catheter. As exerting forces increases carpal tunnel pressure, the range of recommended postures will decrease from those suggested when exerting forces with the hand.

Posture, and more specifically joint angle, affects comfort. Hsiao and Keyserling (1991) described three regions: a "neutral" region that represents minimal discomfort to the joint and adjacent structures, an "effort" region defined as one associated with mild discomfort, and a "maximum range" defined as the limit of the joint ROM. Comfort evaluations are commonly performed with minimal exertion or external load (Kee and Karwowski 2001). Postural loads of long duration lead to static muscle activity, which may result in muscle pain. Short-term static comfort was estimated from Kee and Karwowski (2001) and Diffrient et al. (1985). Kee and Karwowski (2001) rated relative comfort levels across joints and proportions of the joint ROM. However, as they held each posture for 60 s, the rating is for this short duration only. While the data provide good insight into relative comfort of different joints, high discomfort might be expected in a posture rated as good, if held for prolonged periods of time.

The region of maximal strength for power grip was taken from O'Driscoll et al. (1992) as 35° ± 2° wrist extension and 7° ± 2° wrist ulnar abduction from Eastman Kodak Company (1983) and for elbow flexor moment from Kulig et al. (1985) as 65° to 100° elbow flexion. Strength decrements of more than 20% were estimated from Singh and Karpovich (1968), Jørgensen and Bankov (1971), Kulig et al. (1985), Hazelton et al. (1975), Pryce (1980), Eastman Kodak Company (1983), and Miller and Wells (1988).

Local contact stresses should be avoided in the regions shown in Figure 10.8 to avoid pressure on nerves and blood vessels on the inside of elbow, sides of fingers and thumb, and carpal canal region. It must again be emphasized that the presence of external forces must be considered simultaneously with the effects of posture. The ranges presented therefore should be used with this in mind.

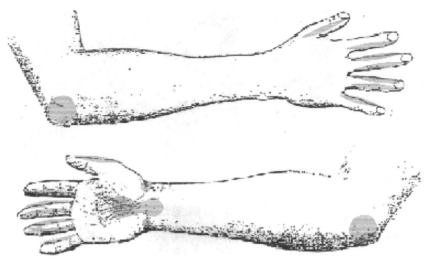

FIGURE 10.8
Areas of the upper limb where local contact stresses should not be applied.

10.2.5 Interaction of Elbow, Wrist, and Finger Postures

The musculoskeletal system distal to the elbow functions synergistically during prehension. The synergism is due to the multiple muscles spanning these joints as well as the demands of the task. There thus exist strong interactions among elbow posture, wrist posture, and finger posture. For example, carpal canal pressures are affected by forearm rotation and meta-carpophalangeal (knuckle) flexion (Rempel et al. 1998). In general, there are insufficient data to fully map these interaction effects. They can perhaps best be documented using modeling approaches: for example, with the fingers straight, even small amounts of wrist extension become uncomfortable (see Figure 10.9). Keir et al. (1996) showed that by using a modeling approach, these interactions could be predicted.

10.2.6 Effects of Movement

The time variation pattern of posture is movement; movement differs from static posture in that muscles shorten and lengthen, tendons slide against their sheaths, nerves slide through their sheaths, articular surfaces slide and roll on each other, and fluid flow is increased in most tissues.

Moore et al. (1991) proposed that the velocity of sliding of tendons in their sheaths under load created frictional work. As the frequency, hand force and wrist deviation from neutral increased so would the frictional work. They showed that under controlled conditions during repetitive work at high and

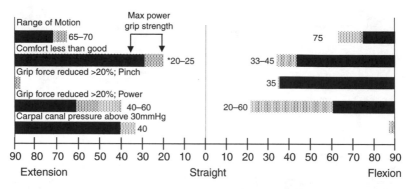

FIGURE 10.9
Postural ranges for wrist flexion/extension. Unshaded (clear central) areas denote healthy and effective posture. The shading is the range of maximum limits in the literature and the start of the dark shading (not recommended area) is the upper bound over all sources. *Postural data for Kee and Karwowski (2001), gathered with straight fingers (Karwowski 2002).

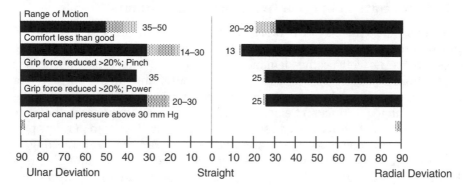

FIGURE 10.10
Postural ranges for wrist ulnar/radial deviation (abduction). Unshaded (clear central) areas denote healthy and effective posture. The shading is the range of maximum limits in the literature and the start of the dark shading (not recommended area) is the upper bound over all sources.

low repetition rates with and without the wrist deviated from neutral, the modeled frictional work matched the injury response observed by Silverstein et al. (1986) in their study of industrial workers.

Movement implies both velocity and acceleration. Marras and Schoenmarklin (1993) in a study of wrist angular accelerations showed that the high peak wrist angular accelerations were related to increased risk of cumulative trauma disorders. They suggested that the increased risk was due to increased loads on tendons and other structures that were involved in accelerating the hand.

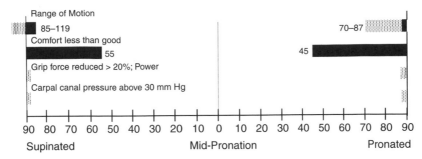

FIGURE 10.11

Postural ranges for forearm pronation/supination. Unshaded (clear central) areas denote healthy and effective posture. The shading is the range of maximum limits in the literature and the start of the dark shading (not recommended area) is the upper bound over all sources.

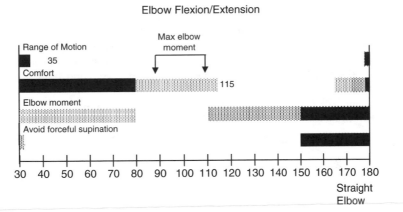

FIGURE 10.12

Postural recommendations for elbow flexion/extension. Unshaded (clear central) areas denote healthy and effective posture. The shading is the range of maximum limits in the literature and the start of the dark shading (not recommended area) is the upper bound over all sources.

Summary

This section has presented reasons that posture of the elbow, forearm, and wrist is important for the effective and healthy performance of manual activities. Evaluation criteria including performance, comfort, and musculo-skeletal injury risk have been presented and data extracted from the litera-ture to quantify their effects over the range of movement of the wrist, forearm, and elbow. Forces in tissues are created by the effects of external forces transmitted through the skeletal linkage together with those due to postural effects. Comfort and injury risk are related to these forces and their time variation pattern and duration. Posture therefore is an important con-tributor to healthy and effective activity.

References

American Academy of Orthopaedic Surgeons, 1965. *Joint Motion: Method of Measuring and Recording*, Chicago: American Academy of Orthopaedic Surgeons.

Amis, A.A., Dowson, D., and Wright, V., 1979. Muscle strengths and musculo-skeletal geometry of the upper limb. *Engineering in Medicine*, 8(1), 41–48.

Anderson, C.T., 1965. Wrist Joint Position Influences Normal Hand Function, Master's thesis, University of Iowa, Iowa City.

Armstrong, T.J. 1986. Ergonomics and cumulative trauma disorders. *Hand Clinics*, 2(3), 553–565.

Åstrand, P.O. and Rodahl, K., 1986. *Textbook of Work Physiology: Physiological Bases of Exercise*, New York: McGraw-Hill.

Ayoub, M.M. and Lo Presti, P., 1971. The determination of an optimum size cylindrical handle by use of electromyography. *Ergonomics*, 4(4), 503–518.

Berg, V.J., Clay, D.J., Fathallah, F.A., and Higginbotham, V.L., 1988. The effects of instruction on finger strength measurements: applicability of the Caldwell regimen, in *Trends in Ergonomics/Human Factors V*, Aghazadeh, F., Ed., Amsterdam: North-Holland, Elsevier Science, 191–198.

Bernard, B., 1997. A Critical Review of Epidemiological Evidence for Work-Related Musculoskeletal Disorders of the Neck, Upper Extremity, and Low Back, DHHS Publ. 97-141, Cincinnati, OH: U.S. Department of Health and Human Services, National Institute of Occupational Safety and Health.

Björing, G. and Hägg, G., 2000. The ergonomics of spray guns — users' opinions and technical measurements on spray guns compared with previous recommendations for hand tools. *International Journal of Industrial Ergonomics*, 25, 405–414.

Bobjer, O., 1989. Ergonomic knives, in *Advances in Industrial Ergonomics and Safety I*, Mital, A., Ed., London: Taylor & Francis, 291–298.

Boone, D.C. and Azen, S.P., 1979. Normal range of motion of joints in male subjects. *Journal of Bone and Joint Surgery*, 61(5), 756–759.

Brand, P.W., 1985, *Clinical Mechanics of the Hand*, St. Louis, MO: C.V. Mosby.

Carey, E.J. and Gallwey, T.J., 2002. Effects of wrist posture, pace and exertion on discomfort. *International Journal of Industrial Ergonomics*, 29, 85–94.

Casanova, J.S. and Grunert, K., 1989. Adult prehension: patterns and nomenclature for pinches. *Journal of Hand Therapy*, 2(6), 231–244.

CEN, 2000. EN 894-3:2000, *Safety of Machinery — Ergonomic Requirements for the Design of Displays and Control Actuators*, Part 3: *Control Actuators*, Brussels: European Committee for Standardization.

Cochran, D.J., Albin, T.J., Bishu, R.R., and Riley, M.W., 1986. An analysis of grasp degradation with commercially available gloves, in *Proceedings of the Human Factors Society 30th Annual Meeting*, Dayton, 29 September–3 October 1986, Santa Monica, CA: Human Factors Society, 852–855.

Crosby, C.A. and Wehbe, M.A., 1994. Hand strength: normative values. *Journal of Hand Surgery*, 19A(4), 665–670.

Delleman, N.J., 1999. Working Postures: Prediction and Evaluation, Ph.D. thesis, University of Amsterdam.

Diffrient, N., Tilley, A.R., and Harman, D., 1985. *Human Scale*, Cambridge, MA: MIT Press.

Eastman Kodak Company, 1983. *Ergonomic Design for People at Work*, Vol. 2, New York: Van Nostrand Reinhold.

Fellows, G.L. and Freivalds, A., 1991, Ergonomics evaluation of a foam rubber grip for tool handles. *Applied Ergonomics*, 22(4), 225–230.

Flatt, A.F., 1961. Kinesiology of the hand. *American Academy of Orthopaedic Surgeons Instructional Course Lectures*, 18: 266–281.

Fransson, C. and Winkel, J., 1991. Hand strength: the influence of grip span and grip type. *Ergonomics*, 34(7), 881–892.

Fransson-Hall, C. and Kilbom, A., 1993. Sensitivity of the hand to surface pressure. *Applied Ergonomics*, 24(3), 181–189.

Grandjean, E., 1985. *Fitting the Task to the Man*, London: Taylor & Francis.

Gunning, J., Callaghan, J., and McGill, S.M., 2001. The role of prior loading history and spinal posture on compressive tolerance and type of failure in the spine using a porcine trauma model. *Clinical Biomechanics*, 16(6), 471–480.

Hagberg, M., Silverstein, B., Wells, R., Smith, R., Carayon, P., Hendrick, H., Perusse, M., Kourinka, I., and Forcier, L., Eds., 1995. *Work-Related Musculoskeletal Disorders (WMSDs): A Reference Book for Prevention*, London: Taylor & Francis.

Hägg, G.M. and Hallbeck, M.S., 2001. Do handle design and hand posture affect pointing accuracy? in *Proceedings of the Human Factors and Ergonomics Society, 45th Annual Meeting*, Santa Monica, CA: Human Factors and Ergonomics Society, 731–735.

Hallbeck, M.S. and McMullin, D.L., 1993. Maximal power grasp and three-jaw chuck pinch force as a function of wrist position, age, and glove type. *International Journal of Industrial Ergonomics*, 11, 195–206.

Hallbeck, M.S., Cochran, D.J., Stonecipher, B.L., Riley, M.W., and Bishu, R.R., 1990. Hand-handle orientation and maximum force, in *Proceedings of the Human Factors Society 34th Annual Meeting*, Orlando, 2–5 October 1990, Santa Monica, CA: Human Factors Society, 800–834.

Hallbeck, M.S., Muralidhar, A., and Balachandran, R., 1994. Effect of interphalangeal separation on grasp strength., in *Proceedings of the 12th Triennial Congress of the International Ergonomics Association*, Toronto, 15–19 August 1994, Mississauga, Ontario: Human Factors Association of Canada, Vol. 2, 83–85.

Härkönen, R., Piirtomaa, M., and Alaranta, H., 1993. Grip strength and hand position of the dynamometer in 204 Finnish adults. *Journal of Hand Surgery*, 18B(1), 129–132.

Hazelton, H.J., Smidt, G.L., Flatt, A.F., and Stephens, R.I., 1975. The influence of wrist position on the force produced by the finger flexors. *Journal of Biomechanics*, 8, 301–306.

Hertzberg, T., 1955. Some contributions of applied physical anthropometry to human engineering. *Annals of New York Academy of Sciences*, 63, 621–623.

Holling, H.E. and Verel, D., 1957. Circulation in the elevated forearm. *Clinical Science*, 16, 197–213.

Hsiao, H. and Keyserling, W.M., 1991. Evaluating posture behavior during seated tasks. *International Journal of Industrial Ergonomics*, 8, 313–334.

Hünting, W., Läubli, T., and Grandjean, E., 1981. Postural and visual loads at VDT workplaces. I. Constrained postures. *Ergonomics*, 24(12), 917–931.

ISO, 1984. *ISO 447, Machine Tools — Direction of Operation of Controls*, Vol. 2, Geneva: International Organization for Standardization.

ISO, 2000. *ISO 11226, Ergonomics — Evaluation of Static Working Posture*, Geneva: International Organization for Standardization.

Jørgensen, K. and Bankov, S., 1971. Strength of elbow flexors with pronated and supinated forearm, in *Biomechanics II*, Jokl, E., Ed., Basel: Karger, 174–180.

Jørgensen, M.J., Riley, M.W., Cochran, D.J., and Bishu, R.R., 1989. Maximum forces in simulated meat-cutting tasks, in *Proceedings of the Human Factors Society 33rd Annual Meeting,* Denver, CO, Vol. 1, Santa Monica, CA: Human Factors Society, 641–645.

Jung, M.-C. and Hallbeck, M.S., 2002. The effect of wrist position, angular velocity, and exertion direction on simultaneous maximal grip force and wrist torque under isokinetic conditions. *International Journal of Industrial Ergonomics*, 29(3), 133–143.

Kadefors, R. and Sperling, L., 1993. Kravspecificationer [Demand specifications] (report from the Swedish Hand Tool Project), Stockholm, Sweden: The Swedish Working Life Fund [in Swedish].

Kamal, A.H., Moore, B.J., and Hallbeck, M.S., 1992. The effects of wrist position/ glove type on maximal peak lateral pinch force, in *Advances in Industrial Ergonomics and Safety IV,* Kumar, S., Ed., London: Taylor & Francis, 701–708.

Kaplan, E.B. and Smith, R.J., 1984. Kinesiology of the hand and wrist, in *Kaplan's Functional and Surgical Anatomy of the Hand,* Spinner, M., Ed., Philadelphia, PA: J.B. Lippincott, 283–349.

Kardborn, A. 1998. Inter-organizational participation and user focus in a large scale product development programme: the Swedish Hand Tool Project. *International Journal of Industrial Ergonomics,* 21, 369–381.

Karwowski, W., 2002. Personal communication.

Kee, D. and Karwowski, W., 2001. The boundaries for joint angles of isocomfort for sitting and standing males based on perceived comfort of static joint postures. *Ergonomics*, 44(6), 614–648.

Keenan, P.M., 1998. A Study of the Factors That May Affect a Subject's Ability to Produce Force with a Simulated Power Tool Trigger, Master's thesis, University of Nebraska, Lincoln.

Keir, P. and Wells, R., 1999. Changes in the geometry of the carpal tunnel contents due to wrist posture and tendon load: an MRI study on normal wrists. *Clinical Biomechanics*, 14(9), 635–645.

Keir, P.J. and Wells, R., 2002. The effect of typing posture on wrist extensor muscle loading. *Human Factors*, 44(3), 392–403.

Keir, P., Wells, R., and Ranney, D., 1996. Passive stiffness of the forearm musculature and functional implications. *Clinical Biomechanics,* 11(7), 401–409.

Keir, P.J., Wells, R., Ranney, D., and Lavery, W., 1997. The effects of tendon load and posture on carpal tunnel pressure. *Journal of Hand Surgery,* 22A(4), 628–634.

Ketchum, L.D., Brand, P.W., Thompson, D., and Pocock, G.S., 1978. The determination of moments for extension of the wrist generated by muscles of the forearm. *Journal of Hand Surgery,* 3(3), 205–210.

Kilbom, Å., Mäkäräinen, M., Sperling, L., Kadefors, R., and Liedberg, L., 1993. Tool design, user characteristics and performance: a case study on plate-shears. *Applied Ergonomics*, 24, 221–230.

Kraft, G.H. and Detels, P.E., 1972. Position of function of the wrist. *Archives of Physical Medicine and Rehabilitation*, 53, 272–275.

Kroemer, K.H.E., 1986. Coupling the hand with the handle: an improved notation of touch, grip, and grasp. *Human Factors*, 28(3), 337–339.

Kulig, K., Andrews, J.G., and Hay, J.G., 1985. Human strength curves. *Exercise and Sport Science Reviews*, 12, 417–466.

Landsmeer, J.M.F., 1962. Power grip and precision handling. *Annals of the Rheumatic Diseases*, 21(2), 164–169.

Lee, Y.H. and Cheng, S.L., 1995. Triggering force and measurement of maximal finger flexion force. *International Journal of Industrial Ergonomics*, 15, 167–177.

Lehman, K.R., Allread, W.G., Wright, P.L., and Marras, W.S., 1993. Quantification of hand grip force under dynamic conditions, in *Proceedings of the Human Factors and Ergonomics Society 37th Annual Meeting*, Seattle, 11–15 October 1993, Santa Monica, CA: Human Factors and Ergonomics Society, 715–719.

Lundborg, G., Gelberman, R.H., Minteer-Covery, M., Lee, Y.H., and Hargens, A.R., 1982. Median nerve compression in the carpal tunnel: functional response to experimentally induced controlled pressure. *Journal of Hand Surgery*, 7(3) 252–259.

Marley, R.J. and Wehrman, R.R., 1992. Grip strength as a function of forearm rotation and elbow posture, in *Proceedings of the Human Factors Society 36th Annual Meeting*, Atlanta, GA, Santa Monica, CA: Human Factors Society, 791–794.

Marras, W.S. and Schoenmarklin, R.W., 1993. Wrist motions in industry. *Ergonomics*, 36(4), 341–251.

McGill, S.M. and Brown, S., 1992. Creep response of the lumbar spine to prolonged full flexion. *Clinical Biomechanics*, 7, 43–46.

Miller, M. and Wells, R.P., 1988. The influence of wrist flexion, wrist deviation and forearm pronation on pinch and power grip strengths, in *Proceedings of the Fifth Biennial Conference of the Canadian Society for Biomechanics*, London, Ontario: Spodym Publishers, 112–113.

Moore, A.E., Wells, R.P., and Ranney, D.A., 1991. Quantifying exposure in occupational manual tasks with cumulative trauma potential. *Ergonomics*, 34(12), 1433–1453.

Napier, J.R., 1956. Prehensile movements of the human hand. *Journal of Bone and Joint Surgery*, 38B, 902–913.

NRC (National Research Council), 2001. Musculoskeletal Disorders and the Workplace, Washington, D.C.: National Academy Press.

O'Driscoll, S.W., Horii, E., Ness, R., Cahalan, T.D., Richards, R.R., and An, K.-N., 1992. The relationship between wrist position, grasp size and grip strength. *Journal of Hand Surgery*, 17A(1), 169–177.

Oh, S. and Radwin, R.G., 1993. Pistol grip power tool handle and trigger size effects on grip exertions and operator preference. *Human Factors*, 35(3), 551–569.

Pryce, J.C., 1980. The wrist position between neutral and ulnar deviation that facilitates the maximum power grip strength. *Journal of Biomechanics*, 13, 505–511.

Putz-Anderson, V., 1988. *Cumulative Trauma Disorders — A Manual for Musculoskeletal Disease of the Upper Limbs*, London: Taylor & Francis.

Radwin, R.G. and Oh, S., 1991. Power tool handle design factors affecting operator grip exertions, in *Proceedings of the 11th Congress of the International Ergonomics Association*, Paris, Vol. 3, 23–24.

Ramakrishnan, B., Bronkema, L.A., and Hallbeck, M.S., 1994. Effects of grip span, wrist position, hand and gender on grip strength, in *Proceedings of the Human Factors and Ergonomics Society 38th Annual Meeting*, Nashville, TN, Santa Monica, CA: Human Factors and Ergonomics Society, 554–558.

Rempel, D., Bach, J.M., Gordon, L., and So, Y., 1998. Effects of forearm pronation/supination on carpal tunnel pressure. *Journal of Hand Surgery*, 23A, 38–42.

Rose, L., Ericson, M., and Örtengren, R., 2000. Endurance time, pain and resumption in passive loading of the elbow. *Ergonomics*, 43(3), 405–420.

Silverstein, B.A., Fine, L.J., and Armstrong, T.J., 1986. Hand wrist cumulative trauma disorders in industry. *British Journal of Industrial Medicine*, 43, 779–784.

Singh, M., and Karpovich, P.V., 1968. Strength of forearm flexors and extensors in men and women. *Journal of Applied Physiology*, 25, 177–180.

Sperling, L., 1989. Kvinnohandens ergonomi [The ergonomics of the female hand. Grip function and demands on hand tools], Internal project report, Ergonoma AB [in Swedish], Gothenborg, Sweden.

Sudhakar, L.R., Schoenmarklin, R.W., Lavender, S.A., and Marras, W.S., 1988. The effects of gloves on grip strength and muscle activity, in *Proceedings of Human Factors Society 32nd Annual Meeting*, Anaheim, 24–28 October, Santa Monica, CA: Human Factors Society, 647–650.

Taylor, C.L. and Schwarz, R.J., 1955. The anatomy and mechanics of the human hand. *Artificial Limbs*, 2(2), 22–35.

Terrell, R. and Purswell, J., 1976. The influence of forearm and wrist orientation on static grip strength as a design criterion for hand tools, in *Proceedings of Human Factors Society 20th Annual Meeting*, College Park, MD, Santa Monica, CA: Human Factors Society, 309.

Tichauer, E.R., 1966. Some aspects of stress on forearm and hand in industry. *Journal of Occupational Medicine*, 8(20), 63–71.

Wang, M., 1982. A Study of Grip Strength from Static Efforts and Anthropometric Measurement, Unpublished Master's thesis, University of Nebraska, Lincoln, NE.

Wells, R., Norman, R., Neumann, P., Andrews, D., Frank, J., Shannon, H., and Kerr, M., 1997. Assessment of physical work load in epidemiologic studies: common measurement metrics for exposure. *Ergonomics*, 40(1), 51–62.

Werner, R., Armstrong, T.J., Bir, C., and Aylard, M.K., 1997. Intracarpal canal pressures: the role of finger, hand, wrist and forearm position. *Clinical Biomechanics*, 12(1), 44–51.

Wikström, L., Byström, S. Dahlman, S., Fransson, C., Kadefors, R., Kilbom, Å., Landervik, E., Liedberg, L., Sperling, L., and Öster, J., 1991. Kriterier vid val och utveckling av handverktyg, Undersökningsrapport [Criteria when choosing and developing hand tools], Research Report 1991:18, Solna, Sweden: National Institute for Occupational Health [in Swedish].

Wilhelm, G.A. and Hallbeck, M.S., 1997. The effects of gender, wrist angle, exertion direction, angular velocity, and simultaneous grasp force on isokinetic wrist torque, in *Triennial Congress of the International Ergonomics Association*, Tampere, Finland, Vol. 4, 126–128.

Winkel, J. and Mathiassen, S.-E., 1994. Assessment of physical load in epidemiologic studies: concepts, issues and operational considerations. *Ergonomics*, 37(6), 979–989.

Woody, R. and Mathiowetz, V., 1988. Effect of forearm position on pinch strength measurements. *Journal of Hand Therapy*, 1(2), 124–126.

Zellers, K., Gandini, C., and Loveless, T., 1992. The effect of arm positions on pinch strength measurements: a pilot study, in *Advances in Industrial Ergonomics and Safety IV*, Kumar, S., Ed., London: Taylor & Francis.

11

Multiple Factor Models and Work Organization

CONTENTS

Introduction

This chapter contains two sections, both stressing the importance of considering multiple risk factors when working on the prevention and reduction of musculoskeletal fatigue, discomfort, and injuries. The first section provides definitions, criteria, and procedures useful to describe and, wherever possible, to assess the risk factors that can represent a physical overload for the different structures and segments of the upper limb. Furthermore, it presents the most widely used synthesized methods and indices for risk assessment: the RULA and REBA methods, the strain index, the CTD risk index, the ACGIH threshold limit value, and the OCRA risk index.

The second section focuses primarily on the effects of changes in posture, termed posture variation, a time-related aspect of work. Methods for assessing posture variation are described, as well as the effects of combining different tasks within an individual's job, for example, through job rotation. Particular emphasis is given to the need for autonomy of operators or groups of operators, that is, their decision latitude with regard to job content and work performance, in reducing the risk of developing musculoskeletal disorders.

11.1   Multiple Factor Models

Daniela Colombini and Enrico Occhipinti

Introduction

This section, while taking into consideration the most recent and significant contributions in the literature, intends to supply a set of definitions, criteria, and procedures useful to describe and, wherever possible, to assess the risk factors that can represent a physical overload for the different structures and segments of the upper limbs. The consequences of physical overload are represented by upper limb work-related musculoskeletal disorders (WMSDs) (Hagberg et al. 1995). This section focuses specifically on identification of risk factors and describes some of the methods that have been developed for evaluating them.

11.1.1   Relevant Studies

This section should not be to be considered a complete overview of the literature on identification of the main risk factors but is intended to direct readers toward studies that represent essential contributions for the operational choices, which are then suggested when presenting the concrete evaluation models.

In 1987, Drury discussed a method for the biomechanical assessment of pathologies due to repetitive movements, and focused on three main factors: force, frequency, and posture. He suggested a description and assessment method that counts the daily number of "hazardous movements" for the body, and particularly for the wrist.

In 1988, Putz-Anderson published a book in which he systematically listed all the practical and theoretical knowledge that was available at the time on the control and management of cumulative trauma disorders (CTDs). Among other things, the book postulates a "risk model" for CTDs, based on the interaction of four main factors: repetitiveness, force, posture, and recovery time. In the previous two years, Silverstein et al. (1986, 1987) highlighted the connection between repetition and force risk factors and CTDs (particularly carpal tunnel syndrome, or CTS). They also threw light on the fact that there is a synergistic mechanism operating between those two factors.

In 1992, Winkel and Westgaard suggested guidelines for the study of occupational risk factors — and of personal individual factors — related to shoulder and neck disorders.

In 1993, a large group of authors who were part of a working group of ICOH (International Commission on Occupational Health, or ICOH), mainly Scandinavian and American, presented a conceptual model for the interpretation of the development of occupational musculoskeletal disorders of the neck and upper limbs (Armstrong et al. 1993).

Again in 1993, Tanaka and McGlothlin presented an analysis of biomechanical determinants of occupational CTS. In their model, the exposure limit is determined by the repetitiveness of movements, by the force used, and by the postural deviations of the joint involved, the wrist in this case.

Guidelines for "practitioners" were presented and discussed by Kilbom in 1994, for the analysis and assessment of repetitive tasks for the upper limbs (Kilbom 1994a,b). This is an extremely important review, both theoretically and practically, which supplies useful suggestions for both the definition of repetitive tasks and the classification of the different issues to consider during analysis. Frequency of movement is pointed out as particularly important for the characterization of risk. For each body region (hands, wrist, elbow, shoulder), indications are given of maximum limits for frequency of similar movements above which there is likely to be a high risk for upper limb injuries. The existence of other overloading factors (high force, high static load, speed, extreme postures, duration of exposure) is considered to amplify the risk level.

In 1995, the contributions to analysis and prevention of WMSDs of a large panel of authors were summarized by Hagberg et al. (1995). Starting from a critical analysis of epidemiological studies on the subject, the book examines the different elements representing occupational risks that could be a

TABLE 11.1

Upper Limb WMSDs: Relevance of Different Risk Factors

Body Part	Risk Factor	Strong Evidence	Evidence	Insufficient Evidence
Neck and neck/	Repetition		×	
shoulder	Force		×	
	Posture	×		
	Vibration			×
Shoulder	Repetition		×	
	Force			×
	Posture		×	
	Vibration			×
Elbow	Repetition			×
	Force		×	
	Posture			×
	Combination	×		
Hand/wrist carpal	Repetition		×	
tunnel syndrome	Force		×	
	Posture			×
	Vibration		×	
	Combination	×		
Hand/wrist	Repetition		×	
tendinitis	Force		×	
	Posture		×	
	Combination	×		

Source: NIOSH (1997).

cause of the various pathologies of the upper limbs, and indicates possible measurement and analysis methods for each of the elements considered.

More recently, the U.S. National Institute for Occupational Safety and Health (NIOSH) published "Musculoskeletal Disorders and Workplace Factors: A Critical Review of Epidemiological Evidence for WMSDs of the Neck, Upper Extremity and Low Back" (NIOSH 1997) providing a critical report of studies that have shown the association between certain risk factors in work (particularly repetitiveness, force, posture, and vibration) and particular upper limb pathologies. Table 11.1 presents a synthesis of the main results of this critical overview.

A consensus document was prepared and published (Colombini et al. 2001) by the Technical Committee "Musculoskeletal Disorders" of the International Ergonomics Association (IEA) with the endorsement of ICOH. This document defines, in a general model, the main risk factors to be considered and presents observational methods for each that can be used in their description, classification, and evaluation. In its conclusions the consensus document underlines the need for an integrated evaluation by means of concise exposure indices. It also reports all the essential literature regarding these aspects. The general assessment model and definitions from this document are reported in Section 11.1.2. A presentation of the most widely used synthesized methods and indices is then given in Section 11.1.3.

11.1.2 General Assessment Model and Definitions

The general model of description and assessment of tasks, concerning all exposed workers in a given situation, is aimed at evaluating four main collective risk factors: repetitiveness, force, awkward posture and movements, and lack of proper recovery periods. Such factors should be assessed as functions of time (mainly considering their respective durations). In addition to these factors, others, grouped under the term *additional factors*, should be considered; these are mechanical factors (e.g., vibrations, localized mechanical compressions), environmental factors (e.g., exposure to cold), and organizational factors (e.g., pace determined by machinery), and for most of them there is evidence of association with WMSDs of the upper limbs.

Each identified risk factor should be properly described and classified (i.e., assessed, even if roughly). This allows, on the one hand, identification of possible requirements and preliminary preventive interventions for each factor and, on the other hand, eventually, the consideration of all the factors contributing to the overall "exposure" within a general and mutually integrated framework. From this viewpoint "numerical" or "categorical" classifications of results may be useful to make management of results easier, even if it is important to avoid the feeling of an excessive objectiveness of methods whose classification criteria may still be empirical. To this end, the following definitions are important:

- Work is composed of one or more *tasks*; these are definite activities that can occur one or more times in one work shift.
- Within a single task, several *cycles* may be identified. Cycles are sequences of technical actions, of relatively short duration, which are repeated over and over, always the same.
- Within each cycle, several *technical actions* may be identified. These are elementary operations that enable the completion of the cycle operational requirements. The action is not necessarily identified with the movement of a single body segment, but rather with a group of movements, by one or more body segments, which enable the completion of an elementary operation.

The suggested procedure for assessing the risk should follow the general phases listed below:

1. Pinpointing the typical tasks of any job, and — among them — those that take place in repetitive and equal cycles for significant lengths of time
2. Finding the sequence of technical actions in the representative cycles of each task

3. Describing and classifying the risk factors within each cycle (repetitiveness, force, posture, additional factors)

4. Combining the data concerning the cycles in each task during the whole work shift, taking into consideration the duration and sequences of the different tasks and of the recovery periods

5. Making a brief and structured assessment of the risk factors for the job as a whole (exposure index)

11.1.3   Methods and Indices for Risk Assessment

This section contains a presentation of the most widely used synthesized methods and indices for risk assessment.

11.1.3.1   *RULA and REBA*

RULA (rapid upper limb assessment) is a survey method developed for use in ergonomics investigations of workplaces where work-related upper limb disorders are reported (McAtamney and Corlett 1993). This tool requires no special equipment in providing a quick assessment of the postures of neck, trunk, and upper limbs along with muscle function and the external loads experienced by the body. A coding system is used to generate an action list that indicates the level of intervention required to reduce the risks of injury due to physical loading on the operator. RULA is mainly based on the classification of working postures, but it gives some importance to repetition and to the use of force. It can be considered a useful tool for a first screening of a job/task in which there is exposure to repetitive/strainful exertions of the upper limbs.

REBA (rapid entire body assessment) was developed on the basis of RULA system and was specifically designed to be sensitive to the type of unpredictable working postures found in health care and other service industries (Hignett and McAtamney 2000). It is appropriate for evaluating tasks where postures are dynamic, static, or where gross changes in position take place. As with RULA, special attention is devoted to the external load acting on trunk, neck, and legs and to the human–load interface (coupling) using the upper limbs. Those aspects are specifically scored and then processed into a single combined risk score using a table provided.

11.1.3.2   *Strain Index*

The strain index (SI; Moore and Garg 1995) is a semiquantitative job analysis method that involves the measurement or estimation of six task variables (intensity of exertion, duration of exertion per cycle, efforts per minute, hand/wrist posture, speed of work, and duration of task per day). The first five task variables were identified from scientific principles. The sixth was included because the authors believed that it was a relevant factor. An ordinal rating is

TABLE 11.2

Rating Criteria for the Strain Index

Rating	Intensity of Exertion	Duration of Exertion per Cycle (%)	Efforts per Minute	Hand/ Wrist Posture	Speed of Work	Duration of Task per Day (h)
1	Light	<10	<4	Very good	Very slow	<1
2	Somewhat hard	10–29	4–8	Good	Slow	1–2
3	Hard	30–49	9–14	Fair	Fair	2–4
4	Very hard	50–79	15–19	Bad	Fast	4–8
5	Near maximal	>79	>19	Very bad	Very fast	>8

Source: Moore, J.S. and Garg, A., *American Industrial Hygiene Association Journal*, 56, 443–458, 1995.

TABLE 11.3

Multipliers Corresponding to Ratings for the Six Task Variables of the Strain Index

Rating	Intensity of Exertion A	Duration of Exertion per Cycle B	Efforts per Minute C	Hand/Wrist Posture D	Speed of Work E	Duration of Task per Day F
1	1	0.5	0.5	1.0	1.0	0.25
2	3	1.0	1.0	1.0	1.0	0.50
3	6	1.5	1.5	1.5	1.0	0.75
4	9	2.0	2.0	2.0	1.5	1.00
5	13	3.0	3.0	3.0	2.0	1.50

Source: Moore, J.S. and Garg, A., *American Industrial Hygiene Association Journal*, 56, 443–458, 1995.

assigned for each variable according to exposure data; then a multiplier value is assigned for each variable. The SI is the product of these six multipliers:

$$SI = A \times B \times C \times D \times E \times F \tag{11.1}$$

Moore and Garg (1995) suggest that the method accurately identifies jobs that are associated with distal upper extremity disorders when compared with jobs that are not. From current evidence, jobs with an SI ≤ 3 are not considered to present a risk, while jobs with an SI ≥ 7 present a clear risk. For work-related upper limb disorders, an SI in the range of 3 to 7 represents a borderline area between risk and low risk.

To calculate the SI each task variable is rated according to five levels. These ratings are presented in Table 11.2. The multipliers for each task variable are related to the ratings and are given in Table 11.3. The foundations, limitations, and assumptions of the SI method are summarized by Moore and Garg (1995) as follows:

- It only applies to the distal upper extremity (hand/forearm).
- It predicts a spectrum of upper limb disorders (disorders of muscle–tendon units as well as carpal tunnel syndrome), not specific disorders.
- It assesses jobs and not individual workers.
- The relationships between exposure data and the multiplier values are not based on explicit mathematical relationships between the task variables and the physiological, biomechanical, or clinical responses.
- Future studies should include multiple task analysis and evaluation of predictive validity for other outcomes (e.g., severity rate, staff turnover, etc.).

11.1.3.3   CTD Risk Index

The CTD risk assessment model for predicting injury incidence rates was developed by Seth et al. (1999). The model is unique in that it uses quantitative data such as hand motion frequencies and forces together to obtain a frequency factor score that is reflective of the strain imposed on the muscles and tendons of the wrist. Gross upper extremity postures are included in a posture factor score and various minor job stressors are included as a miscellaneous factor score. The individual frequency, posture, and miscellaneous scores were regressed against CTD incidence rates for 24 jobs in a garment sewing company and a printing company to yield weightings for a predicted incidence rate or risk index. The miscellaneous score was not found to be a significant predictor in determining incidence rates and thus was dropped from the model. The results showed that the model was best suited for job tasks with cycle times greater than 4 s. The final best-fit model explained 52% of variance in predicted incidence rates and may be a useful tool in identifying jobs at risk for CTDs, using only frequency and posture scores. The final regression equation is

$$\text{CTD risk index} = -7.80 + 5.33F + 3.89P \qquad (11.2)$$

where F is the force-frequency score and P is the posture score.

11.1.3.4   ACGIH TLV

The ACGIH (American Conference of Governmental Industrial Hygienists) TLV (threshold limit value) is based on epidemiological, psychophysical, and biomechanical studies and is intended for assessing "mono-task" jobs (i.e., jobs in which only one repetitive task is present) performed for 4 or more hours per day (ACGIH 2000). The TLV specifically considers average hand activity level (HAL) and peak of hand force (Table 11.4) and identifies

TABLE 11.4

The ACGIH (2000) TLV and Action Limit for Reduction of Work-Related Musculoskeletal Disorders Based on Hand Activity Level (as defined in Table 11.5) and Normalized Peak Hand Force (on a scale of 0–10 corresponding to 0%–100% of the applicable population's reference strength)

HAL (hand activity level)	1	2	3	4	5	6	7	8	9	10
Normalized peak force: TLV	7.2	6.4	5.6	4.8	4.0	3.2	2.4	1.6	0.8	0
Normalized peak force: Action limit	5.4	4.8	4.2	3.6	3.0	2.4	1.8	1.2	0.6	0

Note: For the action limit general controls are recommended. The values in the table are estimated from the original figure in ACGIH (2000).

TABLE 11.5

Determination of Hand Activity Level (0–10) for the ACGIH (2000) TLV from Data on Exertion Frequency and Duty Cycle (% of work cycle where force is greater than 5% of maximum)

Frequency of Hand Exertion (s⁻¹)	Period of Exertion (s)	Duty Cycle (%)				
		0–20	20–40	40–60	60–80	80–100
0.125	8.0	1	1	—	—	—
0.25	4.0	2	2	3	—	—
0.5	2.0	3	4	5	5	6
1.0	1.0	4	5	5	6	7
2.0	0.5	—	5	6	7	8

Note: Dash (—) means that the combination is not possible.

conditions for which nearly all workers may be repeatedly exposed without adverse health effects.

The measure of HAL is based on the frequency of hand exertions and the duty cycle (relative durations of work and rest). HAL can be determined with ratings by a trained observer using the scale of Latko et al. (1997) or calculated by using information on the frequency of exertions and the work/recovery ratio (Table 11.5).

Peak hand force is normalized on a scale of 0 to 10, which corresponds to 0% to 100% of the applicable population's reference strength. Peak force can be determined with ratings by a trained observer, rated by workers, or measured using instrumentation or biomechanical methods. Peak force requirements can be normalized by dividing the force required to perform the job by the maximum strength capability of the work population for that activity, which is determined (when useful) through relevant databases.

Some combinations of force and hand activity are presumed (by the authors) to be associated with a significantly elevated prevalence of musculoskeletal disorders. Therefore, an Action Limit (Table 11.4) is suggested at which point general controls, including surveillance, are recommended.

The ACGIH (2000) TLV method mainly considers two important risk factors: repetition and the use of force. It is also nominally based on epidemiological studies but no mention is made about whether it can be used in a predictive capacity. In presenting this method, ACGIH (2000) underlines the need for work standards that allow workers to pause as necessary, with a minimum of at least one pause (recovery period) per hour.

11.1.3.5   OCRA Risk Index

11.1.3.5.1 Main Approach

None of above approaches for risk assessment indices considers all the risk variables underlined in the consensus document (Colombini et al. 2001), especially those referring to the analysis of work organization. The OCRA (Occupational Repetitive Actions) concise index (Colombini et al. 1998, 2002) has been developed with a full consideration of the concepts, criteria, and methods suggested by this document.

The OCRA exposure index is the ratio of the number of technical actions (performed during repetitive movement tasks) actually carried out during the work shift (ATA), and the number of technical actions that are specifically recommended by the method (RTA). Technical action frequency is the variable that best characterizes exposure whenever upper limb repetitive movements are analyzed (Colombini et al. 1998, 2002). The technical actions should not be identified as the joint movements or as the elementary movements of motion time measurements (MTM) analysis (Barnes 1968). To make action frequency analyses more accessible, a conventional measurement unit has been chosen — the *"technical action"* of the upper limb. This definition is very similar, although not identical, to the MTM-2 description of movement sequences. Table 11.6 lists the criteria for the definition and counting of technical actions. When reviewing the film of a task in slow motion, all the technical actions carried out by the right and left arm must be listed in order of execution.

The overall number of technical actions carried out within the shift (ATA) can be calculated by organizational analysis (to give the number of actions per cycle, the number of actions per minute, and this last number multiplied by the net duration of the repetitive task analyzed to obtain ATA).

The following general formula then calculates the overall number of technical actions recommended within a shift:

Number of technical actions recommended (RTA) =

$$\sum_{i=1}^{n}\left[CF \times \left(Ff_i \times Fp_i \times Fc_i\right) \times D_i\right] \times Fr \times Fd \tag{11.3}$$

TABLE 11.6

Criteria for the Definition and Counting of Technical Actions for Calculating the OCRA Index

Reach	Reach means shifting the hand toward a pre-fixed destination.
Move	Move means transporting an object to a given destination by using the upper limb.
	Reaching an object should be considered as an action exclusively when the object is positioned beyond the reach of the length of the extended arm of the operator and is not reachable by walking. The operator must then move both the trunk and the shoulder to reach the object. If that workplace is used by both men and women, or by women alone, the measurement of the length of the extended arm corresponds to 50 cm (5th percentile for general adult European population), and this length must be used as a reference point.
	Moving an object should be considered as an action exclusively when the object weighs more then 2 kg in grip (or 1 kg in pinch grip) and the upper limb has a wide shoulder movement covering an area >1 m in diameter.
Grasp/Take	Gripping an object with the hand or fingers, to carry out an activity or task, is a technical action.
	Synonyms: Take, hold, grip again, take again
Grasp with one hand Grasp again with other hand	The actions of gripping with the right hand and gripping again with the left (or vice versa) must be counted as single actions and ascribed to the limb that actually carried them out.
Position	Positioning an object or a tool at a preestablished point constitutes a technical action.
	Synonyms: Position, lean, put, arrange, put down; equally, to reposition, put back, etc.

Sources: Colombini et al. (1998, 2002).

where

$i = 1 \ldots n$ are the tasks performed during a shift featuring repetitive movements of the upper limbs;

CF = frequency constant of technical actions per minute recommended under optimal conditions (30);

Ff; Fp; Fc = multiplier factors with scores ranging from 0 to 1, selected according to the behavior of the "force" (Ff), "posture" (Fp), and "additional elements" (Fc) risk factors in each of the i tasks;

D = duration in minutes of each repetitive task i;

Fr = multiplier factor, with scores ranging between 0 and 1, selected according to the behavior of the "lack of recovery period" risk factor during the entire shift;

Fd = multiplier factor, with scores ranging between 0.5 and 2, selected according to the daily duration of tasks with repetitive upper limb movements.

In practice, the procedure to determine the recommended overall number of actions within a shift is as follows:

1. For each repetitive task, start from the maximum recommended technical action frequency (CF = 30 actions/min). This becomes the reference constant for any repetitive task, given that conditions are optimal or not significant for all other risk factors (force, posture, additional factors, lack of recovery periods).
2. For each task, correct the frequency constant for the presence and degree of the following risk factors: force, posture, and additional factors.
3. Multiply the weighted frequency, thus obtained for each task, by the number of minutes of actual performance of each repetitive task.
4. Sum the values obtained for the different tasks (although of course where the repetitive task considered is unique, this is not necessary).
5. Multiply the resulting value by the multiplier factor for recovery periods.
6. Apply the final multiplier factor that considers the total time (in minutes) spent in doing upper limb repetitive tasks during the shift.

The value thus obtained represents the total RTA in the working shift.

11.1.3.5.2 Criteria and Procedures for Determining the Multiplier Factors

A brief illustration and discussion follows reviewing the criteria and procedures involved in the determination of the OCRA index calculation variables.

Frequency constant of technical actions (CF)

The literature (Drury 1987, Kilbom 1994a, Colombini et al. 1998, ACGIH 2000), albeit not explicitly, supplies suggestions of "limit" or threshold action frequency values for similar actions involving hands and wrists, and these range from 10 to 25 actions/min. However, these values mainly refer to occupational movements (flexion–extension, radial-ulnar deviations of the wrists) similar to each other. Obviously, when speaking of upper limb technical actions in general, it is probable that the joint movements will be more varied than this. It is, however, just as obvious that a high technical action frequency (e.g., more than 50 to 60/min) will imply excessively short times for muscle contraction and relaxation. On the basis of the above and practical considerations of the applicability of these proposals in the workplace, the frequency constant of technical actions (CF) is fixed at 30 actions/min (Colombini et al. 1998, 2000).

Force factor (Ff)

Force is a good direct representation of the biomechanical involvement that is necessary to carry out a given technical action or sequence of actions. It is difficult to quantify force in real working environments. Some authors use a semiquantitative estimation of external force by calculating the weights of the various objects handled; others suggest the use of dynamometers. As for the quantification of internal force, the most widely suggested approach is the use of surface electromyography.

One way to overcome this difficulty could be recourse, as already suggested by some authors (Eastman Kodak Company 1983, Putz-Anderson 1988), to a specific method called the Category-Ratio (CR-10) Scale for the Rating of Perceived Exertion, which extends over a 10-point score and was initially conceived by Borg (1982, 1998). This can describe the degree of muscle exertion that is perceived subjectively in a given segment of the body. The results of the application of Borg's CR-10 scale, where it has been used over a sufficient number of individuals, have turned out to be at least superficially comparable to those obtained with surface electromyography (EMG) — with the rating on Borg scale $\times$ 10 = percentage value with respect to the maximum voluntary contraction (MVC) as obtained by EMG (Grant et al. 1994).

The quantification of the exertion of the whole of the upper limb should be made for every single technical action that makes up the cycle. For practical purposes, actions requiring minimal muscle exertion can be identified (Borg scale rating = 0 to 0.5) and ignored. The average weighted score is then calculated for all the actions that are part of a single cycle. Once the actions requiring exertion have been determined, operators are asked to rate each with a score from 1 to 10: very very weak (0.5), very weak (1), weak (2), moderate (3), strong (5), very strong (7), very very strong (10). The observer will match the score for each of the actions indicated with the relative duration and then estimate its percentage value with respect to cycle duration. The calculation of the average exertion weighted over time involves multiplying the Borg scale score ascribed to each action by its percentage duration within the cycle; the partial results must then be added together. Table 11.7 shows how to identify the force multiplier factors corresponding to the average weighted score calculated in a cycle. When choosing the multiplier factor, it is necessary to refer to the average force value, weighted by cycle duration.

Posture factor (Fp)

The postures and movements executed during repetitive tasks by the various segments of the upper limb are among the elements that contribute most to the risk of onset of various musculoskeletal disorders. Currently, there is a sufficient degree of consensus in the literature to define the following as potentially dangerous: extreme postures and movements of each joint; postures maintained for a long period of time or identical movements of the various segments that are repeated for at least 50% of the work time (stereotypy). An accurate description of postures and movements can also be considered as a predictive element of specific pathologies of the different segments of the upper limbs.

TABLE 11.7

Multiplier Factors for the Different Risks Factors in the OCRA Index

Elements for the determination of the force multiplier factor (Ff)

Force factor	Average perceived effort (Borg CR-10 scale)	≥0.5	1	1.5	2	2.5	3	3.5	4	4.5	5
Force factor	Average force (%MVC)	≥5	10	15	20	25	30	35	40	45	50
Multiplier factor		1	0.85	0.75	0.65	0.55	0.45	0.35	0.2	0.1	0.01[a]

Elements for the determination of the posture multiplier factor (Fp)

Postural involvement score	0–3	4–7	8–11	12–15	16
Multiplier factor	1	0.70	0.60	0.50	0.3

Elements for the determination of the additional elements multiplier factor (Fc)

Additional elements score	0	4	8	12
Multiplier factor	1	0.95	0.90	0.80

Elements for the determination of the recovery period multiplier factor (Fr)

Number of hours without adequate recovery	0	1	2	3	4	5	6	7	8
Multiplier factor	1	0.9	0.8	0.7	0.6	0.45	0.25	0.10	0

Elements for the determination of the duration multiplier factor (Fd)

Minutes devoted to repetitive tasks during shift	<120	120–239	240–480	>480
Multiplier factor	2	1.5	1	0.5

[a] If technical actions requiring the exertion of a force exceeding a Borg CR-10 score of 5 are found, lasting at least 10% of cycle time, the multiplier factor to be used is 0.01.

Sources: Colombini et al. (1998, 2002).

The description/assessment of the postures must be made over a representative cycle for each of the repetitive tasks examined, identifying durations of the postures and/or movements of the four main anatomical segments of the upper limbs (both right and left):

- Posture and movements of the arm with respect to the shoulder (flexion, extension, abduction)
- Movements of the elbow and forearm (flexion–extension, pronation–supination of the forearm)

- Postures and movements of the wrist (flexion–extension, radial-ulnar deviation)
- Postures and movements of the hand (mainly the type of grip)

To simplify the posture analysis, only the more relevant joint movements and/or postures are considered. For classification purposes, it is sufficient to see that, within the execution of every action, the joint segment involved reaches an excursion greater than 40 to 50% of joint range (or is in an unfavorable position for gripping with the hand) to classify the involvement as "heavy." High joint involvement is quantified with different scores (from 1 to 4) extrapolated from the data on subjective perception of joint involvement (Genaidy et al. 1994).

As far as the types of handgrip are concerned, some (pinch grip, etc.) are considered less favorable for the hand and fingers than the power grip, and are therefore classified as implying medium/high involvement.

The recording form illustrated in Figure 11.1 shows the main joint risk areas that are included in the analysis. It shows the degrees beyond 40 to 50% of joint excursion range, the different types of grasp, and the relative score, which has been assigned by weighting with respect to subjective perception and the relative duration in the cycle time.

The posture evaluation requires the four operating steps described below:

1. Describe the postures and/or movements separately for the joints on the right and left side of the body.

2. Establish whether there is joint involvement in a risk area, and its duration within the cycle (1/2, 2/3, 3/3 of cycle time).

3. Establish the presence of stereotypy of certain movements, which can be identified by observing technical actions, or groups of technical actions, and noting those which are always repeated the same for at least 50% of cycle time, or by the presence of static postures that are maintained without movement for at least 50% of cycle time or by a very short duration of the cycle (less than 15 s but obviously characterized by the presence of actions of the upper limbs). The presence of stereotypy adds 4 to the score for the joints involved.

4. Calculate (for each joint and for each arm) the overall involvement score within the cycle taken as representative of the task.

Table 11.7 gives the data for calculating the corresponding posture multiplier factor, starting from the postural involvement descriptive score. This table must be used for the elbow, wrist, and hand. To calculate the OCRA index, the highest score must be taken from those referred to each of the previously mentioned segments. The shoulder involvement score will be dealt with separately (and used for preventive redesign purposes), because at the moment it would be necessary to define a different general frequency constant from those found for the other segments.

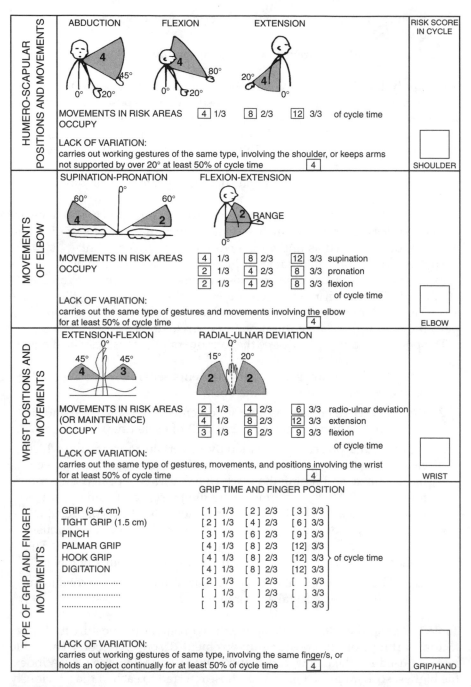

FIGURE 11.1
Evaluation of working postures and movements in OCRA index. (Reprinted from Colombini, D. et al., *Risk Assessment and Management of Repetitive Movements and Exertions of Upper Limbs*, Oxford, U.K.: Elsevier, Copyright (2002) with permission from Elsevier.)

Additional elements factor (Fc)

Side by side with the risk factors that have already been examined, there are other occupational factors to be found in the literature that must be taken into consideration when exposure is assessed. They are defined here as "additional elements." This is not because they are of secondary importance, but because each can, from time to time, be present or absent in the contexts examined. They include the following factors, although the list is not necessarily exhaustive:

- The use of vibrating tools (even if only for part of the actions)
- Requirement for absolute accuracy (tolerance 1 to 2 mm in positioning a piece or object)
- Localized compressions on anatomical structures of the hand or forearm from tools, objects, or working areas
- Exposure to cold or refrigeration
- Use of gloves, which interfere with the handling ability required by the task
- Objects to be handled that have a slippery surface
- Sudden or fast movements including "tearing" or "ripping" movements, hammering, hitting with a pick over hard surfaces, or using the hand as a tool

The list above is only concerned with factors of a physical or mechanical nature; other factors, which are listed under the general term of "psychosocial factors," have also been identified as factors in the onset of WMSDs. Among these, some are concerned with individual characteristics and cannot therefore be included in methods considering a collective and occupational type of exposure of a given group of workers. There are also factors related to the organization of the work (such as working pace determined by a machine, working on a moving object, incentive payment schemes, and insufficient training), which should be taken into consideration.

For each factor indicated, the same risk score is assigned (score 4 when it is present for 1/3 of the cycle time, score 8 for 2/3 of the cycle time, score 12 for 3/3 of the cycle time). It is also possible to assign variable scores to other additional elements factors (scores 1 to 4), according to the type of risk presented. Table 11.7 shows the necessary elements for attributing the multiplier factor for additional elements (Fc), based on the scores proposed above.

Recovery periods factor (Fr)

A recovery period is a time when one or more muscle–tendon groups are basically at rest. The following can be considered recovery periods:

- Pauses from work (both official and unofficial), including the lunch break where it exists
- Periods within the task cycle that leave muscle groups previously employed in the task totally at rest (control/waiting); these rest periods, to be considered significant recovery periods, must be experienced for at least 10 s consecutively almost every few minutes
- Employment in other tasks where the muscle–tendon groups are at rest (e.g., visual control tasks, or tasks that are carried out alternately with one of the two upper limbs)

No unequivocally defined criteria exist in the literature for the evaluation of recovery periods. Byström's (1991) contribution is very important and formulates models for the planning of optimum work/rest ratios in cases where intermittent static muscle actions are involved. Unfortunately, there is still a lack of precise and scientifically proven guidelines concerning recovery periods after repetitive dynamic actions, which represent the majority of working contexts.

Using the indications supplied by governmental or scientific authorities (Victorian Occupational HSC 1988, ACGIH 2000) as a starting point, in the case of repetitive tasks, it is advisable to have a recovery period every 60 min, with a work:recovery ratio of 5:1; the optimal distribution ratio for repetitive tasks then seems to be 50 min "work" and 10 min "recovery." On the basis of this recommended distribution, it is possible to design criteria to evaluate the presence of risk in a concrete situation; the risk may be due to the lack of, or inadequacy of, the distribution of recovery periods.

The easy procedure prepared for an evaluation of this risk factor requires the observation, for each hour, of whether there are repetitive tasks and whether there are adequate recovery periods. The lunch break (if it is present) and the end of the shift are considered to be the recovery periods for the hours preceding these two events. On the basis of the presence or absence of adequate recovery periods within every hour of repetitive work analyzed, each hour is then considered either "risk-free" or "at risk" (lack of recovery periods). The overall risk is determined by the overall number of hours at risk (generally between 0 and 6). Although all the other multiplier factors must be determined by considering each of the possible repetitive tasks that make up the shift, the recovery period factor must be determined by considering the whole of the working shift. For every hour without an adequate recovery period, there is a corresponding multiplier factor: for 1 h in the shift without adequate recovery the multiplier is $Fr = 0.90$, for 2 h in the shift without recovery $Fr = 0.80$, and so forth (as shown in Table 11.7).

Duration factor (Fd)

Within a working shift, the overall duration of repetitive tasks is important in determining overall exposure. The OCRA index calculation model is based on scenarios where repetitive manual tasks continue for a good part (4 to

TABLE 11.8

OCRA Index Classification Criteria with Indications of Recommended Preventive Actions

	OCRA Value	Risk Level	Recommended Preventive Action
Green	≤1	No risk	No action
Yellow/Green	1.1–2.2	Very low risk	No action
Yellow/Red	2.3–3.5	Low risk	Advisable to set up health surveillance
			Advisable to set up improvement actions for exposure conditions, especially for higher scores
Red	>3.5	Presence of risk Index values supply the criteria for priorities of action	Redesign of tasks and workplaces according to priorities
			Health surveillance, training, and information programs to exposed individuals.

8 h) of the shift. In some contexts, however, there may be large differences with respect to this more "typical" scenario (e.g., regularly working overtime, part-time work, repetitive manual tasks for only part of a shift), and that is why the multiplier factor was structured to consider these changes with respect to usual exposure conditions. Table 11.7 supplies the necessary parameters for dealing with the duration factor (where the time indicated in minutes is the total sum of time expended during the shift in repetitive upper limb tasks). The choice of the Fd factor values was made on the basis of data found in the literature (Moore and Garg 1995, ANSI 1995, CEN 2002).

11.1.3.5.3 Application of OCRA

The studies on 23 jobs and a reference group carried out so far (Colombini et al. 1998, 2002) suggest that indications of risk levels (green, yellow, and red lights) can be identified with OCRA scores (Colombini et al. 1998, 2002, Occhipinti and Colombini 2003). OCRA index values equal to or lower than 1 are considered acceptable (green). Values between 1 and 2 are borderline (yellow/green) but are considered to have very low risk. For values higher than this, preventive action is recommended, as indicated in Table 11.8.

Summary

This section, while taking into consideration the most recent and significant contributions in the literature, supplies a set of definitions, criteria, and procedures useful to describe and, wherever possible, to assess the risk factors that can represent a physical overload for the different structures and segments of the upper limb. It focuses specifically on identification of risk factors and describes some of the methods that have been developed for evaluating them.

The general model of description and assessment of tasks, from the Consensus Document of the Technical Committee of the IEA (Colombini et al. 2000), which was endorsed by ICOH, concerning all exposed workers in a given situation, is aimed at evaluating four main collective risk factors: repetitiveness, force, awkward postures and movements, lack of proper recovery periods. Such factors should be assessed as a function of time. In addition to these factors (others, grouped under the term additional factors) should be considered; these are mechanical factors (e.g., vibrations, localized mechanical compression), environmental factors (e.g., exposure to cold), and organizational factors (e.g., pace determined by machinery) for which there is evidence of association with WMSDs of the upper limbs. A brief presentation of the best-known synthesized methods/indices is given.

11.2   Variation and Autonomy

Svend Erik Mathiassen and Marita Christmansson

Introduction

Several reviews discuss the scientific evidence that postures at work are related to the risk of contracting musculoskeletal disorders. Some have focused on specific body regions, for example, the low back (Burdorf and Sorock 1997, Op de Beeck and Hermans 2000) or the upper extremity, including shoulders and neck (Winkel and Westgaard 1992, Buckle and Devereux 1999, Ariëns et al. 2000, Van der Windt et al. 2000, Sluiter et al. 2001). Other publications have provided comprehensive general overviews (Hagberg et al. 1995, Bernard 1997). Relationships between postures and risk have also been discussed in previous chapters of this book.

The scientific community generally agrees that disorders, in particular in the back, shoulders, and wrists, are related to posture descriptors such as "prolonged flexion of the back," "static neck posture," or "repetitive movements of the arms and hand." These three examples attempt to capture important features of a complete recording of posture vs. time from an individual worker (Figure 11.2). In principle, such a recording could be extended to include an entire work career.

Obviously, the three posture descriptors are not strictly defined. Different studies have quantified them (and other expressions of postures at work) in different ways, and most often using measurement methods that are per se sensitive to subjective interpretation, such as self-reports or expert observations (Winkel and Mathiassen 1994, Van der Beek and Frings-Dresen 1998). Poor strategies for workload assessment have been suggested to be one important reason that quantitative relationships between exposure and disorders are largely unknown (Hagberg et al. 1993, Winkel and Mathiassen

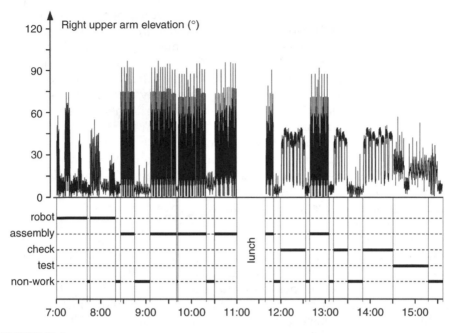

FIGURE 11.2
Elevation of the right upper arm against gravity during an entire working day, excluding lunch. Illustrative example based on occupational data. The day included five major task categories, occurring according to the pattern indicated in the lower part of the figure: robot maintenance, electronics assembly, quality checking, function tests, and non-work activities.

1994, Burdorf and Van der Beek 1999, Hagberg et al. 2001). Uncertain relationships, in turn, lead to problems in determining the risk associated with a particular posture and load pattern and in designing effective interventions against disorders.

Scientists have in general agreed that three fundamental dimensions of physical load are equally important to its effects: the amplitude (level, intensity), the frequency (repetitiveness), and the duration (Winkel and Mathiassen 1994, Van der Beek and Frings-Dresen 1998, Burdorf and Van der Beek 1999, Hagberg et al. 2001). Frequency and duration, which both refer to time-related aspects of work, have traditionally been given less attention by ergonomists than factors related to load amplitudes, such as the presence of "extreme postures" (Winkel and Westgaard 1992). This ergonomic tradition is also reflected in the fact that interventions initiated by ergonomics experts have, to a large extent, focused on technical issues confined to workplaces. Numerous initiatives have concerned, for example, workstation design, tools, or the behavior and capacity of employees, while few have aimed at time-related organizational factors such as the distribution of work among individuals, work hours, or control systems (Westgaard and Winkel 1997).

Still, time aspects of load and posture are known to be important risk factors for musculoskeletal disorders, as shown by the three examples above of posture descriptors related to risk. *Prolonged* indicates a long duration (*in casu* of a flexed back), *static* indicates that posture (of the neck) changes little across time or is rarely neutral, while *repetitive movements* typically means that a certain movement pattern is repeated again and again at a high frequency (Kilbom 1994b). Time aspects in production are normally controlled by management and engineers, and possible positive effects of a "traditional" ergonomic intervention have probably often been suppressed by a subsequent change in organizational factors (Winkel and Westgaard 1996, Westgaard and Winkel 1997).

Time-related aspects of postures and loads have been included in several ergonomic guidelines and standards (CEN 1995, Konz 1998a, Fallentin et al. 2001). Even the current Swedish legal text on prevention of musculoskeletal disorders states: "'The employer shall ensure that work which is physically monotonous, repetitive, closely controlled or restricted does not normally occur.... [T]he risks of ill-health ... shall be averted by job rotation, job diversification, breaks or other measures which can augment the variation at work" (Swedish National Board of Occupational Safety and Health 1998). At the same time, current trends in European production, in industry as well as in the service sector, point toward more time-intensive systems with greater occurrence of standardized, short-cycle tasks with little variation (Vahtera et al. 1997, Landsbergis et al. 1999, Neumann et al. 2002). Thus, a conflict appears in contemporary production systems between work "as it should be" according to standards and guidelines, and work "as it is."

The present section focuses primarily on posture *variation*, suggested to be a particularly important time-related aspect of work. Methods for assessing variation are considered, as well as the available evidence that posture variation — including the occurrence of rest pauses — is an important indicator of ergonomic quality. Ergonomic effects of combining different tasks in a job, for example through job rotation, are also discussed. A particular emphasis is given to the *autonomy* of operators and groups of operators, that is, their decision latitude with regard to job content and work performance. Autonomy is discussed in terms of its possible influence on posture variation, while its psychological effects, for example on job satisfaction and mental variation, are not considered.

11.2.1   What Is "Variation"?

According to the *Oxford Dictionary* (Allen 1990), *vary* means "make different" or "undergo change," and *variation* is described as "the act or instance of varying" or "departure from a former or normal condition." In the context of postures, this would mean that variation is present as soon as a posture changes; varying becomes the opposite of *static* in the basic sense of that word: that conditions do not change at all. However, in customary ergonomic use of variation and static, the distinction is less clear-cut. A job described as implying static neck postures usually does show some posture fluctuations (Bao et al. 1996, Åkesson et al. 1997). In addition, postures probably change when work is interrupted by breaks or personal allowances. Conversely, a job that to a major extent implies unchanging, yet neutral or relaxed, postures is rarely reported to show static postures.

Thus, static is often used in ergonomics to describe a situation where nonneutral postures change only a little during prolonged periods of time. The opposite would be frequently changing postures covering a large angular range, also allowing for neutral positions. Using this as a description of work with variation would be consistent with the notion of variation indicating change. However, it would also imply that repeated performances of, for example, short-cycle assembly tasks would have to be classified as varying work. This contrasts to common ergonomics semantics, regarding repetitions of similar operations again and again as a nonvarying job.

This discussion illustrates that the terms *static* and *variation* are often used loosely in the ergonomics literature. It also shows the need for composite measures of posture variability, which capture the size and frequency of posture changes, as well as the similarity of work sequences. Although both of these characteristics are expressions of the time pattern of postures, it seems appropriate to assess them independently (Moore and Wells 1992, Mathiassen and Winkel 1995). The job illustrated in Figure 11.2 comprises four work tasks with different characteristics as regards variation in upper arm posture (Figure 11.3): robot maintenance with a low similarity of work sequences and occasional high frequencies of posture changes, electronics

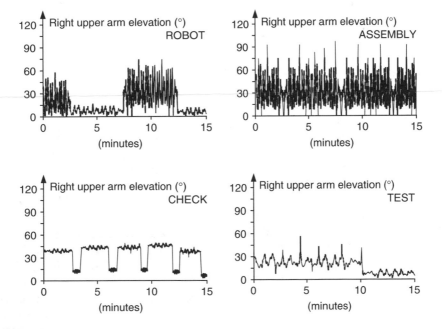

FIGURE 11.3
Magnified posture recordings from each of the four work tasks in the job illustrated in Figure 11.2. Non-work activities are not shown.

assembly showing high frequencies and high similarity between cycles, assembly quality check with a clear cyclic similarity but low frequencies of posture changes, and function tests characterized by low frequencies and low similarity.

11.2.2   Is Variation Related to Health?

There is a general agreement among ergonomists that variation in postures and loads is positive, as reflected in guidelines (Fallentin et al. 2001), standards (CEN 1995), and national legislation (Swedish National Board of Occupational Safety and Health 1998). However, to a major extent, this notion is based on indirect evidence. It has been documented that jobs characterized as monotonous, static, or repetitive are associated with increased risks for a number of upper extremity musculoskeletal disorders, as compared with more varying jobs (Bernard 1997, Sluiter et al. 2001). However, often, postures and loads are only superficially described in the varying reference job(s), and with less accurate methods than in the group considered to have hazardous exposures (e.g., Nordander et al. 1999). Very few studies have shown positively, for example, that jobs requiring workers to alternate between tasks imply smaller risks for musculoskeletal disorders than jobs consisting of only one task (Roquelaure et al. 1997, Jonsson et al. 1988). Thus, ergonomic epidemiology offers only vague suggestions regarding the possible effects on health of more variation in different jobs, as well as on what health-promoting patterns of variation might look like.

Many studies designed to investigate psychosocial factors at work have included recordings of perceived variation in the job. For example, in common assessments of the relationship between job demands and the possibilities of individuals controlling their own situation, the workers are asked to consider whether they can decide for themselves how to carry out the job, whether they can take a break at discretion, and whether the work is monotonous (Karasek and Theorell 1990). Obviously, answers to these questions will reflect physical workloads as well as perceived demands and control. High psychosocial demands have been shown to imply increased risk for musculoskeletal disorders, in particular if they occur together with low decision latitude (Bongers et al. 1993, 2002, Carayon et al. 1999). This may be taken as evidence that variation in the job is positive, in particular if it occurs according to autonomic decisions by the workers themselves. In addition, it stresses the need to consider the interaction of psychological and physiological dimensions of work when assessing risks (Westgaard 1999).

Several musculoskeletal disorders, for example low back pain, can be the result of loads accumulated throughout a long latency period (Norman et al. 1998). From a company point of view, latency can complicate the use of disorder statistics for evaluation of ergonomic conditions. Newly employed operators may develop disorders that are caused primarily by previous jobs, and critical conditions may appear not to cause disorders due to a rapid turnover of exposed employees (Östlin 1989, Punnett 1996). This problem is aggravated if subjects migrate frequently between different jobs, even within a company. Thus, in order for a company to understand the impact of different tasks and combinations of tasks on health and productivity, it seems attractive to analyze loads and/or acute responses, rather than concentrating on disorder outcomes (cf. Winkel and Mathiassen 1994).

This approach accentuates the need for methods that can retrieve and interpret the pattern of variation of work postures across time (Dempsey 1999). A majority of existing tools for posture surveillance and control are designed to identify critical circumstances in an isolated task, e.g., RULA (McAtamney and Corlett 1993), the NIOSH lifting guide (Waters et al. 1993), the strain index (Moore and Garg 1995), OCRA (Colombini 1998, cf. Section 11.1), ErgoSAM (Christmansson et al. 2000). Several of the methods include considerations of time aspects of actions and movements in the task being assessed, but the effect of alternating with tasks of another character is, at the most, assessed in general terms (OCRA).

11.2.3 How Can Variation Be Measured?

Frequency characteristics of postures during work have been measured by a variety of methods in the literature, ranging from simple counting of motions or work cycles during a specified time period (Section 11.1; Colombini 1998), through the number of times per hour that the upper arms move between certain angle sectors (Kilbom and Persson 1987), to key parameters from the frequency spectrum of posture recordings (Ohlsson et al. 1994, Radwin et al. 1994). While all of these procedures isolate the frequency dimension of posture, the exposure variation analysis (EVA; Mathiassen and Winkel 1991) attempts to capture the amplitude and frequency dimensions simultaneously. This is achieved by calculating the proportion of work spent in uninterrupted periods within specified posture categories. The uninterrupted periods are classified according to their duration (Table 11.9). Thus, EVA answers the needs for a method expressing both the size and frequency of posture changes. To date, EVA has mainly been used for analyzing muscle activity during work, but some studies have applied EVA to posture recordings (Torgén et al. 1995, Burdorf and Van der Beek 1999, Jansen et al. 2001, Torén and Öberg 2001). EVA still needs to be developed further and standardized for posture assessments, and the ability of EVA parameters to predict fatigue or disorders has not yet been sufficiently investigated.

In contrast to the reasonable assortment of approaches for quantifying the frequency dimension of posture, similarity has rarely been quantified. A number of studies of repetitive work have reported the task cycle time, under the implicit but uncontrolled assumption that the load pattern is identical from one repetition to the next (Silverstein et al. 1986). Several investigations have included questions to employees of whether they carry out "the same" movements or tasks again and again (e.g., Hansson et al. 2001). However,

TABLE 11.9

Exposure Variation Analysis (Mathiassen and Winkel 1991) of the Recording of Right Upper Arm Elevation Illustrated in Figure 11.2

Posture Interval (°)	Sequence Duration (s)						
	0–1	1–3	3–7	7–15	15–31	31–63	All
0–15	4.3	1.9	0.8	1.1	2.0	8.3	18.4
15–30	5.8	3.9	5.2	5.8	3.2	0.0	23.9
30–45	8.7	9.8	3.4	2.5	7.9	0.0	32.3
45–60	5.2	2.5	5.7	0.0	4.1	0.0	17.5
60–75	3.8	0.0	0.0	0.0	0.0	0.0	3.8
75–90	3.2	0.0	0.0	0.0	0.0	0.0	3.2
90–120	0.9	0.0	0.0	0.0	0.0	0.0	0.9
All	31.9	18.1	15.1	9.4	17.2	8.3	100.0

Note: The value in a cell shows the percentage of total work time spent in a particular posture interval (row), *and* in uninterrupted sequences of a specified duration (column). For example, during 9.8% of the working day, the arm was elevated between 30° and 45° in sequences lasting between 1 and 3 s. "All" indicates marginal distributions of posture amplitudes (right-most column) and frequencies (bottom row). Thus, the arm was elevated less than 15° for, in total, 18.4% of the working day.

studies of routine work tasks have indicated that postures do vary between repetitions, with a coefficient of variation (CV) of 0.05 to 0.30 depending on task and posture parameter (Johansson et al. 1998, Kjellberg et al. 1998, Granata et al. 1999, Van Dieën et al. 2002, Mathiassen et al. 2003b). Thus, the assumption of a near-perfect cycle-to-cycle similarity in repetitive work is not trivial. Probably, the term *repetitive work* has been used in ergonomic studies for a number of situations that have differed considerably with regard to similarity. More data from different kinds of cyclic work in occupational life are needed to obtain a clearer impression of what "high" and "low" similarity might be, and of the associations among similarity, fatigue, and disorders.

The cited studies have quantified similarity through a straightforward calculation of the cycle-to-cycle standard deviation (SD) of one or more key posture parameters, often normalized by the mean to give the CV. This is an attractive and simple option, and the cycle-to-cycle SD can easily be retrieved as part of the descriptive statistics of a study (Mathiassen et al. 2002). It is important to note that similarity can be assessed independently for parameters belonging to any of the three main dimensions of posture: amplitude, frequency, and duration (Möller et al. 2004). For example, consecutive cycles could differ in duration (i.e., cycle time) and frequency of movements, but be similar as regards average postures.

In work consisting of cyclically repeated tasks, the cycle offers an obvious choice for the basic analysis unit of posture data with respect to similarity. In noncyclic tasks, or work containing more than one task, the analysis unit

TABLE 11.10

Task EVA of the Job Illustrated in Figure 11.2

Task	Sequence Duration (min)				
	0–5	5–15	15–35	35–75	All
Robot	0.0	0.0	7.0	8.4	15.4
Assembly	0.0	2.4	15.7	15.0	33.1
Check	0.0	0.0	11.1	8.4	19.5
Test	0.0	0.0	0.0	10.0	10.0
Nonwork	1.3	8.1	12.6	0.0	22.0
All	1.3	10.5	46.4	41.8	100.0

Note: Cells show the percentage of the working day spent in robot maintenance, electronics assembly, quality checking, function tests, and nonwork activities (rows), *and* in sequences of a specified duration (column). The marginal summations shown by "all" indicate the overall pattern of shifts between tasks (bottom row) and the cumulated percent of the day spent at each task (right-most column).

must be defined in terms of a time period, e.g., minutes (Mathiassen et al. 2003a) or hours (Westgaard et al. 2001). For statistical reasons, the size of the period-to-period SD will be expected to decrease as the duration of the analysis unit increases. Moreover, the relationship between analysis unit duration and SD will be sensitive to recurring tasks or patterns of work during the day (Mathiassen et al. 2003a). These circumstances suggest that similarity should be retrieved for different durations of the basic analysis unit. Obviously, this generalized approach also gives data on similarity in cyclic work, as the average cycle time appears as a special case of analysis unit duration. The approach has not been tested on occupational postures, partly because it requires posture registrations of long duration.

A more straightforward assessment of similarity in a job can be obtained through a task EVA (Table 11.10), that is, an EVA (cf. Table 11.9) in which the amplitude categories are replaced by the different tasks occurring in the job. Thus, the analysis expresses similarity in the sense of whether the workday contains prolonged periods at the same task, or frequent shifts between tasks. For this result to be properly interpreted, some information must also be available of posture differences between tasks (so as to evaluate whether changes between tasks do, indeed, imply changes in postures), as well as within tasks (so as to assess the contribution of individual tasks to the overall posture variation). Thus, postures need to be analyzed by task. The task exposure profile (Figure 11.4) of the job illustrated and analyzed in previous figures and tables indicates that the job contains substantial proportions of tasks that differ considerably in upper arm postures. The task EVA concept, as well as its linkages to task-based posture analysis, needs to be explored further on occupational data.

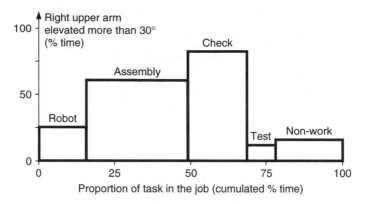

FIGURE 11.4

Task exposure profile of the job illustrated in Figure 11.2. The relative proportion of each task in the job is shown, as well as the percentage of time spent in that task with the right upper arm elevated more than 30°.

A few studies of postures and posture variability have made use of predetermined time systems (PTSs) as practiced in industry (Mital et al. 1987, Bao et al. 1996, Colombini 1998, Christmansson et al. 2000, Laring et al. 2002). A number of PTS systems exist, with the common denominator that work tasks are split into sequences of movements according to a predetermined set of movement categories. Each category has a standard duration assigned to it, so that the total duration of the task can be estimated. The primary engineering application is in production planning, for example, when balancing an assembly line, and when setting work pace standards. From an ergonomic viewpoint, the real-time sequence of movements generated by a PTS seems to provide a basis for assessing occurrence of critical postures, movement frequencies, and even similarity within and between tasks. The first two applications have been investigated (Bao et al. 1996, Christmansson et al. 2000, and Colombini 1998, respectively), while similarity information in PTS has not been explored so far. Further adaptation of PTSs and the use of their outputs to ergonomics assessments could be a promising line of research and development.

11.2.4   Variation within an Occupational Task

At the level of the individual worker, three principally different sources of posture variation in the job can be identified: variability associated with the performance of each specific task in the job (within-task variability), variability due to different tasks having different posture profiles (between-task variability), and variability related to the time schedule for changing between tasks (schedule variability). In this nomenclature, even nonwork activities such as breaks must be interpreted as tasks, since they contribute to the overall pattern of posture variation across the day.

A proper understanding of the posture variation *within* a work task will require data on the average amplitude and frequency of postures occurring during the task and, when the task is repeated, on the cycle-to-cycle variability around this average, i.e., the similarity as defined above. Average patterns of posture have been shown in numerous studies to be associated with technical factors such as product and workstation design. This is thoroughly discussed in previous chapters of this book. Factors influencing cycle-to-cycle similarity, on the other hand, have received much less attention, although they are equally important for the overall posture variability.

Even strictly controlled tasks with a cycle time of few seconds exhibit a marked cycle-to-cycle CV in posture parameters (Kjellberg et al. 1998, Granata et al. 1999, Van Dieën et al. 2002, Mathiassen et al. 2003b), as well as in muscle activation (Bao et al. 1995, Nussbaum 2001). In less constrained tasks with cycle times of 1 to 2 min, both postures and EMG parameters seem to vary more (Harber et al. 1992, Johansson et al. 1998, Christensen et al. 2000, Mathiassen et al. 2002, Möller et al. 2004). Thus, it seems reasonable to assume that longer cycles and less standardized performance requirements will lead to more within-task variability, although closer relationships cannot be established so far.

A common ergonomic intervention with the specific purpose of increasing posture variability is to provide the worker with an adjustable chair or table, allowing easy changes between sitting and standing during work. In a study of computer operators by Winkel and Oxenburgh (1990), this intervention was shown to increase the exposure variability of the back and legs, but not of the shoulders and neck. It also reduced discomfort. Other studies indicate that the opportunity to change posture may be used only rarely if the worker lacks motivation or is driven by other incentives into a restricted type of behavior (Neumann et al. 2002).

In industry, some production models emphasize the need to strictly standardize the way that tasks are carried out, for example, by means of detailed manuals, or by marking on the floor where workers are supposed to place their feet. The purpose can be to minimize balancing losses on an assembly line, or to strive for perfect product quality. The standardized process may be designed with ergonomics in mind so as to avoid extreme or constrained postures, but the very goal of the standardization is to eliminate deviation in work techniques. Thus, minimal cycle-to-cycle posture variability will be interpreted by the production engineer as a success.

An intriguing hypothesis is that workers repeating a task very stereotypically will be more prone to develop musculoskeletal disorders than workers managing to vary postures and loads slightly between task cycles. Some evidence suggests that individuals may, indeed, differ in consistency, even when repeating a strictly controlled task (Mathiassen et al. 2001, 2003b). Whether this is a protective ability when doing cyclic occupational work remains to be investigated.

11.2.5   Variation between Tasks — Breaks in Static and Dynamic Work

Obviously, increased variability *between* tasks in the job of an individual can be achieved by introducing new tasks that differ in postures from those already present. Job enlargement and job enrichment are examples of interventions with this purpose, which will be discussed in a following section on group-level reorganizations. Introducing more breaks is an additional way of increasing variation in the job, which does not require work tasks to be reallocated among workers. This may be one of the explanations of why the use of rest pauses is one of the most frequently recommended interventions against musculoskeletal disorders in ergonomics textbooks, for repetitive tasks with low loads as well as for jobs with high metabolic demands (Konz 1998b). The strong belief in rest pauses is also reflected by the fact that commercial software has been developed to help workers design proper rest schedules (Asundi 1995).

When rest pauses are introduced in a constant-force muscle contraction in a fixed posture, muscle fatigue develops gradually over time, but at a lower rate than in exercise without rest (Mathiassen 1993a). Fatigue is always attenuated by rest pauses, even when the reduction in average load caused by the inactive pauses is compensated for by increasing the load in active periods (Mathiassen 1993a). Thus, the overall time-integrated load that can be accomplished up to the point of exhaustion is nearly doubled when the contraction is interrupted by regular pauses as compared to a continuous activity. The rate of fatigue development is influenced by the duration of the complete activity/rest cycle as well as by the duty cycle, i.e., the active proportion of the cycle. Frequent shifts between activity and rest (short cycle time) lead, in general, to less fatigue than rare shifts (long cycle time), at a maintained load and duty cycle. The effects of changing the duty cycle at a maintained cycle time and average load seem to be more complex. Thus, in the cited study by Mathiassen (1993a), fatigue developed comparatively fast in a protocol with a 50/10 s contraction/rest schedule at an average load corresponding to about 15% of the maximal voluntary capacity. A similar increase in fatigue was found in a 20/40 s schedule with the same average load. However, when activity and rest were combined in 30/30 s and 40/20 s protocols, fatigue was significantly less pronounced. Thus, an optimal duty cycle, if a certain average load has to be produced, seems to be a medium alternative with neither too much nor too little of the cycle occupied by rest.

The general findings described above apply to a number of important signs of fatigue, e.g., changes in the EMG signal, decreased force-generating capacity, increased heart rate, perceived fatigue, and time until exhaustion. Several models have been presented, relating some of these fatigue indicators to the extent and timing of pauses (e.g., Rohmert 1960, Pottier et al. 1969, Milner et al. 1986). Other responses, e.g., loss of potassium from active muscles, seem to be related more to the total exercise duration, and show larger deviations from the nonfatigued state after exhaustive exercise with intermittent rest pauses than without (Sjøgaard 1996). This has led to the theory that appropriate biological brakes acting in continuous isometric exercise are circumvented by introducing regular rest periods, and thus that rest is not necessarily beneficial in a long-term perspective (Mathiassen 1993b). It is also important to notice that rest for prolonged periods can have negative biological effects, which are best counteracted by activity. This applies, for example, to leg swelling (Winkel 1985) and on a longer timescale to the capacity for doing physical work in general (Erikssen 2001).

The reason rest alleviates fatigue in low-level isometric contractions has been the subject of much debate. An attractive hypothesis states that the rest pauses allow for metabolites to be cleared from the active muscles, and for oxygen and energy substrates to be replenished. However, rest pauses have a large effect even if they occur for very short periods at rare intervals (Byström et al. 1991), and even if they are introduced at load levels where blood flow through the active muscles does not seem to be severely restricted (Byström and Kilbom 1990). Thus, it has instead been suggested that pauses exert their effect by modifying sensory responses to exercise (Byström et al. 1991), or by triggering a substitution in which different subparts of the active muscle(s) deliver the force output at different points in time. Several studies have supported that alternating recruitment may, indeed, occur between muscles in a synergy (Sjøgaard et al. 1986, Van Dieën et al. 1993, Mathiassen and Aminoff 1997), and even between motor units within a specific muscle (Forsman et al. 1999, Westgaard and De Luca 1999, Forsman et al. 2002, Thorn et al. 2002). The later studies also showed, however, that some motor units can be continuously active for long periods according to a Cinderella principle, at least during contractions with little or no variation in force and posture.

The physiological effects of introducing variation in a static posture by means other than rest has gained less attention. Again, the effects to be expected depend on the recruitment principle of the active muscles, as well as on possible peripheral or central sensory responses (Enoka 1995). According to the Cinderella principle, rest is the only intervention that will ensure that all muscle fibers relax. If an alternating recruitment principle applies, other sources of variation may also be effective. The latter possibility is supported by observations that short periods of increased load in a maintained static shoulder elevation can trigger a changed recruitment pattern in the trapezius muscle (Westad and Westgaard 2001) and can alleviate fatigue (Mathiassen and Turpin-Legendre 1998). Activation of other body regions, and even mental efforts, has been shown to accelerate recovery from fatiguing exercise (Asmussen and Mazin 1978). This suggests that alternative work tasks or active pauses would be a more effective source of variation than total rest in reducing fatigue. Still, the concern that a suppression of fatigue perceptions may represent a risk in the long run must be borne in mind.

Dynamic exercise implies, by definition, movements and will therefore always, to some extent, contain posture variation in the frequency sense of that word. Some results from the laboratory indicate that slow movements around a certain mean posture may cause similar fatigue responses to those when keeping that posture static (Hagberg 1981). In both cases, muscles are kept constantly active. However, many dynamic activities, such as walking, bicycling, and repeated lifting, are probably accomplished through periods of intense muscle activity alternating with periods of very low activity or even rest (Chaffin and Andersson 1991). The literature on the physiological results of scheduled rest pauses in this kind of dynamic exercise is abundant, particularly within the sports sciences, and the reader is referred to other overviews (e.g., Billat 2001). As with static contractions, the effect of pauses on fatigue depends on the exercise intensity as well as on the proportion and distribution of rest. In addition, movement speed in the exercise bouts influences fatigue development (Mathiassen 1989). Rest pauses do alleviate fatigue, in the sense that exercise intensity in the work periods can be kept higher with than without intermittent rest.

11.2.6   Breaks in Occupational Work

Numerous mathematical models have been presented of how to optimize rest schedules in an occupational setting (Rohmert 1973a,b, Rose et al. 1992, Fisher et al. 1993, Byström and Fransson-Hall 1994, Dul et al. 1994, Wood et al. 1997; reviews in Price 1990a,b, Mital et al. 1991, Konz 1998b). Several of the models have been derived directly from experiments on constant-force muscle contractions in fixed postures. Some of these models have been cited in ergonomics textbooks, even though the validity of isometric exercise as a general model for occupational loads can be seriously questioned (Mathiassen and Winkel 1996). Quantitative work/rest models have also been developed in management science, with the primary purpose of maximizing the daily work output of individuals (Bechtold et al. 1984, Janaro and Bechtold 1985). Thus, determination of appropriate relaxation allowances is considered an important element in industrial engineering (Lund and Mericle 2000).

Studies of interventions in simulated or real occupational work provide, however, a more ambiguous impression of the effect of pauses. In an investigation of repetitive meat-processing work, 36 min of breaks were added on top of the normal schedule of 30-min lunch and one 15-min break per half-day (Dababneh et al. 2001). The total daily production was shown to be maintained in spite of the added breaks, and discomfort in the lower leg decreased. In a field study of data-entry workers, Galinsky et al. (2000) found that discomfort and eyestrain were attenuated, yet not eliminated, by four additional 5-min breaks distributed across the workday. However, total production decreased with added breaks.

Secretaries doing computer work were shown by Hagberg and Sundelin (1986) to prefer a protocol with a cycle of 15 s of rest and 5.75 min of work to a protocol without scheduled pauses. McLean et al. (2001) investigated the development of discomfort and EMG changes in office workers during a nonbreak session at their desk as compared to work with discretionary rest breaks or with regular 30 s breaks every 20 or 40 min as signaled by the computer. All of the protocols with added breaks reduced discomfort as compared to continuous work, but no clear indications were found of any of the break protocols influencing the EMG. The authors suggested that the work per se was so varied in terms of muscle activity that additional variation provided by rest pauses had no effect on the pattern of muscle activation (McLean et al. 2001).

In a laboratory study of simulated short-cycle assembly work, Sundelin (1993) found no conclusive differences in EMG signs of muscle fatigue during 1 h of continuous work as compared to 1 h of work with a 1-min rest for every 5 min of work. Work pace was adjusted to reach the same hourly production in the two protocols. Discomfort did not differ between work with and without pauses.

Vague or absent effects of breaks on physiological fatigue indicators was a major finding also in a laboratory study of an industrial assembly task by Mathiassen and Winkel (1996). Fatigue developed gradually during the workday in a protocol with breaks according to industrial regulations, and 20 min of added rest breaks every 2 h did not reduce fatigue significantly, irrespective of whether the production loss caused by the breaks was compensated for by an increased work pace or not.

An obvious prerequisite for breaks to have a physiological effect is that postures and loads during breaks differ from those during work. Only a few studies have actually checked whether this is the case during normal occupational work. In a study of cleaners and office workers, the proportion of time with shoulder muscles in a resting state was as much as three times larger during breaks than during cleaning and office tasks (Nordander et al. 2000, Mathiassen et al. 2003a). Another study on office workers by Fernström and Åborg (1999) confirmed that shoulder load amplitudes were less during coffee breaks than during work, although the difference corresponded to, at the most, about 3% of the maximal contraction capacity. Observations of the pattern of discretionary breaks among computer operators (Hagberg and Sundelin 1986) showed that individuals did rise to leave the workstation about twice per hour, but that most breaks only consisted in withdrawing the hands from the keyboard. These results suggest that postures and loads are, in general, in a more rested state during discretionary breaks than during work. However, the difference does not seem to be particularly pronounced if the work per se is characterized by low-level, steady activity in a seated posture.

11.2.7 Scheduled or Discretionary Breaks?

Field studies show that discretionary breaks occur, sometimes to a considerable extent, outside the periods of scheduled breaks and lunch. This applies to jobs allowing the individual a substantial control of his or her own work schedule (McGehee and Owen 1940, Hagberg and Sundelin 1986, Nordander et al. 2000, Mathiassen et al. 2003a), but also to short-cycle work controlled by strict productivity requirements where some individuals may work ahead of the predetermined pace (Christensen et al. 2000). Introduction of scheduled breaks can lead to a reduction in the occurrence of discretionary breaks (McGehee and Owen 1940, Hagberg and Sundelin 1986). This can be one reason that scheduled breaks seem to have a minor impact on fatigue development, as discussed above.

Another explanation may be that a break schedule that is forced upon the worker is perceived as an annoyance because it interrupts work at inconvenient moments. This was clearly demonstrated in a study on office workers by Henning et al. (1997), in which only about half of the group complied with 4.5 min of additional breaks per hour, appearing in a predetermined schedule. The scheduled breaks were considered particularly unattractive by a subgroup engaged in prolonged tasks involving customer contacts. Forced and unavoidable breaks can be perceived as stressful, as shown in a study of packing machine operators (Veiersted 1994). In that study, unexpected machine stops offered an opportunity to take a rest, but some subjects were tense even during these stops.

As an alternative to short, scheduled breaks, which seem to be of limited success in occupational settings, an increased allowance for discretionary breaks could lead to the desired variation in postures. One obvious prerequisite for this to happen is that the break allowance is used. In a study by Genaidy et al. (1995), meatpackers were offered breaks according to their own choice in addition to the normal 15 min of scheduled breaks every 3 h of work. The discretionary breaks were constrained to last at the most 2 min each, and 24 min in total. Workers were told by the researchers that more breaks would be beneficial to their health. However, the average worker took only two additional breaks, lasting in total 48 s, i.e., less than 5% of the allowance. Apparently, stronger drivers superseded the possible wish of the individuals to protect their long-term health.

The study by Genaidy et al. (1995) illustrates that a formal autonomy may not be realized if other factors in the production system act against it. The results also suggest that the perception of weariness during occupational work can be so weak that it is easily overruled by an urge to keep on working. This observation has been confirmed in studies showing that physiological signs of fatigue can occur even in subjects allowed to take self-administered rest pauses without limitations (Rohmert and Luczak 1973, Henning et al. 1989). In a laboratory study (Henning et al. 1996), computer operators were given continuous feedback on their current total of break time, and they were reminded by the computer to take a break of 30 s every 10 min in case they had not done so on their own. The break account information resulted in more spontaneous pauses and less back discomfort than if no feedback was provided. Again, this suggests that the biological drive to take a rest is weak, at least in low-level steady work. Thus, a necessary, but perhaps not sufficient, prerequisite for a self-administered break allowance to be used effectively seems to be that the organizational context strongly supports it. This applies to technical factors in the system, e.g., buffer sizes between workstations, as well as to administrative issues, e.g., the salary system.

11.2.8 Variation through Alternating Activities

For long-lasting work at low load levels, for example, in front of a computer or at an electronics assembly line, it has been proposed that physiological needs for variation are better met by introducing more activity than by introducing more rest (Winkel and Oxenburgh 1990). The idea has been based on studies of leg swelling, showing that even very little activity, i.e., small leg movements, improved cardiovascular responses as compared to total rest (Winkel 1985). A similar situation may apply to spinal shrinkage (Van Dieën and Oude Vrielink 1998). Active breaks, i.e., arranged activities intended to be more vigorous than normal work, have been investigated in office work (Sihvonen et al. 1989, Sundelin and Hagberg 1989, Henning et al. 1997) and assembly work (Silverstein et al. 1994, Mathiassen and Winkel 1996).

In the field study by Sundelin and Hagberg (1989), short gymnastics bouts — 15 to 20 s every 6 min — were shown to alter the amplitude of shoulder muscle activity as intended. A similar, but weaker, effect was seen when the subjects walked in the corridor during the breaks. Subjects preferred the active conditions to an equivalent amount of rest, but physiological responses were not assessed. The preference is in agreement with the results of Henning et al. (1997), who applied stretching exercise.

In the study by Sihvonen et al. (1989), pause gymnastics was reported to decrease the amplitude of shoulder muscle activity during subsequent computer work, as compared to levels observed immediately before the gymnastics. This effect was not found by Sundelin and Hagberg (1989). Pause gymnastics lasting 10 to 15 min per day did not reduce discomfort in a study by Silverstein and co-workers (1988), and a drop-out of 20% occurred among participants within a year. When 20 min of active breaks were introduced in the light assembly task studied by Mathiassen and Winkel (1996), fatigue developed in a similar fashion to that observed for work without breaks and for work with breaks consisting of seated rest. In this study, the active breaks comprised administrative work and manual handling of boxes weighing 8 to 15 kg in a stockroom. Although the cited studies do not agree completely, the general impression is that the effect of short active breaks on fatigue and discomfort is weak, and similar to that of rest.

Likewise, physical activity programs at the workplace have been suggested to be of limited success in improving job satisfaction and reducing absenteeism (Proper et al. 2002). A major endeavor in such programs is to promote fitness in general rather than to introduce more variation into the job, but they often include active breaks during one or more workdays per week.

11.2.9 Variation through Changed Schedules

Some attention has been paid to the schedule variability of rest, that is, the issue of identifying the most effective distribution across time of work and rest, given that their overall proportions in the job are fixed. In this case, the real-time pattern of tasks and postures is manipulated, but not the overall posture distribution and average. Inspired by results from constant-force experiments, it is often suggested that short and frequent micropauses give a better recovery than longer but rarer rest pauses. However, for some important responses, e.g., leg swelling during standing work (Van Dieën and Oude Vrielink 1998), experiments have shown that micropauses can be a less effective preventive measure than accumulated long breaks. Furthermore, laboratory studies of repeated gripping (Moore 2000) and intermittent maximal shoulder extensions (Elert and Gerdle 1989) suggest that if rest pauses are too short, some individuals may fail to actually relax their muscles.

The supremacy of the micropause principle does not have firm support from occupational field studies either. In a study of seated work in front of a video screen, Rohmert and Luczak (1973) found that a work/rest cycle of 25/5 min led to better productivity than a 100/20 min cycle, but that both led to signs of fatigue. In contrast, a 9-min break in the middle of every 2-h block of work was preferred to 3 min of break every 30 min by workers engaged in repetitive meat processing (Dababneh et al. 2001). Beynon et al. (2000) found signs among hospital porters that perceived exertion and average energy expenditure during 4 h of work were larger if 25 min of breaks were distributed as two 12.5-min breaks than if they appeared as one 15-min and two 5-min breaks.

Over an extended time perspective, the distribution of tasks, and even of jobs for the individual, leads into questions concerning, for example, shift-work patterns, the proper arrangement of vacations, and life-time occupational careers (Konz 1998b). These aspects lie beyond the scope of the present section.

As it appears, little evidence exists concerning the effects of different sources of variation in different types of job. Thus, focal questions concerning what a sufficient or optimal posture variation might be cannot be answered today: What characterizes complementary or symbiotic tasks that will promote recovery from one another when combined in a job? Which are the optimal proportions of complementary tasks or of work and rest? What is the optimal schedule of changing between tasks? Still, the discussion above suggests that the most effective way to obtain posture variation in a job is to incorporate substantial proportions of productive tasks that differ significantly in postures.

11.2.10 Autonomy at the Individual Level

As reflected, for example, by the discussion above of scheduled vs. discretionary breaks, a crucial determinant of variation in the job of an individual is the extent of autonomy, that is, the decision latitude with respect to scheduling work and determining the procedures to carry it out (Hackman and Oldham 1980). Numerous definitions of autonomy have been proposed (see, for example, Christmansson 1997), and several examples are available of expressions of autonomy at different companies, at individual as well as group levels (cf. Section 11.2.12; Christmansson 1997). The examples include autonomy to decide on work hours, work allocation within teams, employment issues, and the setting of goals for production.

The autonomy is to a considerable extent the result of the production process, especially the material flow and the size of buffers, and the degree to which workers are controlled by the design and function of machinery and equipment. Together, these factors have been summarized in the term *technical autonomy* (Karlsson 1979). *Administrative autonomy,* on the other hand, covers factors in the social system at the company, e.g., control systems, organizational culture, staff, manufacturing strategies, and work design. Thus, administrative autonomy concerns whether a worker has the authority to make decisions. Administrative autonomy cannot be realized in the absence of technical autonomy, as the worker would then be primarily controlled by the equipment or the production process (Karlsson 1979).

An extended autonomy is generally considered an important element of a "good" job in the literature on occupational ergonomics and psychology. However, different aspects of autonomy are likely to differ with respect to their impact on physical loads and postures. Some aspects of autonomy are directly related to work posture patterns, for example, scheduling of breaks, deciding on work amount and work hours, and selection of work sequences (Christmansson 1997). Other aspects may influence postures indirectly by changing the work context, for example, salary setting, and the opportunity for competence development. It is important to note that an increased autonomy at the individual level can also result in impaired load and posture patterns, for example, if the work pace or the duration of work is increased. As stated above, a formal autonomy needs support from other control systems as well as from the organizational culture in order to be practiced according to intentions.

11.2.11 Variation through Work Reorganizations

Job rotation, job enlargement, or job enrichment have been suggested frequently during the past decades as ergonomic interventions reducing the risk for musculoskeletal disorders. All terms refer to the idea of introducing new work tasks into the job of an individual to change load and posture patterns in a favorable direction. In contrast to technical changes in tasks or increased break allowances as discussed in the preceding section, implementation of job rotation usually requires tasks to be reorganized between individuals, at the level of groups or teams of workers. Whether this is a viable source of increased variation depends basically on the availability of tasks at the group level.

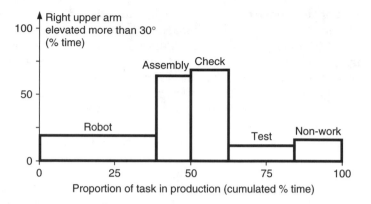

FIGURE 11.5
Product cycle exposure profile, illustrating the average proportions of tasks associated with manufacturing one product, as well as the percentage of time spent in each task with the right upper arm elevated more than 30°.

The concept of product cycle exposure (Bao et al. 1996, Mathiassen and Winkel 1997) has been proposed to capture loads at the level of production systems involving several workers. Product cycle exposure denotes the loads associated with producing average items in a production system. Thus, it includes all tasks carried out by humans during production, including allowances for nonproductive activities. Tasks are represented in proportions as occurring during production. The product cycle exposure profile (Figure 11.5) is thus a group-level equivalent of the individual job exposure profile of postures analyzed by task (Figure 11.4).

On the basis of the product cycle exposure profile (Figure 11.5), different ways of allocating tasks to workers can be contemplated, for example, with respect to their effects on average postures and posture variation for individual workers. Thus, Figure 11.6 illustrates two alternatives for distributing the tasks in the system represented by Figure 11.5 among three workers. The upper alternative results in workers becoming specialists. In this case, the left-most worker has a varying job, yet with a large overall proportion of the job spent with the arm elevated more than 30°, while the two other workers have jobs in more neutral postures, but with less variation. If tasks are instead distributed equally among the workers, all become generalists, enjoying a job composed of tasks with widely varying posture characteristics. The product cycle exposure profile (Figure 11.5) also allows for analysis of the consequences of changing the proportions of tasks. For example, a reduction of 'test' in the system shown in Figure 11.5 would imply an increased occurrence of arm postures greater than 30°, as well as decreased variability between tasks.

Thus, the product cycle exposure profile can reveal whether the system has a potential to offer loads and postures of a good ergonomic quality to the workers. The concept of exposure latitude has been suggested as a description of this potential (Mathiassen and Winkel 1997). Because variation

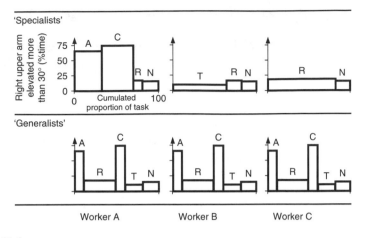

FIGURE 11.6

Two hypothetical distributions of tasks among three workers in the system shown in Figure 11.4. The abscissa and ordinate axes of an individual profile show task proportions and postures in tasks, respectively, as indicated in the upper left diagram. Task codes A, C, R, T, and N refer to the tasks electronics Assembly, quality Checking, Robot maintenance, function Tests, and Nonwork activities (cf. Figures 11.2 through 11.5).

is believed to be important to ergonomic quality, the exposure latitude must include consideration of posture variability within and between tasks. Systems containing only very similar tasks, each characterized by a low within-task variation, will be considered to have a low exposure latitude, as postures will vary little in an individual worker's job, irrespective of how tasks are distributed among workers. Exposure latitude can be increased by introducing more tasks into the system under consideration, or decreased by removing tasks from the system. Even though the exposure latitude is high in a particular system, running production may show a low ergonomic quality. This can happen, for example, if most workers are required to carry out only one limited task, as in a typical line-based industrial layout. Current trends in industry favor short-cycle, line-based solutions, even though several studies indicate that parallel, long-cycle systems can perform equally well, while at the same time being of better ergonomic quality (Engström et al. 1995, 1996a,b, Christmansson 1997),

The expectation that job rotation within a group will reduce disorders is based on the belief that subjects working at low loads will benefit from periods at higher loads, and that the opposite is also true. Thus, the low- and high-load tasks are assumed to be complementary in the sense that they promote recovery from each other. Less similarity is assumed to imply less risk for disorders. In contrast, if the probability of contracting a disorder has a straightforward linear relationship with cumulated or average loads, reallocation of tasks within a group cannot be expected to change the occurrence of disorders. Some workers will experience an increased risk, while risk will decrease for others, provided that all are equally susceptible to the disorder, but because the product cycle exposure does not change, neither does the risk at the group level.

Surprisingly little research has focused on the consequences of job rotation in terms of loads, physiological responses, or risk. Jonsson (1988) discussed whether high shoulder muscle loads in different jobs could be reduced by including less strenuous tasks. He was able to show that tasks with relatively low loads were, indeed, available in some occupations, but he did not proceed to estimate the possible effects of a realistic job rotation on the overall job load. Carnahan et al. (2000) developed a mathematical model of how to distribute lifting tasks optimally among workers, based on the same endeavor to reduce the relative load of individuals. Thus, the basic aim in these two studies was not to increase variation.

In a field study of refuse collectors, Kuijer et al. (1999) were able to show that workers who shifted within a day between two of the three major work tasks of truck driving, street sweeping, and refuse collection perceived less fatigue and exertion than workers solely carrying out any one of the tasks. Amplitudes of arm and trunk posture and energetic loads in the rotation groups were intermediate between those observed in the two constituent tasks, as could be expected. In a later prospective study, the prevalence of musculoskeletal disorders was shown to be similar for a group only collecting refuse and a group rotating between collecting and truck driving, with the exception that the rotating group reported more low-back complaints (Kuijer 2002). However, this group also reported less need for recovery when going home from work.

Icelandic fish processing plants were studied by Ólafsdottir and Rafnsson (1998) before and after a rationalization. Prior to the change, jobs were dominated by repetitive short-cycle tasks, but they also included periods of less controlled weighing and packing tasks with other posture and load patterns. The rationalization consisted in introducing a line-based system where some workers performed a specific short-cycle task all day long while weighing and packing were allocated to other workers. The prevalence of musculoskeletal disorders in elbow, wrist, and fingers increased after the change, while it decreased for the ankles. No significant effects were observed for other body parts. The changed disorder pattern was explained by the work being less varied but the workstations better at the flow-line than in the previous system, although variation was not assessed in quantitative terms.

A reorganization of an office department studied by Fernström and Åborg (1999) implied that workers became less specialized on single tasks, and that the proportion of desk tasks, as opposed to computer work, was generally increased. Shoulder load amplitudes did not change due to the reorganization, nor did the proportion of muscular rest or the frequency of rest periods. One probable explanation could be that all tasks in the production system were of the same character.

In a study of supermarket cashiers (Hinnen et al. 1992), rotation between seated cashiers' work and standing tasks such as filling up shelves or selling at the delicatessen counter was shown to imply a smaller prevalence of musculoskeletal complaints than doing cashiers' work alone. However, this positive effect of job rotation was only seen with scanner-type counters, not with traditional checkout counters requiring manual entry of prices. The latter finding was explained by alternative tasks being perceived as an annoyance when working with the traditional counters, but not when working with laser scanners (Hinnen et al. 1992).

Another explanation is implicitly suggested in a study by Christmansson et al. (1999). An assembly workshop with a clear distinction between the workers assembling the product and those planning work and supplying the workstations was replaced by a team-based system where all workers had considerable opportunities to change between assembly and administrative tasks according to their own decisions. The jobs of the former assembly workers were, indeed, shown to be enriched, for example, by materials handling and administration, although assembly still occupied the major part of the workday. However, little changed in workers' perception of their psychosocial work environment, and musculoskeletal health did not improve. As an explanation, the authors suggest that the reorganization was not accompanied by sufficient education, and that social support diminished. Thus, increased task variation was perceived as stressful rather than stimulating by the workers (Christmansson et al. 1999).

Obviously, a reallocation of tasks within a group will have an effect on the posture variation of individual workers only if tasks differ in postures. As suggested above, common office tasks may not differ much with respect to upper extremity loads (Fernström and Åborg 1999, Mathiassen et al. 2003a), although both seated and standing tasks may occur. Studies of tasks in assembly workshops indicate that administrative tasks may differ in load from assembly tasks (Christmansson et al. 2002) but, on the other hand, that different assembly tasks can be very similar (Möller et al. 2004). In the latter study, it was even shown that rotating between three assembly stations may imply less overall posture variability, as measured through the cycle-to-cycle variance, than working at one station only. This can happen when average posture amplitudes differ little between stations, while, at the same time, one station is characterized by a large variation between cycles as compared to the other stations.

11.2.12 Autonomy of Groups

At the individual level, autonomy was described as the latitude to schedule work and determine the procedures to carry it out. Autonomy can be viewed as comprising an administrative and a technical dimension also at the group level (Bailey and Adiga 1997). The former covers decisions concerning group processes and management, and the latter includes decisions regarding products, equipment, and the production process. Instruments have been developed for measuring autonomy, taking into consideration operational, tactical, and strategic items (Bailey and Adiga 1997). Obviously, autonomy at the individual level interacts with autonomy at the group level.

As in the individual case, a formal autonomy in a group may not be practiced. This can happen because control systems restrict realistic decisions or because of social interactions or pressures inside the group. For example, specialization of workers inside a group may develop in spite of authority of the group to distribute tasks equally among workers (cf. Figure 11.6). Thus, formal autonomy must be supported by adequate control systems in order to be realized, for example, concerning incentives, production planning (Christmansson et al. 1999), attendance (Parenmark et al. 1993), and communication (Klein 1991). Factors in the organizational culture, such as the company's attitude toward change initiatives from workers, can also promote or obstruct the realization of formal autonomy. The relationships are, to a large extent, unknown between, on the one hand, task availability, formal autonomy, and control system constraints and, on the other, posture variation in the jobs of individual workers in a group.

11.2.13   Company Strategies, Variation, and Autonomy

The discussion above on posture variation and autonomy at the levels of individuals and groups has shown that the posture patterns of an individual worker are determined by decisions made by the individual as well as by the team or group in which the individual is working, and even by strategic decisions made at a management level.

Strategic decisions may change the assortment of tasks at the company, for example, through an outsourcing of certain production segments or support functions, or through an increase of the general level of automation (Hayes and Wheelright 1984). Also, changes to the design of a product will affect the tasks associated with its production (Helander and Nagamachi 1992). These examples illustrate changes in exposure latitude at the company level, i.e., changes in the potential for designing production systems with a good ergonomic quality.

Strategic decisions can also constrain the distribution of tasks among employees, for example, by subdividing the company into departments or by favoring a particular production model, as exemplified in the previously mentioned discourse concerning serial or parallel assembly systems. These choices influence whether the prevailing exposure latitude is used to its full potential. Finally, management decisions may directly affect the autonomy of the individual worker, for example, by requiring standardized assembly methods as part of a quality control initiative.

Management-level decisions are often implemented through formal control systems, for example, regarding work standards, quality check routines, and incentives, but also through attitudes and company culture. Although it may seem obvious that strategic decisions have ergonomic consequences, even when they are not explicitly intended to do so, little empirical research exists to support this notion, or to identify the impact of different decisions at different levels (Neumann 2001). A number of studies have indicated that comprehensive company-level initiatives with ergonomic objectives, including, for example, job rotation, workstation redesign, and more breaks, can lead to a reduction in musculoskeletal disorders in industries as well as in office environments (Ohara et al. 1976, Itani et al. 1979, Silverstein et al. 1994). It has been suggested that the effectiveness of these interventions relies on their integration into the formal and informal control systems of the company (Westgaard and Winkel 1997). Thus, interventions initiated and driven by external experts are less likely to be successful and sustainable (Winkel and Westgaard 1996, Westgaard and Winkel 1997, Langaa Jensen 2002).

At the individual and group levels, autonomy described the latitude for designing products and production according to discretion. A definition of autonomy at the company level would focus on the extent to which management is constrained by external factors in their decision making. Thus, a number of issues, some of which occur frequently even in the public debate, relate to company autonomy. Examples with an obvious association to ergonomics include rules and regulations on work environment and terms of employment, prevailing conditions in the market, and available production technology. As with individual and group autonomy, modifying factors may promote or obstruct the practice of a formal autonomy.

11.2.14   A General Model of Variation and Autonomy

Factors supporting posture variation in the jobs of individual workers are summarized in Figure 11.7. In line with the previous discussion, this variation is suggested to be dependent on the available tasks as well as on the extent of formal autonomy and on factors influencing whether this autonomy is practiced. Important aspects of tasks, autonomy, and contextual promoters are present at the level of individuals, groups, and companies. At the individual level, the aspects of tasks and autonomy listed refer to the assignment of the particular worker. At the group level, they represent factors across a number of individuals. For example, autonomy with respect to work/break schedule at the individual level concerns the worker's right to decide his or her own workday. The equivalent group-level term includes planning of presence and absence of the collective of workers, which could in fact obstruct the autonomy of the individual.

The figure expresses the authors' advice to company stakeholders, including occupational health staff, concerning factors to be aware of when designing a production facility that will meet the ergonomic quality demands of the future.

Thus, good jobs are most likely to occur in companies where the available work comprises substantial proportions of tasks that differ in exposure and that can be carried out using flexible methods. This, in turn, requires that companies practice their autonomy as regards, for example, manufacturing strategies, incentive systems, and within-company departmentalization in a way that allows the exposure latitude to be fully used. Groups of workers

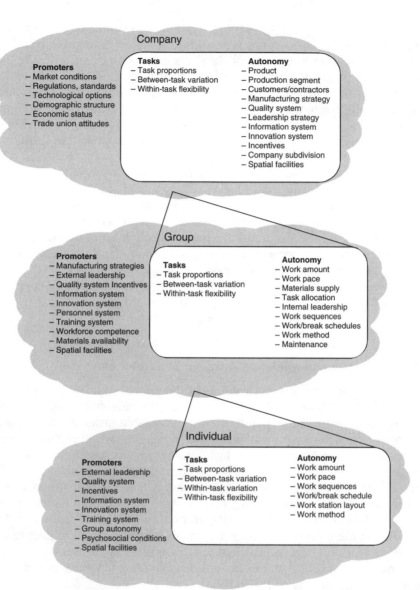

FIGURE 11.7

Factors at company, group, and individual levels that influence whether the job of an individual worker will be varying with respect to physical loads in general, and postures in particular. At each level, the figure shows factors associated with the available tasks, factors associated with aspects of autonomy, and contextual factors, which can act as promoters when formal autonomy is to be practiced.

in the company should also preferably be assigned a selection of tasks with varying loads and postures, and they should be given autonomy with respect to, for example, how tasks are distributed among workers in the group and which work methods could be used to perform the tasks. Important promoters at the company supporting the realization of formal autonomy are, for example, systems for supply of materials, training of new workers, and the setting of salaries. The individual worker should be given a technical autonomy, for example, as regards workstation layout, as well as an administrative autonomy, for example, with respect to the timing of work and breaks, but should also be offered work tasks allowing for posture variation within and between workdays. A job situation with good variation and autonomy requires that systems for incentives and competence development, just to take two examples, promote the realization of formal autonomy, and also requires that the psychosocial work environment be supportive.

The authors believe that a high ergonomic quality is best obtained by establishing a large exposure latitude at the company and group levels, that is, a wide variety of tasks in combination with control systems ensuring that groups and individuals have substantial freedom to structure their own work. They also believe that this is compatible with high demands of productivity, quality, and effectiveness, and in certain aspects even a prerequisite for successful production.

Summary

This section has discussed the scientific evidence that variation in posture is an important determinant of risk for musculoskeletal disorders. It suggested principles for assessing variation and emphasized that posture variation in the job of an individual is a consequence of the variation within and between different tasks in the job, including nonwork activities. Thus, variation is, to a large extent, dependent on the work organization, including break allowances and the distribution of available tasks between workers. In addition, variation is influenced by the extent of autonomy for individual workers and groups, i.e., the latitude to make decisions concerning technical and administrative issues. A production system with a high ergonomic quality is suggested to be characterized by containing a variety of tasks with differing posture patterns, in combination with control systems encouraging individuals and groups to use the possibilities offered to change between tasks.

References

ACGIH, 2000. Statement on work related musculoskeletal disorders – hand activity level, in 2000 Threshold Limit Values for Chemical Substances in the Work Environment, Cincinnati, OH: American Conference of Governmental Industrial Hygienists, 117–121.

Åkesson, I., Hansson, G.-Å., Balogh, I., Moritz, U., and Skerfving, S., 1997. Quantifying work load in neck, shoulders and wrists in female dentists. *International Archives of Occupational and Environmental Health*, 69, 461–474.

Allen, R.E., 1990. *The Concise Oxford Dictionary of Current English,* Oxford: Clarendon Press.

ANSI, 1995. Draft ANSI Z-365 Control of Work-Related Cumulative Trauma Disorders, Washington, D.C.: American National Standards Institute.

Ariëns, G.A., Van Mechelen, W., Bongers, P., Bouter, L.M., and Van der Wal, G., 2000. Physical risk factors for neck pain. *Scandinavian Journal of Work, Environment & Health,* 26, 7–19.

Armstrong, T.J., Buckle, P., Fine, L.J., Hagberg, M., Jonsson, B., Kilbom, A., Kuorinka, I., Silverstein, B.A., Sjøgaard, G., and Viikari-Juntura, E., 1993. A conceptual model for work-related neck and upper-limb musculoskeletal disorders. *Scandinavian Journal of Work, Environment & Health,* 19, 73–84.

Asmussen, E. and Mazin, B., 1978. A central nervous component in local muscular fatigue. *European Journal of Applied Physiology,* 38, 9–15.

Asundi, P., 1995. Rest reminders and multimedia. *Ergonomics,* 38, 2131–2133.

Bailey, D.E. and Adiga, S., 1997. Measuring manufacturing work group autonomy. *IEEE Transactions on Engineering Management,* 44, 158–174.

Bao, S., Mathiassen, S.E., and Winkel, J., 1995. Normalizing upper trapezius EMG amplitude: comparison of different procedures. *Journal of Electromyography and Kinesiology,* 5, 251–257.

Bao, S., Mathiassen, S.E., and Winkel, J., 1996. Ergonomic effects of a management-based rationalization in assembly work — a case study. *Applied Ergonomics,* 27, 89–99.

Barnes, R.M., 1968. *Motion and Time Study. Design and Measurement of Work,* 6th ed., New York: Wiley.

Bechtold, S.E., Janaro, R.E., and DeWitt, L.S., 1984. Maximization of labor productivity through optimal rest-break schedules. *Management Science,* 30, 1442–1458.

Bernard, B., 1997. Musculoskeletal Disorders and Workplace Factors. NIOSH Publication 97-141, 2nd ed., Cincinnati, OH: U.S. Department of Health and Human Services.

Beynon, C., Burke, J., Doran, D., and Nevill, A., 2000. Effects of activity–rest schedules on physiological strain and spinal load in hospital-based porters. *Ergonomics,* 43, 1763–1770.

Billat L.V., 2001. Interval training for performance: a scientific and empirical practice. Special recommendations for middle- and long-distance running. Part II: Anaerobic interval training. *Sports Medicine,* 31, 75–90.

Bongers, P., De Winter, C.R., Kompier, M.A., and Hildebrandt, V.H., 1993. Psychosocial factors at work and musculoskeletal disease. *Scandinavian Journal of Work, Environment & Health,* 19, 297–312.

Bongers, P., Kremer, A.M., and Ter Laak, J., 2002. Are psychosocial factors, risk factors for symptoms and signs of the shoulder, elbow, or hand/wrist? A review of the epidemiological literature. *American Journal of Industrial Medicine,* 41, 315–342.

Borg, G.A.V., 1982. A category scale with ratio properties for intermodal and interindividual comparison, in *Psychophysical Judgement and the Process of Perception,* Geissler, H.G. and Petzold, A., Eds., Berlin: VEB Deutscher Verlag der Wissenschaften, 25–34.

Borg, G.A.V., 1998. *Borg's Perceived Exertion and Pain Scales,* Leeds, U.K.: Human Kinetics.

Buckle, P. and Devereux, J., 1999. Work-Related Neck and Upper Limb Musculoskeletal Disorders. Report for European Agency for Safety and Health at Work, Luxembourg: Office for Official Publications of the European Communities.

Burdorf, A. and Sorock, G., 1997. Positive and negative evidence of risk factors for back disorders. *Scandinavian Journal of Work, Environment & Health*, 23, 243–256.

Burdorf, A. and Van der Beek, A.J., 1999. Exposure assessment strategies for work-related risk factors for musculoskeletal disorders. *Scandinavian Journal of Work, Environment & Health*, 25(Suppl. 4), 25–30.

Byström S., 1991. Physiological response and acceptability of isometric intermittent handgrip contractions, Arbete och Halsa 38, Stockholm: National Institute of Occupational Safety and Health.

Byström, S. and Fransson-Hall, C., 1994. Acceptability of intermittent handgrip contractions based on physiological response. *Human Factors*, 36, 158–171.

Byström, S. and Kilbom, Å., 1990. Physiological response in the forearm during and after isometric intermittent handgrip. *European Journal of Applied Physiology*, 60, 457–466.

Byström, S., Mathiassen, S.E., and Fransson-Hall, C., 1991. Physiological effects of micropauses in isometric handgrip exercise. *European Journal of Applied Physiology*, 63, 405–411.

Carayon, P., Smith, M.J., and Haims, M.C., 1999. Work organization, job stress, and work-related musculoskeletal disorders. *Human Factors*, 41, 644–663.

Carnahan, B.J., Redfern, M.S., and Norman, B., 2000. Designing safe job rotation schedules using optimization and heuristic search. *Ergonomics*, 43, 543–560.

CEN, 1995. *EN 614-1. Safety of Machinery — Ergonomic Design Principles*, Part 1: *Terminology and General Principles*, Brussels: European Committee for Standardization.

CEN, 2002. *EN 1005-3 Safety of Machinery. Human Physical Performance*, Part 3: *Recommended Force Limits for Machinery Operation*. Brussels: European Committee for Standardization.

Chaffin, D.B. and Andersson, G.B.J., 1991. *Occupational Biomechanics*, 2nd ed., New York: John Wiley & Sons.

Christensen, H., Søgaard, K., Pilegaard, M., and Olsen, H.M., 2000. The importance of the work/rest pattern as a risk factor in repetitive monotonous work. *International Journal of Industrial Ergonomics*, 25, 367–373.

Christmansson, M., 1997. Production Systems for Manual Repetitive Work. Effects of Autonomy and Variety of Work and Prevalence of Musculoskeletal Disorders in Upper Limbs, Ph.D. thesis, Gothenborg: Department of Operations Management and Work Organization, Chalmers University of Technology.

Christmansson, M., Falck, A.-C., Amprazis, J., Forsman, M., Rasmusson, L., and Kadefors, R., 2000. Modified method time measurements for ergonomic planning of production systems in the manufacturing industry. *International Journal of Production Research*, 38, 4051–4059.

Christmansson, M., Fridén, J., and Sollerman, C., 1999. Task design, psycho-social work climate and upper extremity pain disorders — effects of an organisational redesign on manual repetitive assembly jobs. *Applied Ergonomics*, 30, 463–472.

Christmansson, M., Medbo, L., Hansson, G.-Å., Ohlsson, K., Unge Byström, J., Möller, T., and Forsman, M., 2002. A case study of a principally new way of materials kitting — an evaluation of time consumption and physical workload. *International Journal of Industrial Ergonomics*, 30, 49–65.

Colombini, D., 1998. An observational method for classifying exposure to repetitive movements of the upper limbs. *Ergonomics*, 41, 1261–1289.

Colombini, D., Grieco, A., and Occhipinti, E., Eds., 1998. Occupational musculoskeletal disorders of the upper limbs due to mechanical overload. *Ergonomics*, Special Issue, 41(9), 1251–1397.

Colombini, D., Occhipinti, E., and Grieco, A., 2000. *La Valutazione e la Gestione del Rischio da Movimenti e Sforzi Ripetuti degli Arti Superiori [Assessment and Management of the Risk Connected to Repetitive Exertions of the Upper Limbs]*. Milan: Franco Angeli.

Colombini, D., Occhipinti, E., Delleman, N., Fallentin, N., Kilbom, Å., and Grieco, A., 2001. Exposure assessment of upper limb repetitive movements: a consensus document, in *International Encyclopaedia of Ergonomics and Human Factors*, Part 1, Karwowski, W., Ed., London: Taylor & Francis, 52–66.

Colombini, D., Occhipinti, E., and Grieco, A., 2002. *Risk Assessment and Management of Repetitive Movements and Exertions of Upper Limbs*, Oxford: Elsevier Science.

Dababneh, A.J., Swanson, N., and Shell, R.L., 2001. Impact of added rest breaks on the productivity and well being of workers. *Ergonomics*, 44, 164–174.

Dempsey, P., 1999. Utilizing criteria for assessing multiple-task manual materials handling jobs. *International Journal of Industrial Ergonomics*, 24, 405–416.

Drury, C.G., 1987. A biomechanical evaluation of the repetitive motion injury potential of industrial jobs. *Seminars on Occupational Medicine*, 2, 41–49.

Dul, J., Douwes, M., and Smitt, P., 1994. Ergonomic guidelines for the prevention of discomfort of static postures based on endurance data. *Ergonomics*, 37, 807–815.

Eastman Kodak Company, 1983. *Ergonomic Design for People at Work*. Vols. 1 and 2, New York: Van Nostrand Reinhold.

Elert, J. and Gerdle, B., 1989. The relationship between contraction and relaxation during fatiguing isokinetic shoulder flexions. An electromyographic study. *European Journal of Applied Physiology*, 59, 303–309.

Engström, T., Johansson, J.Å., Jonsson, D., and Medbo, L., 1995. Empirical evaluation of the reformed assembly work at the Volvo Uddevalla plant: psychosocial effects and performance aspects. *International Journal of Industrial Ergonomics*, 16, 293–308.

Engström, T., Jonsson, D., and Johansson, B., 1996a. Alternatives to line assembly: some Swedish examples. *International Journal of Industrial Ergonomics*, 17, 235–245.

Engström, T., Jonsson, D., and Medbo, L., 1996b. Production model discourse and experiences from the Swedish automotive industry. *International Journal of Operational Production Management*, 16, 141–158.

Enoka, R.M., 1995. Mechanisms of muscle fatigue: central factors and task dependency. *Journal of Electromyography and Kinesiology*, 5, 141–149.

Erikssen, G., 2001. Physical fitness and changes in mortality: the survival of the fittest. *Sports Medicine*, 31, 571–576.

Fallentin, N., Jørgensen, K., and Simonsen, E.B., 1993. Motor unit recruitment during prolonged isometric contractions. *European Journal of Applied Physiology*, 67, 335–341.

Fallentin, N., Viikari-Juntura, E., Waersted, M., and Kilbom, Å., 2001. Evaluation of physical workload standards and guidelines from a Nordic perspective. *Scandinavian Journal of Work, Environment & Health*, 27(Suppl. 2), 1–52.

Fernström, E.A.C. and Åborg, C.M., 1999. Alterations in shoulder muscle activity due to changes in data entry organisation. *International Journal of Industrial Ergonomics*, 23, 231–240.

Fisher, D.L., Andres, R.O., Airth, D., and Smith, S.S., 1993. Repetitive motion disorders: the design of optimal rate-rest profiles. *Human Factors*, 35, 283–304.

Forsman, M., Kadefors, R., Zhang, Q., Birch, L., and Palmerud, G., 1999. Motor-unit recruitment in the trapezius muscle during arm movements and in VDU precision work. *International Journal of Industrial Ergonomics*, 24, 619–630.

Forsman, M., Taoda, K., Thorn, S., and Zhang, Q., 2002. Motor-unit recruitment during long-term isometric and wrist motion contractions: a study concerning muscular pain development in computer operators. *International Journal of Industrial Ergonomics*, 30, 237–250.

Galinsky, T.L., Swanson, N.G., Sauter, S.L., Hurrell, J.J., and Schleifer, L.M., 2000. A field study of supplementary rest breaks for data-entry operators. *Ergonomics*, 43, 622–638.

Genaidy, A., Barkawi, H., and Christensen, D., 1994. Ranking of static non-neutral postures around the joints of the upper extremity and spine. *Ergonomics*, 38(9), 1851–1858.

Genaidy, A.M., Delgado, E., and Bustos, T., 1995. Active microbreak effects on musculoskeletal comfort ratings in meatpacking plants. *Ergonomics*, 38, 326–336.

Granata, K.P., Marras, W.S., and Davis, K.G., 1999. Variation in spinal load and trunk dynamics during repeated lifting exertions. *Clinical Biomechanics*, 14, 367–375.

Grant, A.K., Habes, D.J., and Putz-Anderson, V., 1994. Psychophysical and EMG correlates of force exertion in manual work. *International Journal of Industrial Ergonomics*, 13, 31–39.

Hackman, J.R. and Oldham, G.R., 1980. *Work Redesign*, Boston: Addison-Wesley.

Hagberg, M., 1981, Muscular endurance and surface electromyogram in isometric and dynamic exercise. *Journal of Applied Physiology*, 51, 1–7.

Hagberg, M. and Sundelin, G., 1986. Discomfort and load on the upper trapezius muscle when operating a word processor. *Ergonomics*, 29, 1637–1645.

Hagberg, M., Kilbom, Å., Buckle, P., Fine, L.J., Itani, T., Läubli, T., Riihimäki, H., Silverstein, B., Sjøgaard, G., Snook, S.H., Viikari-Juntura, E., and Kolare, S., 1993. Strategies for prevention of work-related musculoskeletal disorders: consensus paper. *International Journal of Industrial Ergonomics*, 11, 77–81.

Hagberg, M., Silverstein, B., Wells, R., Smith, M.J., Hendrick, H.W., Carayon, P., and Pérusse, M., 1995. *Work Related Musculoskeletal Disorders (WMSDS): A Reference Book for Prevention*, London: Taylor & Francis.

Hagberg, M., Punnett, L., Bergqvist, U., Burdorf, A., Härenstam, A., Kristensen, T.S., Lillienberg, L., Quinn, M., Smith, T.J., and Westberg, H., 2001. Broadening the view of exposure assessment. *Scandinavian Journal of Work, Environment & Health*, 27, 354–357.

Hansson, G.-Å., Balogh, I., Unge Byström, J., Ohlsson, K., Nordander, C., Asterland, P., Sjölander, S., Rylander, L., Winkel, J., Skerfving, S., and Malmö Shoulder-Neck Study Group, 2001. Questionnaire versus direct technical measurements for assessment of posture and movements of head, upper back, arms and hands. *Scandinavian Journal of Work, Environment & Health*, 27, 30–40.

Harber, P., Bloswick, D., Peña, L., Beck, J., Lee, J., and Baker, D., 1992. The ergonomic challenge of repetitive motion with varying ergonomic stresses. *Journal of Occupational Medicine*, 34, 518–528.

Hayes, R.H. and Wheelright, S.C., 1984. *Restoring Our Competitive Edge — Competing through Manufacturing*, New York: John Wiley & Sons.

Helander, M. and Nagamachi, M., 1992. *Design for Manufacturability — A Systems Approach to Concurrent Engineering and Ergonomics*, London: Taylor & Francis.

Henning, R.A., Sauter, S.L., Salvendy, G., and Kreig, E.F., 1989. Microbreak length, performance, and stress in a data entry task. *Ergonomics*, 32, 855–864.

Henning, R.A., Callaghan, E.A., Ortega, A.M., Kissel, G.V., Guttman, J.I., and Braun, H.A., 1996. Continuous feedback to promote self-management of rest breaks during computer use. *International Journal of Industrial Ergonomics*, 18, 71–82.

Henning, R.A., Jacques, P., Kissel, G.V., Sullivan, A.B., and Alteras-Webb, S.M., 1997. Frequent short rest breaks from computer work: effects on productivity and well-being at two field sites. *Ergonomics*, 40, 78–91.

Hignett, S. and McAtamney, L., 2000. Rapid entire body assessment (REBA). *Applied Ergonomics*, 31, 201–205.

Hinnen, U., Läubli, T., Guggenbühl, U., and Krueger, H., 1992. Design of checkout systems including laser scanners for sitting work posture. *Scandinavian Journal of Work, Environment & Health*, 18, 186–194.

Itani, T., Onishi, N., Sakai, K., and Shindo, H., 1979. Occupational hazard of female film rolling workers and effects of improved working conditions. *Archiv Za Higijenu Rada I Toksikologiju*, 30, 1243–1251.

Janaro, R.E. and Bechtold, S.E., 1985. A study of fatigue impact on productivity through optimal rest break scheduling. *Human Factors*, 27, 459–466.

Jansen, J.P., Burdorf, A., and Steyerberg, E., 2001. A novel approach for evaluating level, frequency and duration of lumbar posture simultaneously during work. *Scandinavian Journal of Work, Environment & Health*, 27, 373–380.

Johansson, A., Johansson, G., Lundqvist, P., Åkesson, I., Odenrick, P., and Akselsson, R., 1998. Evaluation of a workplace redesign of a grocery checkout system. *Applied Ergonomics*, 29, 261–266.

Jonsson, B., 1988. The static load component in muscle work. *European Journal of Applied Physiology*, 57, 305–310.

Jonsson, B.G., Persson, J., and Kilbom, Å., 1988. Disorders of the cervicobrachial region among female workers in the electronics industry. *International Journal of Industrial Ergonomics*, 3, 1–12.

Karasek, R. and Theorell, T., 1990. *Healthy Work. Stress, Productivity, and the Reconstruction of Working Life*, New York: Basic Books.

Karlsson, U., 1979. Alternativa Produktionssystem till Lineproduktion [Alternative Production Systems to Line Production], Ph.D. thesis, Gothenborg, Sweden: University of Gothenborg [in Swedish].

Kilbom, Å., 1994a. Repetitive work of the upper extremities. Part I: Guidelines for the practitioner. *International Journal of Industrial. Ergonomics*, 14, 51–57.

Kilbom, Å., 1994b. Repetitive work of the upper extremity: Part II: The scientific basis (knowledge base) for the guide. *International Journal of Industrial Ergonomics*, 14, 59–86.

Kilbom, Å. and Persson, J., 1987. Work technique and its consequences for musculoskeletal disorders. *Ergonomics*, 30, 273–279.

Kjellberg, K., Lindbeck, L., and Hagberg, M., 1998. Method and performance: two elements of work technique. *Ergonomics*, 41, 798–816.

Klein, J.A., 1991. A re-examination of autonomy in light of new manufacturing practices. *Human Relations*, 44, 21–38.

Konz, S., 1998a. Work/rest. Part I: Guidelines for the practitioner. *International Journal of Industrial Ergonomics*, 22, 67–71.

Konz, S., 1998b. Work/rest. Part II: The scientific basis (knowledge base) for the guide. *International Journal of Industrial Ergonomics*, 22, 73–99.

Kuijer, P.P.F.M., 2002. Effectiveness of Interventions to Reduce Workload in Refuse Collectors, Ph.D. thesis, Amsterdam: Academic Medical Centre, University of Amsterdam.

Kuijer, P.P.F.M., Visser, B., and Kemper, H.C.G., 1999. Job rotation as a factor in reducing physical workload at a refuse collecting department. *Ergonomics*, 42, 1167–1178.

Landsbergis, P.A., Scnall, P. and Cahill, J., 1999. The impact of lean production and related new systems of work organisation on worker health. *Journal of Occupational Health Psychology*, 4, 108–130.

Langaa Jensen, P., 2002. Human factors and ergonomics in the planning of production. *International Journal of Industrial Ergonomics*, 29, 121–131.

Laring, J., Forsman, M., Kadefors, R., and Örtengren, R., 2002. MTM-based ergonomic workload analysis. *International Journal of Industrial Ergonomics*, 30, 135–148.

Latko, A., Armstrong, T.J., Foulke, J.A., Herrin, G.D, Rabourn, R.A., and Ulin, S.S., 1997. Development and evaluation of an observational method of assessing repetition in hand tasks. *American Industrial Hygienist Association Journal*, 58, 278–285.

Lund, J. and Mericle, K.S., 2000. Determining fatigue allowances for grocery order selectors. *Applied Ergonomics*, 31, 15–24.

Mathiassen, S.E., 1989. Influence of angular velocity and movement frequency on development of fatigue in repeated isokinetic knee extensions. *European Journal of Applied Physiology*, 59, 80–88.

Mathiassen, S.E., 1993a. The influence of exercise/rest-schedule on the physiological and psychophysical response to isometric shoulder–neck exercise. *European Journal of Applied Physiology*, 67, 528–539.

Mathiassen, S.E., 1993b. Variation in shoulder-neck activity – physiological, psychophysical and methodological studies of isometric exercise and light assembly work, Arbete och Hälsa 7, Stockholm: National Institute of Occupational Health.

Mathiassen, S.E. and Aminoff, T., 1997. Motor control and cardiovascular responses during isoelectric contractions of the upper trapezius muscle: evidence for individual adaptation strategies. *European Journal of Applied Physiology*, 76, 434–444.

Mathiassen, S.E. and Turpin-Legendre, E., 1998, Reduction of shoulder elevation fatigue by periods of increased load, in *Proceedings of the PREMUS-ISEOH'98 conference*, Helsinki: Finnish Institute of Occupational Health, 2.

Mathiassen, S.E. and Winkel, J., 1991. Quantifying variation in physical load using exposure-vs-time data. *Ergonomics*, 34, 1455–1468.

Mathiassen, S. and Winkel, J., 1995. Assessment of "repetitive work," in *Proceedings from the 2nd International Conference on Prevention of Musculoskeletal Disorders (PREMUS)*, Montreal, 232–234.

Mathiassen, S.E. and Winkel, J., 1996. Physiologic comparison of three interventions in light assembly work: reduced work pace, increased break allowance and shortened working days. *International Archives of Occupational and Environmental Health*, 68, 94–108.

Mathiassen, S.E. and Winkel, J., 1997. Ergonomic exposure assessment adapted to production system design, in *Proceedings of the 13th Triennial Congress of the International Ergonomics Association*, Seppälä, P., Luopajärvi, T., Nygård, C.-H., and Mattila, M., Eds., Helsinki: Finnish Institute of Occupational Health, 195–197.

Mathiassen, S.E., Hansson, G.-Å., and Balogh, I., 2001. Individual differences in motor consistency when repeating a simple, stereotyped task, in *Proceedings of the Fourth International Scientific Conference on Prevention of Work-Related Musculoskeletal Disorders (PREMUS)*, Amsterdam, 245.

Mathiassen, S.E., Burdorf, A., and Van der Beek, A.J., 2002. Statistical power and measurement allocation in ergonomic intervention studies assessing upper trapezius EMG amplitude. A case study of assembly work. *Journal of Electromyography and Kinesiology*, 12, 27–39.

Mathiassen, S.E., Burdorf, A., Van der Beek, A.J., and Hansson, G.-Å., 2003a. Efficient one-day sampling of mechanical job exposure data — a study based on upper trapezius activity in cleaners and office workers. *American Industrial Hygiene Association Journal,* 64, 146–211.

Mathiassen, S.E., Möller, T., and Forsman, M., 2003b. Variability in mechanical exposure within and between individuals performing a highly constrained industrial work task. *Ergonomics,* 46, 800–824.

McAtamney, L. and Corlett, E.N., 1993. RULA: a survey method for the investigation of work-related upper limb disorders. *Applied Ergonomics,* 24, 91–99.

McGehee, W. and Owen, E.B., 1940. Authorized and unauthorized rest pauses in clerical work. *Journal of Applied Psychology,* 24, 605–613.

McLean, L., Tingley, M., Scott, R.N., and Rickards, J., 2001. Computer terminal work and the benefit of microbreaks. *Applied Ergonomics,* 32, 225–237.

Milner, N.P., Corlett, E.N., and O'Brien, C., 1986, A model to predict recovery from maximal and submaximal isometric exercise, in *The Ergonomics of Working Postures,* Corlett, N., Manenica, I., and Wilson, J., Ed., London: Taylor & Francis, 126–135.

Mital, A., Asfour, S., and Aghazadeh, F., 1987. Limitations of MTM in accurate determination of work standards for physically demanding jobs, in *Trends in Ergonomics — Human Factors IV,* Asfour, S., Ed., Amsterdam: Elsevier Science, 979–985.

Mital, A., Bishu, R.R., and Manjunath, S.G., 1991. Review and evaluation of techniques for determining fatigue allowances. *International Journal of Industrial Ergonomics,* 8, 165–178.

Möller, T., Mathiassen, S.E., Franzon, H., and Kihlberg, S., 2004. Job enlargement and mechanical exposure variability in cyclic assembly work. *Ergonomics,* 47, 19–40.

Moore, A.E., 2000. Effect of cycle time and duty cycle on muscle activity during a repetitive manual task, in *Proceedings of the 14th Triennial Congress of the International Ergonomics Association,* Santa Monica, CA: Human Factors and Ergonomics Society, 461–464.

Moore, J.S. and Garg, A., 1995. The strain index: a proposed method to analyze jobs for risk of upper extremity disorders. *American Industrial Hygiene Association Journal,* 56, 443–458.

Moore, A.E. and Wells, R., 1992. Towards a definition of repetitiveness in manual tasks, in *Computer Applications in Ergonomics, Occupational Safety and Health,* Mattila, M. and Karwowski, W., Eds., New York: Elsevier, 401–408.

Neumann, W.P., 2001. On Risk Factors for Musculoskeletal Disorder and Their Sources in Production System Design, Licenciate thesis, Lund: Department of Design Sciences, Lund University.

Neumann, W.P., Kihlberg, S., Medbo, P., Mathiassen, S.E., and Winkel, J., 2002. A case study evaluating the ergonomic and productivity consequences of partial automation strategies in the electronics industry. *International Journal of Production Research,* 40, 4059–4075.

NIOSH, 1997. Musculoskeletal Disorders and Workplace Factors. A Critical Review of Epidemiologic Evidence for WMSDs of the Neck, Upper Extremity and Low Back, Cincinnati, OH: National Institute for Occupational Safety and Health.

Nordander, C., Ohlsson, K., Balogh, I., Rylander, L., Pålsson, B., and Skerfving, S., 1999. Fish processing work: the impact of two sex dependent exposure profiles on musculoskeletal health. *Occupational and Environmental Medicine,* 56, 256–264.

Nordander, C., Hansson, G.-Å., Rylander, L., Asterland, P., Unge Byström, J., Ohlsson, K., Balogh, I., and Skerfving, S., 2000. Muscular rest and gap frequency as EMG measures of physical exposure: the impact of work tasks and individual related factors. *Ergonomics*, 43, 1904–1919.

Norman, R., Wells, R., Neumann, P., Frank, J., Shannon, H., Kerr, M., and OUBPS Group, 1998. A comparison of peak vs cumulative physical work exposure risk factors for the reporting of low back pain in the automotive industry. *Clinical Biomechanics*, 13, 561–573.

Nussbaum, M.A., 2001. Static and dynamic myoelectric measures of shoulder muscle fatigue during intermittent dynamic exertions of low to moderate intensity. *European Journal of Applied Physiology*, 85, 299–309.

Occhipinti, E. and Colombini, D. 2003. Risk assessment of upper limbs repetitive movements: overview of OCRA methods and new criteria for OCRA index classification, in *Proceedings of the 27th ICOH Conference* (SPS 61.1), Iguasu Falls, Brazil, 23–28 February 2003.

Ohara, H., Aoyama, H., and Itani, Y., 1976. Health hazard among cash register operators and the effects of improved working conditions. *Journal of Human Ergology*, 5, 31–40.

Ohlsson, K., Hansson, G.-Å., Balogh, I., Strömberg, U., Pålsson, B., Nordander, C., Rylander, L., and Skerfving, S., 1994. Disorders of the neck and upper limbs in women in the fish processing industry. *Journal of Occupational and Environmental Medicine*, 54, 826–832.

Ólafsdottir, H. and Rafnsson, V., 1998. Increase in musculoskeletal symptoms of upper limbs among women after introduction of the flow-line in fish-fillet plants. *International Journal of Industrial Ergonomics*, 21, 69–77.

Op de Beeck, R. and Hermans, V., 2000. Work Related Low Back Disorders, Luxembourg: European Agency for Safety and Health at Work, issue 204.

Östlin, P., 1989. The "health-related selection effect" on occupational morbidity rates. *Scandinavian Journal of Social Medicine*, 17, 265–270.

Parenmark, G., Malmkvist, A.-K., and Örtengren, R., 1993. Ergonomic moves in an engineering industry: effects on sick leave frequency, labor turnover and productivity. *International Journal of Industrial Ergonomics*, 11, 291–300.

Pottier, M., Lille, F., Phuon, M., and Monod, H., 1969. Étude de la contraction statique intermittente. *Le Travail Humain*, 32, 271–284.

Price, A.D.F., 1990a. Calculating relaxation allowances for construction operatives. Part 1: Metabolic cost. *Applied Ergonomics*, 21, 311–317.

Price, A.D.F., 1990b. Calculating relaxation allowances for construction operatives. Part 2: Local muscle fatigue. *Applied Ergonomics*, 21, 318–324.

Proper, K., Staal, B.J., Hildebrandt, V.H., Van der Beek, A.J., and Van Mechelen, W., 2002. Effectiveness of physical activity programs at worksites with respect to work-related outcomes. *Scandinavian Journal of Work, Environment & Health*, 28, 75–84.

Punnett, L., 1996. Adjusting for the healthy worker selection effect in cross-sectional studies. *International Journal of Epidemiology*, 25, 1068–1076.

Putz-Anderson, V., 1988. *Cumulative Trauma Disorders — A Manual for Musculoskeletal Disease of the Upper Limbs*, London: Taylor & Francis.

Radwin, R.G., Lin, M.L., and Yen, T.Y., 1994. Exposure assessment of biomechanical stress in repetitive manual work using frequency-weighted filters. *Ergonomics*, 37, 1984–1998.

Rohmert, W., 1960, Ermittlung von Erholungspausen für statische Arbeit des Menschen [Determining rest allowances in human static work]. *Internationale Zeitschrift für Angewandte Physiologie, Einschliessend Arbeitsphysiologie,* 18, 123–164.

Rohmert, W., 1973a. Problems in determination of rest allowances. Part 2: Determining rest allowances in different human tasks. *Applied Ergonomics,* 4, 158–162.

Rohmert, W., 1973b. Problems in determining rest allowances. Part 1: Use of modern methods to evaluate stress and strain in static muscular work. *Applied Ergonomics,* 4, 91–95.

Rohmert, W. and Luczak, H., 1973. Ergonomische Untersuchung von Teilzeit-Schichtsystemen und Pausen bei informatorischer Arbeit [An ergonomic investigation of part-time shift systems and breaks in information tasks]. *Internationale Archive für Arbeitsmedizin,* 31, 171–191.

Roquelaure, Y., Mechali, S., Dano, C., Fanello, S., Benetti, F., Bureau, D., Mariel, J., Martin, Y.-H., Derrienic, F., and Penneau-Fontbonne, D., 1997. Occupational and personal risk factors for carpal tunnel syndrome in industrial workers. *Scandinavian Journal of Work, Environment & Health,* 23, 364–369.

Rose, L., Ericson, M., Glimskär, B., Nordgren, B., and Örtengren, R., 1992. Ergo-index. Development of a model to determine pause needs after fatigue and pain reactions during work, in *Computer Applications in Ergonomics, Occupational Safety and Health,* Mattila, M. and Karwowski, W., Ed., Amsterdam: Elsevier Science, 461–468.

Seth, V., Weston, R.L., and Freivalds, A., 1999. Development of cumulative trauma disorder risk assessment model for the upper extremities. *International Journal of Industrial Ergonomics,* 23, 281–291.

Sihvonen, T., Baskin, K., and Hänninen, O., 1989. Neck-shoulder loading in word-processor use. *International Archives of Occupational and Environmental Health,* 61, 229–233.

Silverstein, B., Armstrong, T., Longmate, A., and Woody, D., 1988. Can in-plant exercise control musculoskeletal symptoms. *Journal of Occupational Medicine,* 30, 922–927.

Silverstein, B.A., Fine, L.J., and Armstrong, T.J., 1986. Hand-wrist cumulative trauma disorders in industry. *British Journal of Industrial Medicine,* 43, 779–784.

Silverstein, B.A., Fine, L.J., and Armstrong, T.J., 1987. Occupational factors and carpal tunnel syndrome. *American Journal of Industrial Medicine,* 11, 343–58.

Silverstein, B.A., Hughes, R., Burt, J., and Kaufman, J., 1994. Reducing musculoskeletal disorders in data entry operators, in *Proceedings of the 12th Triennial Congress of the International Ergonomics Association,* McFadden, S., Innes, L., and Hill, M., Eds., Toronto: Human Factors Association of Canada, 217–218.

Sjøgaard, G., 1996. Potassium and fatigue: the pros and cons. *Acta Physiologica Scandinavica,* 156, 257–264.

Sjøgaard, G., Kiens, B., Jørgensen, K., and Saltin, B., 1986. Intramuscular pressure, EMG and blood flow during low-level prolonged static contraction in man. *Acta Physiologica Scandinavica,* 128, 475–484.

Sluiter, J.K., Rest, K.M., and Frings-Dresen, M.H.W., 2001. Criteria document for evaluating the work-relatedness of upper-extremity musculoskeletal disorders. *Scandinavian Journal of Work, Environment & Health,* 27(Suppl. 1), 1–102.

Sundelin, G., 1993. Patterns of electromyographic shoulder muscle fatigue during MTM-paced repetitive arm work with and without pauses. *International Archives of Occupational and Environmental Health,* 64, 485–493.

Sundelin, G. and Hagberg, M., 1989. The effects of different pause types on neck and shoulder EMG activity during VDU work. *Ergonomics*, 32, 527–537.

Swedish National Board of Occupational Safety and Health, 1998, Ergonomics for the prevention of musculoskeletal disorders. AFS 1998:1, Stockholm: Swedish National Board of Occupational Safety and Health.

Tanaka, J. and McGlothlin, J.D., 1993. A conceptual quantitative model for prevention of work related carpal tunnel syndrome. *International Journal of Industrial Ergonomics*, 11, 181–193.

Thorn, S., Forsman, M., Zhang, Q., and Taoda, K., 2002. Low-threshold motor unit activity during a 1-h static contraction in the trapezius muscle. *International Journal of Industrial Ergonomics*, 30, 225–236.

Torén, A. and Öberg, K., 2001. Change in twisted trunk postures by the use of saddle seats — a conceptual study. *Journal of Agricultural Engineering Research*, 781, 25–34.

Torgén, M., Nygård, C.-H., and Kilbom, Å., 1995. Physical work load, physical capacity and strain among elderly female aides in home-care service. *European Journal of Applied Physiology*, 71, 444–452.

Vahtera, J., Kivimäki, M., and Pentti, J., 1997. Effect of organisational downsizing on health of employees. *Lancet*, 350, 1124–1128.

Van der Beek, A.J. and Frings-Dresen, M.H.W., 1998. Assessment of mechanical exposure in ergonomic epidemiology. *Occupational and Environmental Medicine*, 55, 291–299.

Van der Windt, D.A.W.M., Thomas, E., Pope, D.P., De Winter, A.F., MacFarlane, G.J., Bouter, L.M., and Silman, A.J., 2000. Occupational risk factors for shoulder pain: a systematic review. *Occupational and Environmental Medicine*, 57, 433–442.

Van Dieën, J.H. and Oude Vrielink, H.H.E., 1998. Evaluation of work-rest schedules with respect to the effects of postural workload in standing work. *Ergonomics*, 41, 1832–1844.

Van Dieën, J.H., Oude Vrielink, H.H.E., and Toussaint, H., 1993. An investigation into the relevance of the pattern of temporal activation with respect to the erector spinae muscle's endurance. *European Journal of Applied Physiology*, 66, 70–75.

Van Dieën, J.H., Hoozemans, M.J.M., Van der Beek, A.J., and Mullender, M., 2002. Precision of estimates of mean and peak spinal loads in lifting. *Journal of Biomechanics*, 35, 979–982.

Veiersted, K.B., 1994, Sustained muscle tension as a risk factor for trapezius myalgia. *International Journal of Industrial Ergonomics*, 14, 333–339.

Victorian Occupational HSC, 1988. Draft Code of Practice. Occupational Overuse Syndrome, Melbourne, Australia: Victorian Occupational HSC.

Waters, T.R., Putz-Anderson, V., Garg, A., and Fine, L.J., 1993. Revised NIOSH equation for the design and evaluation of manual lifting tasks. *Ergonomics*, 36, 749–776.

Westad, C. and Westgaard, R.H., 2001. Reducing the risk of motor unit overexertion: methods of inducing motor unit substitution, in *Proceedings of the Fourth International Scientific Conference on Prevention of Work-related Musculoskeletal Disorders (PREMUS)*, Amsterdam, 61.

Westgaard, R.H., 1999. Effects of physical and mental stressors on muscle pain. *Scandinavian Journal of Work, Environment & Health*, 25(Suppl 4), 19–24.

Westgaard, R.H. and De Luca, C.J., 1999. Motor unit substitution in long-duration contractions of the human trapezius muscle. *Journal of Neurophysiology*, 82, 501–504.

Westgaard, R.H. and Winkel, J., 1997. Ergonomic intervention research for improved musculoskeletal health: a critical review. *International Journal of Industrial Ergonomics,* 20, 463–500.

Westgaard, R.H., Vasseljen, O., and Holte, K.A., 2001. Trapezius muscle activity as a risk factor for shoulder and neck pain in female service workers with low biomechanical exposure. *Ergonomics,* 44, 339–353.

Winkel, J., 1985. On Foot Swelling during Prolonged Sedentary Work and the Significance of Leg Activity, Ph.D. thesis, Stockholm: National Board of Occupational Health and Safety.

Winkel, J. and Mathiassen, S.E., 1994. Assessment of physical work load in epidemiologic studies: concepts, issues and operational considerations. *Ergonomics,* 37, 979–988.

Winkel, J. and Oxenburgh, M., 1990. Towards optimizing physical activity in VDT/ office work, in *Promoting Health and Productivity in the Computerized Office: Models of Successful Ergonomic Intervention,* Sauter, S.L., Dainoff, M., and Smith, M., Eds., London: Taylor & Francis, 94–117.

Winkel, J. and Westgaard, R., 1992. Occupational and individual risk factors for shoulder-neck complaints. Part II: The scientific basis (literature review) for the guide. *International Journal of Industrial Ergonomics,* 10, 85–104.

Winkel, J. and Westgaard, R.H., 1996. A model for solving work related musculoskeletal problems in a profitable way. *Applied Ergonomics,* 27, 71–77.

Wood, D.D., Fisher, D.L., and Andres, R.O., 1997. Minimizing fatigue during repetitive jobs: optimal work-rest schedules. *Human Factors,* 39, 83–101.

12

Force Exertion

Christine M. Haslegrave

CONTENTS

Introduction

The exertion of force is a component of many jobs, including such diverse tasks as repetitive carrying and stacking of heavy sacks, impact loading on crowbars or other tools to release rusted fixings during maintenance work, pushing open a heavy door, opening a screw top container, delicate manipulations on teeth during dental work, among many others. Where significant force is involved, it is a major contributor to musculoskeletal stress and the risk of injury. The muscular effort required in a forceful task is easily apparent in Figure 12.1. Much of the research on this topic has been concerned with manual materials handling, and particularly with lifting and carrying, but

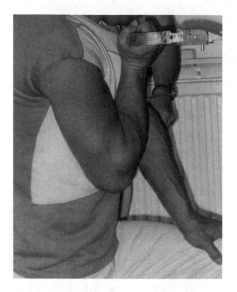

FIGURE 12.1
Exerting a maximal isometric lifting force.

the range of forceful actions at work is very much broader than that. Although precision operations and the use of tools are among these, they are addressed more fully in Chapter 10. However, the general principles discussed below apply to any force exertion.

The examples of tasks quoted above show that very different groups of muscles may be involved. While some of the tasks involve whole-body exertion, others (such as dental work) mainly involve small and relatively weak muscles for exerting the force, although other groups of muscles may be used as stabilizers to hold the arm and body steady while the force is exerted. The consequences of the exertion in terms of performance (effectiveness, physiological efficiency, or accuracy), fatigue, strain, and risk of injury will depend on the degree of loading on each of the muscles involved *relative to* the maximum capacity of that muscle. Thus, the pushing open of a door, while requiring significant force, may be well within the maximum strength of the muscles of the arm and shoulder but dental probing, although requiring much lower forces, may still involve a maximal exertion for the muscles of the dentist's fingers. It also must be remembered that other parts of the body (with stronger muscles) are sometimes used for force exertion instead of the hands and arms, as when someone pushes a door open with the shoulder or back.

The level of force exerted and the loading on the muscles involved therefore depend on both task and technique adopted and, as a result, posture is crucial to performance and strain during force exertion. To understand these relationships, it is necessary to explain some basic principles of muscle physiology and of the biomechanics of the musculoskeletal system. Variation in posture can be used to alter the mechanical advantage of the leverage

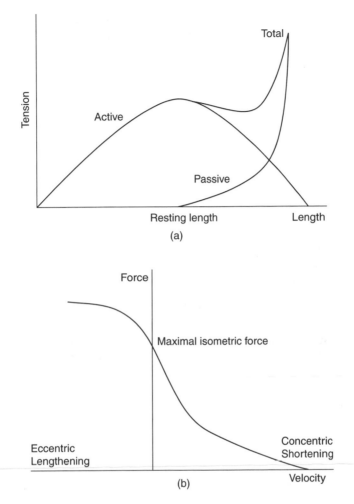

FIGURE 12.2
Muscle characteristics in force exertion: (a) variation in muscle tension with length; (b) variation in muscle force with velocity.

applied by the muscles acting around any particular joint in the body. However, muscle length (and thus its force output as shown in Figure 12.2a) also varies with joint position, so that the maximum torque that can be generated varies throughout the range of movement about the joint (as can be seen for measurements of back extensor strength in Figure 12.3). Joint positions for maximum muscle tension and maximum mechanical advantage rarely coincide. According to Williams and Stutzman (1959), most isometric (static) joint torques show an increase in strength as the joint angle increases and the muscle elongates (as for the back extensor muscle strength shown in Figure 12.3) but exceptions are elbow flexion, shoulder adduction, and knee extension. These are all important for common work tasks and, for these, exertions are likely to be stronger when the joint is in the mid-range of its

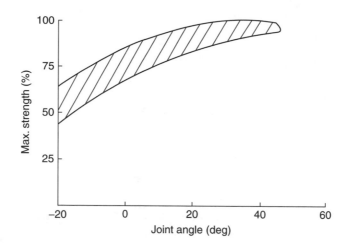

FIGURE 12.3
Maximum back extensor strength as a function of trunk angle (based on data from Svensson 1987). (From Tracy, M.F., in *Evaluation of Human Work*, 2nd ed., Wilson, J.R. and Corlett, E.N., Eds., London: Taylor & Francis, 1995, 714–748. With permission from Taylor & Francis (http://www.tandf.co.uk).)

movement. Data quoted by Williams and Stutzman (1959) and by Svensson (1987) suggest that shoulder flexion torque is relatively level with the upper arm angle between 30° and 90° forward from the vertical, although it is strongest when the arm is extended slightly rearward. Shoulder extension torque is strongest between 30° and 150°. Elbow flexion strength is maximal with the elbow joint angle at about 90°, but this varies with shoulder angle. A summary of these relationships and of the maximum voluntary joint torque strengths for major joints in the body can be found in Tracy (1995, p. 730).

Adoption of such an optimal joint posture, when it is possible, means that the muscle acting across the joint can exert the highest torque or, alternatively, can exert a given submaximal torque with the least muscle effort. Conversely, when a more extreme joint posture has to be adopted, the muscle has less effective leverage (and/or is working at a less effective length) and so has to produce a higher muscle tension to give the same torque output, resulting in increased demand on that muscle, more rapid fatigue, and increased risk of injury.

The type of exertion also has to be considered. A sustained force is held by isometric muscle contraction, where no movement occurs and the muscle fibers are under continuous contraction. In such a situation, muscle fatigue will limit the length of time for which the force can be maintained. In a dynamic force exertion, the muscle will normally exert the force through concentric contraction in which the muscle shortens, although counteracting or restraining forces (usually used, in the work situation, to provide precise control of a movement or to decelerate a moving load) can be exerted through eccentric contraction as the muscle lengthens. Both isometric and eccentric contractions may be used in maintaining the stability of a joint or even of the whole-body posture while a force is being exerted and in this case they are often referred to as co-contraction forces. Strength varies with the type of exertion, as shown for an individual muscle in Figure 12.2b. As a general rule, isometric strength is likely to be greater than concentric strength, although a high force can be generated by an impulsive concentric force exertion, particularly when body (or body segment) weight can be used to generate momentum in addition to the muscle force.

The analysis of dynamic strength is complex because it is velocity dependent, as shown by the muscle strength curve in Figure 12.2b, and because there are inertial effects due to acceleration of body segments. As a result, dynamic strength is highly task dependent and has received little study for most work activities. Two alternative ways of measuring dynamic strength are usually employed in laboratory studies to ensure reproducibility and repeatability; these tests are performed under either isotonic (constant force) or isokinetic (constant velocity) conditions, but neither reproduces exactly the conditions under which force is exerted in real work activities.

Although research has shown that dynamic strength can differ considerably from isometric strength, studies of isometric strength probably provide a good indication of the postures in which force can most effectively be exerted for most work tasks (at least for those that do not involve rapid movements) and of the negative effects imposed by workplace constraints on free choice of posture (such as may occur as a result of barriers restricting access). However, the isometric strength data should not be used to set design criterion limits for maximum force demands in dynamic tasks because this could lead to significant overloading (Kumar 1991).

Finally, it must be recognized that few work tasks demand maximal force exertions. Apart from studies of muscular endurance, as when a posture must be supported for an extended period of time, there have been only a few studies of postures adopted for submaximal exertions. It seems likely that posture has less effect on performance for submaximal than for maximal exertions, and that the posture adopted will be more variable the lower the force required (relative to the strength of the muscles involved). The main focus of concern in this chapter is therefore on postures adopted for tasks in which high forces have to be exerted and on those in which more moderate forces have to be sustained or exerted repetitively.

12.1 Postural Behavior

The above considerations lead to the conclusion that posture is critical in the exertion of a high force. It influences (1) the maximal level of torque that an individual muscle (or group of muscles) is capable of exerting about a joint in the given situation, (2) the degree to which muscles in other parts of the body are capable of supporting the body while performing the task, (3) the level of strain on any of the muscles involved in the exertion, and (4) the risk of injury due to loading on muscles, joints, tendons, ligaments, and other tissues.

Force exertion in any work task will involve many muscles, some acting as prime movers in generating the force and others acting to stabilize the joints in the rest of the body. For example, depressing the brake or accelerator pedal in a car will require plantarflexion of the ankle, extension of the knee, extension of the hip, and stabilization of the spinal joints to maintain the posture of the pelvis and trunk with respect to the seat. The force exerted on the pedal will be reacted against the seat back and seat surface and against the floor surface through contact with the other foot, so that forces will be transmitted through the trunk and both legs while the force is being exerted on the pedal. Given such a "kinetic chain" through which the force is transmitted across the musculoskeletal system, the limiting factor in the maximum force that can be exerted is most likely to be determined by the weakest link — the most highly stressed muscle (Dempster 1955, Grieve and Pheasant 1981, Rohmert et al. 1987). When a person is forced to adopt an awkward posture for a force exertion, it is likely that some of the muscles in the kinetic chain will be attempting to exert torque under less than optimal conditions, with either muscle length or moment arm sub-optimal, and there may be a risk of injury if the muscle force required is close to the capacity for that muscle.

With the complexity of such kinetic chains, it is obvious that there will be wide individual variations in strength and risk of injury due to anatomical variability (such as muscle geometry), differences in anthropometric (body segment) dimensions, and differences in physiological condition (muscle fiber composition, strength, and fitness). Nevertheless, common principles can be established for postural strategies that assist in allowing an individual to exert maximum force when performing a particular task — or, for submaximal tasks, to exert sufficient strength with least effort and lowest risk of injury. These principles can be used in forming guidelines for the design of tasks and workplaces when high forces must be exerted.

Measurements of strength have been made in many work situations and it is clear that strength varies with the general posture adopted — whether standing, sitting, or kneeling (or even lying prone or supine in constrained work situations). The effects of these broad types of posture are discussed in more detail later but, generally speaking, high forces should be exerted in a standing posture wherever possible. The strongest muscle groups of the thighs and lower trunk cannot be used when seated, and a kneeling posture is generally less stable that a standing posture. The other most influential factors are the direction in which the force must be exerted and the layout of the workplace (particularly the task height, reach distance, and presence of any constraints on the free choice of posture), although many other factors are important, as shown in Figure 12.4.

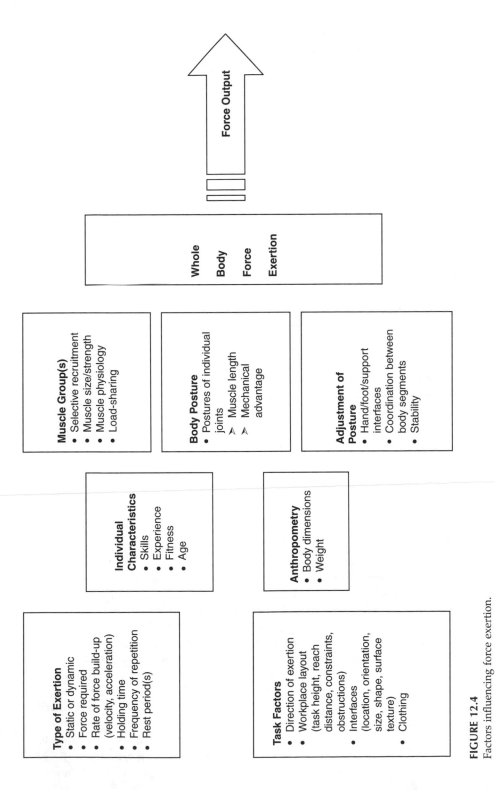

FIGURE 12.4

Factors influencing force exertion.

TABLE 12.1

Results of Rohmert's (1966) Measurements of the Variation in Single-Handed Strength with Direction of Force Exertion (for male subjects standing and with constrained foot positions, and with a vertical grip handle)

Direction of Exertion	Arm Angle (°) (to horizontal)	Force (as % pushing force at 0° arm angle and 100% maximum grip distance) Grip Distance (as % maximum)		
		50%	75%	100%
Push forward, horizontal	30	40	61	80
	0	75	88	100
	−30	70	76	80
	−60	70	80	90
Pull backward, horizontal	30	48	55	65
	0	57	65	72
	−30	70	75	78
	−60	57	70	85
Push to left, horizontal	30	88	76	60
	0	105	83	60
	−30	105	85	65
	−60	83	76	65
Push to right, horizontal	30	60	55	52
	0	76	62	50
	−30	83	67	55
	−60	62	57	52
Lift, vertical	30	70	60	48
	0	85	65	45
	−30	125	100	70
	−60	157	128	102
Press down, vertical	30	190	145	102
	0	140	100	83
	−30	88	83	76
	−60	97	90	80

12.1.1   Effects of Force Direction

The first consideration is probably the direction in which the force is to be exerted. This is clearly illustrated by the variation in strength measured for standing subjects by Rohmert (1966), and these results are summarized in Table 12.1. In general, exerting force in lateral directions (to left or right of the midsagittal plane of the body) should be avoided, because strength in these directions is weak (Pinder et al. 1995, Haslegrave et al. 1997a,b). Whether standing or sitting, the body is also less stable in controlling the twisting forces that result from trying to exert force sideways, so that additional muscle effort is expended in trying to hold the body steady.

Vertical (lifting/lowering) forces and pushing/pulling forces are much stronger than lateral exertions. Rohmert and Jenik measured single-handed strength in six orthogonal directions (lift, press, push, pull, and laterally to left and right) for a large range of hand positions, testing small groups of male and female subjects. Their results can found in Rohmert (1966) for male and in Rohmert and Jenik (1971) for female subjects. They concluded that vertical forces were the highest and pulling forces the lowest (apart from lateral exertions), while strength in pronation (inward rotation of the fore-arm) was generally greater than in supination (outward rotation). However, Rohmert and Jenik constrained the foot positions that their subjects were instructed to adopt (with feet side by side and 30 cm apart) and their con-clusions on the relative strengths of lifting, pressing, pushing, and pulling should not be generalized to other situations.

Pheasant and Grieve (Pheasant and Grieve 1981, Pheasant et al. 1982) explored the effect of direction of exertion in greater detail for two-handed exertions in the sagittal plane with measurements on groups of 10 to 20 men and women, testing directions of exertion intermediate between the six orthogonal directions tested by Rohmert and Jenik. They also systematically varied the foot positions that their subjects adopted. Pheasant and Grieve found a consistent peak force in the downward pressing direction, although this never exceeded body weight. They also found a peak in force exerted when simultaneously pushing upward and forward, approximately along the line between the feet and the hands. Adjustment of foot position could be used to improve strength in some orientations. For example, broadening the foot base increased the range of directions in which the high upward pushing force could be exerted. By contrast, pulling upward and backward was strongest when the feet could be positioned below or just in front of the handhold, when body weight could be used to generate a higher rearward torque, but it must be remembered that this posture increases the risk of foot slipping, because a narrow foot base offers little stability.

Pheasant and Grieve therefore identified two strategies that can be used to maximize strength (as shown by Figure 12.5): to use body weight with the most advantageous mechanical advantage (moment arm) and to use muscular strength to the best advantage by thrusting in a line between the hands and the point at which the force is reacted (which is the foot–ground interface for a free-standing person). To use these strategies, however, the feet must be positioned to support the postures and to provide the necessary stability. The ability to use the two strategies varies, of course, with the direction in which the force is exerted. A further requirement is that the handhold(s) provided must be appropriate for the direction in which the force is to be exerted.

Interestingly, Grieve and Pheasant (1981) found that force is often directed at an angle to the main axis of the thrust, and they showed that such off-axis components of a force exertion can serve to increase the magnitude of the desired component of the force. Once again, it is important that the handhold allow sufficient adjustment of grip and wrist angle to take advantage of this strategy. A corollary of this is that, to assess the risk of musculoskeletal injury in a particular task, the force demands should be measured in all three axes; it is not sufficient simply to measure the force in the desired direction of exertion.

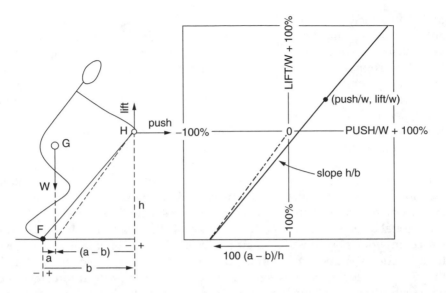

FIGURE 12.5

Strategies identified by Pheasant and Grieve (1981) for maximizing strength. The figure illustrates the terms used in their equation of static exertion and the representation of that equation on a postural stability diagram (where LIFT is the vertical component of force at the hands, PUSH is the horizontal component of force at the hands, W is body weight, h is the height of the bar H against which the force is exerted, b is the horizontal distance of the bar in front of the center of pressure of the feet, and a is the horizontal distance of the center of gravity G in front of the center of pressure). (From Pheasant, S.T. and Grieve, D.W., *Ergonomics*, 24(5), 327–338, 1981. With permission from Taylor & Francis (http://www.tandf.co.uk).)

12.1.2   Effects of Workplace Layout and Environment

The two most important aspects of the layout of a workplace are the height at which force is exerted and the distance of the hand from the body (reach distance) because these impose significant constraints on the posture that can be adopted. Rohmert and Jenik showed that horizontal pushing and pulling strengths tend to increase with reach distance, but that strength in other directions decreases (Rohmert 1966, Rohmert and Jenik 1971). These findings reflect the ability to use the postural strategies discussed above, but they can be considerably affected by other factors (some of which are discussed in Sections 12.1.3 and 12.1.4) and should not be regarded as universal principles. The optimum layouts for particular tasks are described in more detail in Section 12.2.

Workplace layout is important for dynamic exertions as well as isometric exertions. Imrhan and Ayoub (1990) found that isokinetic pull strength depended strongly on arm posture at the starting point of the pull and on velocity. The strongest starting position for their seated subjects was with the arm extended horizontally, which, they suggest, is a posture in which the shoulder extensor muscles are somewhat stretched and in which good mechanical advantage can be obtained. With the arm raised farther, the benefit of increased stretching of the shoulder extensors is negated by a reduced mechanical advantage, so that the pull that can be generated is not as strong. When the arm was below the horizontal, both mechanisms were less effective. Peak force decreased as the velocity of the pull increased. At each velocity, there appeared to be an optimum arm posture at the moment of peak force, although the biomechanical reasons are not yet clear. For example, at 15 cm/s the upper arm had moved through 43° from its starting position and the elbow angle was 99°, while at 45 cm/s the upper arm had moved through 59° and the elbow angle was 71°, with relatively little variation among their ten subjects.

If the layout of the workplace imposes even apparently minor constraints on the posture that can be adopted, strength can be severely limited. For example, foot position is very important (Chaffin et al. 1983, Daams 1993, Haslegrave et al. 1997a) as is the floor surface. Perhaps this should not be surprising, as the force exerted at the hands must be generated against a stable base and reacted against the foot–floor interface. The studies performed by Pheasant and Grieve (1981) showed that, for a stable posture, the center of gravity of the body must be within the base of support, which is usually defined by the positions of the feet. A broader base of support (or stability area) is achieved by separating the feet longitudinally or laterally, and this stance will be adjusted according to the direction in which the desired force is to be exerted. However, Holbein and Chaffin (1997) pointed out that the functional stability area is usually smaller than the theoretical base of support because it may be limited by muscle strength needed to maintain the posture, by internal postural control, or by the need to control an unstable external load.

Foot–floor friction is in fact often the limiting factor in determining maximum pushing or pulling strength (Kroemer and Robinson 1971). Conversely, the pushing of a heavy load creates a risk of foot slipping, especially for a person of light weight who will generate a relatively low force normal to the ground and so will not be able to generate high shearing forces at the foot to react against the pushing force and resist slipping. High friction floor surfaces and shoe soles are essential when high pushing or pulling forces are required in a job.

As with any skill, experienced workers adjust their posture according to the environmental conditions as well as according to the task. Lavender et al. (1998) observed the difference in posture for a pulling task between two floor conditions. On a non-slippery, dry surface their subjects leaned backward, extending their legs at the knee and their arms at the elbow, with considerable trunk flexion. In contrast, when the subjects were asked to perform the same task on a slippery, wet surface, they chose to adopt much more upright trunk postures.

Fothergill et al. (1992) demonstrated that the hand–handle interface can also be a weak link in the kinetic chain in addition to muscles at particular body joints. An inadequate handle can impede the transmission of strength, particularly when it imposes an awkward or stressful deviation at the wrist. The handles that gave the largest degree of hand–handle contact and that permitted a power grip to be employed enabled the largest pulling forces to be produced.

Another very common situation in the workplace is the presence of obstacles that restrict access, often increasing reach distance, as when having to lean across a barrier fence or stretch across a work surface to lift a component from a conveyor belt. As a *general* principle, any obstructions that increase reach distance will reduce the strength that can be exerted and will often at the same time increase the moment arm of the force, increasing the torque exerted on arm joints, trunk, and spine and therefore increasing the risk of musculoskeletal injury.

There are some exceptions to this principle that should be noted — some structures within the workplace may provide support, although only if they are strategically placed (and their effect on any individual will usually vary with their anthropometry). Kroemer and Robinson (1971) showed that pushing or pulling strength can be greatly enhanced if the foot, shoulder, or back can be braced against a rigid structure. Taking the lifting of components from a high-sided container as a simple example, it is possible to lean on the side of the container while stretching to pick up and lift the component with the other arm. If the container is rigid and relatively shallow, body weight can be supported and the component lifted without undue loading on the musculoskeletal system. If the box container is deep, the degree of support is immaterial and the component can only be lifted with the spine extremely flexed, thus risking injury. However, Ferguson et al. (2002) showed that container design can also influence the kinematics of a lift and that such lifts are often carried out while standing on one leg. Lifting style, container height, and position of the object within the bin all affect spinal loading and the interactions between these factors are complex due to "trade-offs" between the possibility of bracing or supporting part of the body weight on the side of the container and the biomechanical disadvantages of stooped postures, asymmetry (through one-handed lifting or standing on one leg), or long reach distances.

Taking a different situation of working overhead, strength in the vertical direction will be maximized with the arm extended (at a longer reach distance) because the elbow joint is braced (Haslegrave et al. 1997a). It can therefore be a distinct disadvantage to be too close to the task when working close to a ceiling or underneath a vehicle in a garage.

Another important factor in the layout of the workplace is whether the force can be exerted on a handhold in the midsagittal plane or whether the person has to twist. Twisting generally results in poorer strength capability, at least for two-handed exertions, but the effect does vary with the direction of exertion of the force. Warwick et al. (1980) showed that twisting the trunk to the right by 90° or 135° had little effect on two-handed pulling strength at shoulder height but reduced pushing, pressing downward, and lifting strength by 18%, 28%, and 38%, respectively. At knee height, when twisting through 90°, lifting strength was reduced by over 50% while strength when pressing down was only reduced by 10% and pushing strength was increased. This clearly shows the significant effects on strength of working in non-optimal postures but equally shows that the effects differ with direction of exertion and with the ability (or otherwise) to use body weight to increase the force exerted. It also has to be remembered that stability will be significantly reduced in twisted postures (particularly when foot positions are not freely chosen, as in Warwick et al.'s study) and that muscle activity and consequently loads in the spine will be greater.

Warwick et al. (1980) and Haslegrave et al. (1997a,b) investigated strength for single-handed exertions in awkward, twisted postures and their results showed that strength varied by more than a factor of three within the task height, reach distance, and orientation conditions tested. However, the studies differed in the degree of control placed on the postures adopted by their subjects, as discussed later, and asymmetric postures may not always affect strength to this degree. Haslegrave et al. (1997a) found that the most notable effect of the degree of body twist was in pushing, where there was a moderate reduction in capability as the subjects twisted 90° or 135° rearward. Reach distance had the greatest effect on lifting force capability, where there was a marked advantage in working at a short reach distance. Nevertheless, the risk of injury in twisted postures may be significant and the biomechanical loadings in such postures should be evaluated.

12.1.3 Work in Confined Spaces

Cramped or obstructed workspaces are a common source of constraints on posture. Seminara and Parsons (1982) provided some graphic illustrations of these in their report of maintenance work in power plants, where the poor siting of machinery and control panels had led to extended reach distances, overhead exertions, and work while standing precariously on pipes or ladders.

Lack of space to position the feet can significantly restrict ability to exert vertical forces and alters the posture in which loads can be lifted. Burgess-Limerick and Abernethy (1998) showed that, when reach distance (ankle to handhold) was restricted to 20 cm, ankle and knee flexion were restricted and a stooped posture had to be adopted, with trunk inclination 10° greater than when the load was at a distance of 40 or 60 cm. They confirmed that this effect was due to the lack of space and not induced by the change in load moment, despite that the muscle forces and joint torques required during lifting will be significantly affected by the load moment. The 20 cm distance was well outside the range of reach distances chosen freely by subjects when lifting loads in an earlier experiment; their chosen ankle-handhold distances were 24 to 40 cm (Burgess-Limerick et al. 1995).

If work has to be done close to a wall (say, at 40 cm), only downward pressing is unaffected; strength in all other directions is reduced unless the wall can be used for bracing (Pheasant et al. 1982). The reduction in strength is most marked for pulling, because the presence of the wall restricts foot placement and limits the moment arm for developing torque from body weight to assist in pulling.

Miners, aircraft maintenance workers, and baggage handlers loading aircraft cargo holds (among others) often have to work stooping, crouching, or kneeling in areas with restricted headroom. Strength is significantly reduced and the risk of injury is increased under such circumstances. Gallagher and colleagues (Gallagher et al. 1988, Gallagher and Unger 1990, Gallagher 1991) at the U.S. Bureau of Mines compared stooping and kneeling postures for lifting under a roof height of 120 cm, and showed that capacity for repetitive lifting was significantly less when kneeling than when adopting a stooped posture, as different muscle groups are used in the two situations and the muscles of the low back are more active in the kneeling posture. When kneeling, lifting is primarily an upper body exertion and many strong groups of muscles, which might otherwise assist, cannot be used. The biomechanical loadings on the spine in the two situations were not evaluated, but it is clear that handling tasks in such confined headroom should be designed in accordance with the more limited lifting capacity in the kneeling posture.

There is also an indication from studies by Haslegrave et al. (1997b) that, for single-handed exertions, less force can be exerted when kneeling on two knees than when kneeling on one knee (although a direct comparison was not made on the same group of subjects). This was most marked for pulling forces but the difference was slight when pressing downward. In either posture, the influence of reach distance and direction of exertion was found to be very complex but task height had surprisingly little effect. The conclusions drawn from the results are that kneeling on two knees is probably a less stable posture than kneeling on one knee, that the kneeling posture allows some freedom to adjust the relative positioning of arm, legs, and trunk to bring strong thigh and trunk muscles into play and to orient the hand, arm, and shoulder advantageously, and that careful positioning in terms of reach distance is particularly important when working in a kneeling posture.

Pheasant et al. (1982) studied the effect of a low ceiling on pushing downward and found that strength in the push–press direction can be greatly increased if the ceiling is so low that the person is able to brace against it. They noted that both this effect and the reduction in lifting strength depended more on the distance between the handhold and ceiling than on the ceiling height per se. So these findings apply to work on a ladder, as well as to work when stooping or kneeling.

In very confined spaces, work may have to be carried out while lying on the ground. In this situation, strength in the horizontal direction (pushing or pulling) becomes weak. However, when lying supine, a high force can be generated when pushing upward because the shoulders and trunk can be braced against the ground, provided there is room to lie with the shoulder directly underneath the point of application of the force and that this is between 50% and 100% of maximum arm reach (Haslegrave et al. 1997a). Strength pulling downward is also high because it can be counteracted by body weight. These advantages are lost, of course, if the shoulder cannot be positioned underneath the point of exertion or if the person is lying prone (Hertzberg 1972).

12.1.4 Freely Chosen Postures for Maximal Force Exertion

As has been discussed earlier, foot position and arm posture are critical in generating and supporting a high force exertion. Chaffin and Erig (1991), using a three-dimensional strength prediction model, estimated that small deviations of less than 10° in the angle of the limiting joint for a particular task could potentially cause strength values to change by 10% or more (and by up to 30% for pressing downward, for which strength is probably limited by knee and ankle strengths).

Haslegrave et al. (1997a) replicated the experimental conditions from Warwick et al.'s (1980) study of strength while twisting rearward, with the exception of control on foot position. To define the reach distance, Haslegrave et al. (1997a) instructed their subjects to position the ball of the foot corresponding to their dominant hand at a given distance from the handhold. Otherwise the subjects had a free choice of posture, in contrast to Warwick et al.'s (1980) subjects who were instructed to place their feet symmetrically and 30 cm apart. Comparison of the results showed that the subjects having a free choice of posture were able to exert very much greater forces, especially in the vertical and lateral directions. One important conclusion from this is that strength measurements reported from laboratory studies in standardized postures may well underestimate the true strength capability of workers in industry — but equally may underestimate the potential resultant musculoskeletal loading and risk of injury in industrial tasks.

Relatively few studies have reported the freestyle postures that are chosen when no constraints are placed on subjects exerting a maximal force. Chaffin et al. (1983) demonstrated the large influence of foot position on pulling and pushing capability. The preferred pushing and pulling postures are shown in Figure 12.6. For pushing, their subjects placed their feet farther apart (by approximately 43 cm on average for a handle height of 109 cm) in freely chosen postures and were able to exert a much greater pushing strength than when their feet were side by side. This allowed them to lean forward, pivoting about the rear foot (which was about 103 cm from the handholds) and using the forward leg to increase the forward turning moment and to catch themselves if they slipped. A similar but closer stance was chosen for pulling, with the feet separated by 48 cm and the rearward foot 78 cm from the handholds (again, for a handle height of 109 cm). This allowed them to lean backward, pivoting about the forward foot and producing a greater turning moment than when placing the feet side by side. In this case the rear foot appeared to be positioned to maintain balance if a slip occurred.

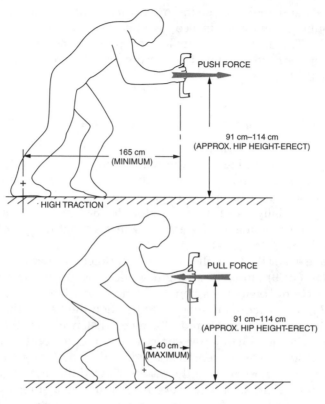

FIGURE 12.6
Freely chosen postures for maximal pushing and pulling in a study by Chaffin et al. (1983). (Illustration from *Occupational Biomechanics*, 2nd ed., Chaffin, D.B. and Andersson, G.B.J., Copyright (1991). John Wiley & Sons, Inc. This material is used by permission of John Wiley & Sons, Inc.)

As can be seen in Figure 12.6, the postures were compared for two-handed and single-handed exertions. Similar postures were adopted in the two cases for pulling but the single-handed exertions were about 25% weaker than the two-handed exertions, which Chaffin et al. (1983) attributed to the reduced upper extremity and shoulder strength capability. The postures adopted for pushing differed in the two situations. When pushing with two hands, the subjects stood farther from the handhold and leaned forward to a greater extent, with a more acute ankle angle, so that they were able to obtain greater advantage from the use of body weight; the force was then 42% greater than the force exerted single-handedly.

Daams (1993) also showed that the forces that could be exerted in constrained postures were very much lower than those when the postures could be chosen freely. The postures she recorded for single-handed pushing and pulling are shown in Figure 12.7. Postures recorded by one of Fothergill et al.'s (1991) subjects for single-handed and two-handed exertions are shown in Figure 12.8, illustrating the greater freedom to adopt an advantageous posture when one hand is free, which permits better use of body weight.

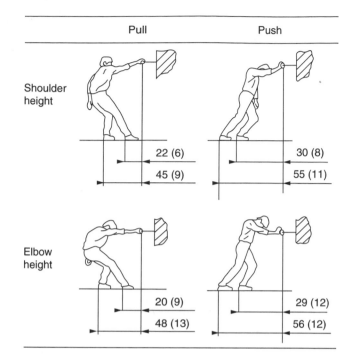

FIGURE 12.7
Positions of feet (mean and standard deviation as a percentage of body height) in freely chosen postures. (From Daams, B.J., *Ergonomics*, 36(4), 397–406, 1993. With permission from Taylor & Francis (http://www.tandf.co.uk).)

Fothergill et al. caution that this will result in higher loading on the shoulder and arm than would occur in a two-handed exertion, a concern that needs further investigation for high force exertions.

Arm posture appears to be very important in exerting a high force because the force must be transmitted across the complex structure and joints of the arm and shoulder. In analyzing postures from various experiments in which subjects had performed single-handed maximal exertions (including situations where they were working overhead, working at shoulder height while standing, twisting toward the side or rear, and kneeling), Haslegrave (1990) found that two elbow postures predominated — with elbow angles between 20° and 50° or 140° and 180°. These corresponded to the arm nearly fully extended and to the arm tightly bent with the shoulder close to the hand (which can be seen in Figure 12.1 and Figure 12.9). These elbow postures did not seem to be influenced by reach distance, because the two postures were found when reach distances of either 50% and 100% maximum arm reach were imposed and when the posture could be chosen freely. Moreover, with reach distance imposed, considerable adjustments to the posture of the whole body had to be made to achieve the preferred arm posture. The two preferred elbow postures do not correspond to peak torque strength for either the elbow flexor or extensor muscles, so that these postures must be

FIGURE 12.8
Freestyle postures adopted for single-handed and two-handed force exertion when pulling downward and backward (pull/press). (From Fothergill, D.M. et al., *Ergonomics*, 34(5), 563–573, 1991. With permission from Taylor & Francis (http://www.tandf.co.uk).)

FIGURE 12.9
Freestyle postures adopted for exerting single-handed force vertically upward (left) and downward (right). (From Haslegrave, C.M., *Ergonomics*, 37(4), 781–799, 1994. With permission from Taylor & Francis (http://www.tandf.co.uk).)

influenced by other factors, and are probably chosen to stabilize the elbow joint and minimize the torques at the elbow and shoulder joints. Other postural strategies for bracing the skeletal framework were also observed during the experiments, including supporting the trunk or shoulder with the free hand braced against knee, chest, or shoulder.

In dynamic tasks, postural strategies can involve coordinated movement patterns. A variety of postures can be observed when loads are lifted from a low level, for example, a back (or derrick) lift and a leg lift (which can be separated into a squat lift and a semisquat lift). In a back lift, a stooped posture is adopted in which the knee joints are extended and the hip joints and spine are flexed to reach the load. In a full squat lift, the knee joints are fully flexed and the trunk is held as vertically as possible. A semisquat lift is intermediate between the two and involves a moderate range of knee flexion with an inclined trunk. Burgess-Limerick et al. (1995), in a study of 39 volunteers lifting loads from close to floor level, found that the posture most commonly adopted at the start of the lift was a semisquat posture with a moderate range of knee flexion (on average 92°), a similar amount of hip flexion (on average 87°), and 45° flexion in the lumbar spine. However, they pointed out that these lifting postures only define the starting point for the lift and that later stages of the transfer, and the ways in which the movements are coordinated, may be equally important.

The resulting biomechanical loads vary between postures and, as with static postures, the musculoskeletal loads depend on task height, reach distance (which may be determined by the size or depth of the load), and asymmetry of posture. Moreover, inertial forces are very important in lifts that are performed rapidly, so that biomechanical loads will be influenced by speed of lift as well as by lifting technique. Although the biomechanical effects of lifting have been analyzed, relatively little is known about the range of movement trajectories among people handling loads in industry.

An observational study by Burgess-Limerick and Abernethy (1997) showed that peoples' freely chosen technique for lifting can change with the weight of the load (the range of weights tested was 2.5 to 10.5 kg). Most of their subjects adopted a semisquat posture at the start of lifting the load (from the floor) and a few adopted a stooped posture. However, some of the subjects used both techniques interchangeably when the load was relatively light but only the semisquat posture when the load weight was increased. Interestingly, the subjects who adopted stooped postures in some or all of the trials were male and, on average, taller, heavier, and stronger than the other subjects. Burgess-Limerick and Abernethy suggest that the alternative postures demonstrate a trade-off between the costs and benefits of different movement patterns. The extent of knee flexion when a squat or semisquat posture is adopted permits greater coordination of the whole body throughout the lift and can reduce muscular effort. They found that when subjects adopted this posture, knee extension occurred rapidly early on in the lifting movement, followed by extension at the hip, and only later by lumbar vertebral extension. This coordinated postural behavior allows the quadriceps muscles to contribute to hip extension and uses the hamstring and erector spinae muscles at a more extended length (and these coordination strategies are discussed in greater depth in Burgess-Limerick et al. 1995). When their subjects started the lift in a stooped posture, this had an advantage of not lowering the center of gravity as far as in a semisquat lift so that less energy was expended while raising the body mass. However, the strength advantage of the semisquat posture was lost because the hamstring muscles shorten rapidly and the quadriceps muscles cannot be recruited. Burgess-Limerick and Abernethy postulate that for some subjects (and particularly for strong subjects lifting light loads) the two alternative movement patterns may be equally sustainable, but that the balance of costs (presumably in terms of muscular effort) changed with a heavy load so that the semisquat posture became predominant.

Although exerting force in postures in which the back is both flexed and twisted or flexed and bending laterally is generally recognized as presenting a risk of back injury, such postures are frequently observed in workplaces, even among workers who have received manual handling training. Gagnon et al. (1993) offered a plausible explanation of such behavior, noting that a twist in the direction opposite to the exertion of the force can serve to increase trunk muscular strength for the exertion (because of the length–tension relationship of the muscle fibers). The workers' behavior is therefore probably influenced by a desire to minimize energy expenditure or muscular fatigue, which overrides their training advice to move or pivot on their feet rather than twist their body. A similar type of response is seen in the change in lifting technique over a workshift as workers tend to use a back lift more frequently than a leg (or squat) lift toward the end of the shift.

12.2 Strength Capability and Guidelines for Force Exertion

Extensive measurements of the effect of posture on isometric strength capability come from studies carried out by Rohmert and colleagues (Rohmert 1966, for men, and Rohmert and Jenik 1971, for women), providing an "atlas" of strength in single-handed, standing force exertions at three handgrip distances and various arm angles (relative to the shoulder joint) within the three-dimensional reach envelope around the body. The main limitations of Rohmert's database are that the handle on which the forces were exerted was vertical (which is not optimal when exerting vertical forces), subjects' foot positions were constrained (placed parallel and 300 mm apart), and the data were collected from two groups of only five young male and ten young female subjects. Some caution therefore has to be taken in applying these strength data. Nevertheless, they give an indication of strength variation with direction of exertion and reach distance. When exerting a right-handed force in the midsagittal plane, horizontal forces (pushing and pulling) increased as the reach distance increased (between 50% to 100% of arm length) but did not change greatly as the direction of exertion changed from 0° (horizontal) to 60° below the horizontal. Lateral force exertions reduced as the reach distance increased. Vertical forces (lifting and pressing) decreased with reach distance, and lifting force increased considerably as the direction of exertion changed from 0° to 60° below the horizontal while the downward pressing force similarly decreased but only at the closest reach distance.

Summary data on strength from this and other sources are given in several textbooks and design handbooks, including Hertzberg (1972), Chaffin et al. (1999), and Peebles and Norris (1998). Kroemer (1968, 1974) and Laubach (1976a,b) also provide extensive sources for strength data for standing subjects in many different task situations. Kroemer (1974) showed that very much higher forces can be generated if there is a wall or footrest conveniently placed against which the body can be braced.

Other sources of data on strength in standing exertions include Fothergill et al.'s (1991) study of isometric strength capabilities during single-handed and two-handed exertions in all directions in the fore–aft plane at 1 m and 1.75 m heights. They showed that single-handed strength could equal two-handed strength under some circumstances, the single-handed to two-handed strength ratio ranging from 0.64 to 1.04. Differences were less at the higher task height, and at this height one-handed exertions in the pull–press direction often exceeded two-handed forces. Sanchez and Grieve (1992) developed gender-free predictive equations for isometric strengths when lifting symmetrically and asymmetrically, covering positions of the hand(s) between head height and 15 cm above floor level within the zone between the midsagittal plane and 90° to the right for two-handed exertions and within the zone from 90° to the left to 90° to the right for single-handed exertions (using the right hand). Chaffin et al. (1999, p. 333) have summarized the force data available to give design limits for pushing and pulling activities on various friction surfaces. Kumar and Garand (1992) have compared static and dynamic lifting strengths in symmetrical and asymmetrical postures.

Strength data are also available for some other task situations. Strength while working overhead (both standing and lying supine) and while kneeling can be found in Haslegrave et al. (1997a) and (1997b), respectively. Laubach (1978) provided strength data for seated pilots, but the subjects wore shoulder and lap safety belts, which would have provided bracing when exerting forces in some directions. Mital and colleagues have provided torque strength data for use of a variety of hand tools (Mital 1986, Mital and Sanghavi 1986) and Peebles and Norris (2003) data on strength for various actions involved when using handles or controls on consumer products (pushing, pulling, gripping, and twisting). Woldstad et al. (1995) measured strength capabilities for the turning of large-diameter vertical handwheels, which is a common task in heavy industries.

In addition various standards and guidelines have been developed to specify maximum force limits for safety or acceptability. The European Standards EN 1005-2 and EN 1005-3 (CEN 2002, 2003) provide guidance on recommended force limits for several common actions concerned in the operation of machinery. The work of Snook and Ciriello (1991) has led to guidelines for maximum force exertion based on the criterion of acceptability to experienced workers (from psychophysical testing). Also using a psychophysical methodology, Smith et al. (1992) have produced lifting, lowering, and carrying capability data (for both women and men) for a wide range of nonstandard postures, representing tasks carried out twisting, stretching, sitting, kneeling, crouching, lying supine, prone, or sideways, and with a restricted ceiling height — in all 99 situations for materials handling tasks that are not two-handed, symmetric, and in the sagittal plane. Guidance on maximum acceptable forces for a variety of types of task, based on epidemiological, physiological, biomechanical, or psychophysical criteria, has also been compiled by Mital et al. (1993).

It is apparent that the strength database is extensive, but there are many gaps and it is not always easy to find data relevant to a particular design situation. The influence of posture on force exertion in some common work situations is discussed in more detail below.

12.2.1   Dynamic Tasks

During the development of skilled performance at a physical task, movements and coordination will be refined so that speed and efficiency are improved while effort (muscular load or energy expenditure) and discomfort are minimized. Initial training to develop manual skills commonly takes several weeks and gradual improvement continues, possibly for years. The early work of Frederick W. Taylor and Frank and Lillian Gilbreth in the first quarter of the 20th century (Barnes 1968) led to motion analysis and time studies, which still influence the current predetermined motion–time systems used in industry. However, most of this work is directed at minimizing the time required for completion of a task, whereas, to minimize musculoskeletal stress, the choice of method (posture, movements, and speed) should also include consideration of the biomechanical and physiological effects. A review of modern work analysis methods can be found in Chaffin et al. (1999).

Models have been developed to predict movement time according to the precision required (or positioning tolerance), using Fitt's law, which defines movement as a function of an Index of Difficulty. Drury (1975) applied this to foot motion when operating foot pedals and Woldstad (1985) to the operation of materials handling manipulators. However, other variables have been found to affect the relationship, including the body segment moved (Drury 1975) and the plane of movement (Li et al. 1995).

There may well be optimum speeds at which tasks can be performed. Hettinger and Müller (1953), for example, showed that any individual has his or her own optimum speed of walking, which minimizes energy expenditure, but that this optimum speed varies with conditions such as the weight of shoes worn. This is the result of the natural frequency of the leg's pendulum-like movement during steady walking (the natural frequency varying with the weight and weight distribution of the leg); gravity can then be used to minimize the muscular force required to decelerate and accelerate the leg through the gait cycle.

Such effects should also be considered when designing repetitive manual tasks that have to be performed in conjunction with conveyors or other machinery. Performance rate and energy expenditure during arm movements vary depending on whether the task can be performed with forearm motion alone or necessitates the movement of the whole arm from the shoulder joint (Konz 1990). Tichauer (1968) showed that slight abduction of the upper arm (8° to 23°) was also desirable, perhaps because this posture permits lateral movement of the hand with minimal rotation of the forearm. The complexity of the anatomy of the arm and shoulder is such that small changes in the layout of a workplace can have a disproportionate effect on efficiency and biomechanical loading. Acceleration and deceleration should be avoided in the movements because these interrupt natural rhythms and involve considerable muscular effort and energy expenditure. This can be achieved by appropriate arrangement of the workplace, and particularly by facilitating circular or elliptical motions without abrupt changes in direction (Konz 1990).

The maximum strength that can be exerted in most dynamic industrial tasks is likely to be lower than static (isometric) strength because of the tension–velocity relationship in concentric muscle contractions. Kumar et al. (1988), for example, showed that lifting strength for either a back lift or an arm lift steadily reduced as the speed of lifting increased.

Some tasks require short-duration dynamic strength in rapid pushing or pulling actions. Mital and Faard (1990) showed that standing pulls are about 37% stronger than seated pulls and that pull strength is greater with a long reach distance. Imrhan and Ramakhrishnan (1992) showed that isokinetic pull strength varies considerably with direction and speed of pull as well as with arm elevation and that, under the best conditions, strength can be greater by a factor of 2.1 than under the worst conditions. Pulling toward the body was found to be 9% stronger than pulling across the body.

Postures have not been much studied for dynamic work tasks (unlike sports activities and gait analysis) and, when evaluating their biomechanical effects, work postures tend to be represented as static "snapshots" at a moment in time. Identifying the critical moments in a dynamic task is more difficult than with static exertions, as peak forces are influenced by inertial forces of both the body (or body segment) and external objects, while these peak forces may not coincide with the most extreme or awkward postures. Usually, several "snapshots" will be chosen for further biomechanical analysis to estimate the points of greatest stress and risk of injury during a task.

12.2.2 Lifting and Carrying

As a general principle, the posture for lifting should be one in which the back is held in a stabilized neutral position with bending taking place at the hips rather than in the lumbar region of the spine. Movements should be smooth and any jerking, lateral bending, or twisting should be avoided. Asymmetric lifting is much more stressful than sagittal plane lifting from the evidence of both biomechanical and psychophysical studies reviewed by NIOSH (Waters et al. 1994). Lifting should be confined to transfers at approximately waist level (i.e., between mid-thigh and mid-chest level). If it is necessary to lift over a greater height range than this, it is likely that the grip on the handhold or object will have to be adjusted, which is undesirable; it is better to have an intermediate surface on which the load can be rested, so that arm postures do not have to be altered in mid-lift.

Marras and Davis (1998) measured spinal loading in compression and shear during asymmetric lifting tasks (with a load of 13.7 kg) and their results indicate that the effect of asymmetry is complex. Both compression and lateral shear forces in the spine increase with asymmetry of posture (i.e., twisting due to the position of the load relative to the sagittal plane through the mid-position of the feet), but anterior–posterior shear force decreases as the posture becomes more asymmetric. The levels of spinal loading calculated by Marras and Davis support the body of literature that has shown that task asymmetry should be minimized in lifting tasks because lateral loading and combined plane loading are hazardous. However, their results did indicate that single-handed lifting at the side of the body might be preferable to two-handed lifting when the task is asymmetric, as compression and anterior–posterior forces in the spine were lower than for either single-handed lifting across the body or two-handed lifting.

Gagnon et al.'s (1993) study showed that pivoting with the feet is a good compromise strategy when transferring loads, instead of twisting sideways. The moments at the lumbar spine are considerably lower than when the trunk is rotated and flexed without moving the feet, and it avoids hazardous eccentric muscle contraction. They note, however, that this handling strategy is technically difficult and requires practice to use effectively, as well as high friction soles on the work shoes (such as those designed for basketball players).

Team lifting is often employed to lift loads that are heavier than considered acceptable for a single person. The maximum load that should be lifted by each member of a two-person team is generally considered to be two thirds of the load that would be acceptable if they were lifting alone. Karwowski and Pongpatanasuegsa (1988) measured the maximum lifting strengths of two-woman and three-woman teams and found that isometric strength of either a two- or a three-woman team was approximately 83% of the sum of the two (or three) individual strengths. The isokinetic strength was approximately 68% of the sum of the individual strengths. These results were similar to those found earlier for two-man and three-man teams (Karwowski and Mital 1986).

The same postural considerations as in single-person lifts apply to the members of the team. However, good postures can be more difficult to achieve, particularly if the load is wide or bulky or if handholds on the load are not appropriate for a team lift. The members of the team should be closely matched in height; otherwise, one member of the team is likely to have to carry a greater proportion of the load than the other(s). The lifting also needs to be carefully timed and coordinated within the team to avoid unequal loading or unexpected changes in speed or direction.

An important question is whether posture and handling techniques can be modified by training. In a study involving warehouse workers, Chaffin et al. (1986) showed that it is possible to change behavior to a certain extent with carefully designed training. The training program tested included having the participants lift different sized boxes in situations representative of their normal work, discussion of why the lifting was performed in the way it was, presentation of basic biomechanical principles of lifting and their application in work situations, and a final practice in lifting the boxes. The lifting techniques used by participants were evaluated in their workplaces just over a month (35 to 51 days) after the training session. The training was found to have had a beneficial effect in reducing the jerkiness of their lifting and in the adoption of a secure grip on the load, but they had not improved significantly on keeping the loads close to their body, keeping their trunk upright, or avoiding twisting. There is some evidence that workers can be trained to adopt more upright trunk postures (evaluated at 3 months after the training) with training spread over a longer period, with feedback and reinforcement of the principles and with active involvement of supervisors (Hultman et al. 1984). It is even more difficult to ensure that the improved techniques are maintained in the long term, and no training will be adequate if it is not supported by appropriate workplace design and organization.

However, Burgess-Limerick et al. (1995) urge caution in designing training to change postural behavior. Their study of postures adopted when lifting from close to floor level suggested that people discover successful strategies of neuromuscular control to minimize muscular effort during load handling. They concluded that normal coordination between knee, hip, and lumbar vertebral joints should not be disrupted by imposed changes in lifting techniques.

12.2.3 Handling Live Loads

The handling of live loads, and particularly of humans in the health-care setting and by the emergency services, poses additional problems and risks. There have been various studies of patient-handling techniques and the resultant biomechanical loadings (for example, Gagnon et al. 1987, Takala and Kukkonen 1987, Elford et al. 2000, Lavender et al. 2000) and certain transfer techniques have been shown to be more hazardous than others. Nevertheless, even with team lifting, the loads involved in lifting an adult are very high and policies adopted now in many health-care facilities are to avoid lifting patients manually.

Hignett and Richardson (1995) have pointed out that manual materials-handling guidelines which have been developed for inanimate loads in industry may not be suitable for the handling of human loads. Their exploratory study of the perceptions of nursing staff has shown that a generic risk assessment approach is not adequate and that further research is needed in this field. The mobility and dependency level of the patient have an important influence on the way that they are handled: they may move during the transfer, and demands for immediate attention from patients or relatives may put additional pressure on staff. All these factors make it difficult for staff to adopt and maintain recommended postures and handling techniques.

The best approach is to assess each transfer of a patient, use the most appropriate of the variety of handling aids that are now available, and transfer in a way in which the patient can assist himself or herself as much as possible. In other words, the focus should be on techniques that assist patients in adopting the best postures for moving themselves (perhaps with the aid of patient-centered assistive equipment) and avoiding heavy lifting by the caregiver, which would inevitably have to be done in a less than optimum posture. The National Back Pain Association/Royal College of Nursing's (1998) guidelines on the handling of patients and Hignett et al.'s (2003) evidence-based review offer advice on methods that can be used in different situations, for example, when transferring a patient from a wheelchair to a chair or turning a patient in bed, and on the use of various forms of handling equipment.

Similar difficult problems occur with the handling of large animals on farms and in veterinary practice, and these are likely to be even more unpredictable than human patients and to require restraint during handling. Again, the policy is to avoid attempting to move or restrain them manually. Specialized structures and equipment are needed in these circumstances, as they are for some other stock-handling tasks that involve work for long periods in very awkward postures, such as shearing sheep (HSE 1996, 1999, 2000).

12.2.4   Using Manual Materials Handling Aids

Several studies have investigated behavior when using materials-handling aids. Such aids include the most common forms of mechanical assistance — hoists, four-wheeled carts, and two-wheeled trolleys — as well as more specialized equipment as diverse as articulated arms used to support components during vehicle assembly and human-powered rickshaws. Handling aids rarely remove all force exertion from the task. Most commonly they replace the vertical forces involved in supporting and transporting the load weight with horizontal pushing or pulling forces (for example, a hoist that supports the weight of the load but does not assist with pushing the load into place), although some powered "manipulator" devices may also assist with horizontal force exertion. In addition to changing the nature of the task from a vertical to a horizontal force exertion, the force required may be increased by adding the inertial load of the assist device itself to the forces involved in manipulating the load. Poor siting at installation may exacerbate the problems, perhaps by increasing reach distances or changing the height of the force exertion, and this often leads to the failure of operators to use assist devices even when they are provided. Moreover, the use of handling aids such as hoists and articulated arms is likely to involve single-handed operation and/or asymmetric postures as well as increasing the time needed to complete the task, as shown, for example, by Resnick and Chaffin (1996) and Nussbaum et al. (2000).

Because the nature of the handling task is changed when a handling aid is introduced, it is important to reassess the user's posture and force capability — and more fundamentally the design of the handling aid itself. The first consideration in the design of handling aids is the location of the hand–load interface, which has a major influence on both posture and force capability. For two-handed pushing and pulling of carts, trolleys, or loads being handled manually, it is clear that strength will vary with height of handholds. Martin and Chaffin (1972) used biomechanical analysis to predict that maximum strength could be applied in either pushing or pulling with hands around 50 to 90 cm from the floor, because this allowed the person to use body weight by leaning forward when pushing or backward when pulling. However, other factors have to be taken into account, including task-related factors such as the need for a clear view as well as the musculoskeletal loading, which may create a risk of injury.

Lee et al.'s (1989, 1991) reports of investigations of cart pushing and pulling (with handle heights between 66 and 152 cm) indicated that pushing appeared to result in lower spinal compression than pulling due to tensing of the anterior abdominal muscles. They also showed that subjects adopted greater torso inclination when pushing than when pulling, thus using body weight more effectively to assist in generating the pushing force. This would not, however, be possible on a low-friction floor surface. Handle height and speed of movement both had a significant effect on spinal compression. Taking the various considerations into account, Chaffin et al. (1999) recommend a handhold height of between 91 and 114 cm (i.e., at approximately hip height) for both pushing and pulling, in conjunction with a high-friction floor surface. The postures that people would adopt would still differ, however, between pushing and pulling; a more rearward foot position (with the rear foot at least 165 cm from the handhold) would be desirable for pushing, while for pulling the leading foot would need to be close to the handhold (with a maximum distance of 40 cm) but with a wider foot base to allow recovery in the event of a foot slip. In this posture, the pulling force would be maximized by leaning backward and pivoting on the front foot.

However, Resnick and Chaffin (1995) pointed out that the initial research based on static strength may not be directly applicable to dynamic use of carts or other handling aids. Lee et al. (1991) showed that subjects assume different postures in dynamic tasks from those that they adopt for static force exertions and Resnick and Chaffin (1995) showed that peak velocity of movements varied with the weight or inertia of the load, but they concluded that static approximations are useful predictions when loads are high and speeds of movement (and accelerations) are consequently low.

The loads involved in the use of other types of transporters such as barrows and trolleys can be very high and, in a study of the use of two-wheeled refuse containers, Frings-Dresen et al. (1995) observed stressful postures with the trunk flexed to an angle that was sometimes greater than 45° or even bent backward. The use of wheeled devices involves more than straightforward pushing and pulling when their loads need to be steered around corners, obstacles, and delivery points. These complex maneuvers have been little studied, but it is clear that the forces that have to be exerted will have a lateral as well as a longitudinal component. It is possible therefore that the best position for the handholds will not be exactly as recommended above for the pushing of carts but will be a compromise to satisfy the three-dimensional components of the force exertion.

A related situation arises with one-wheeled barrows and two-wheeled trolleys, as these require the user to exert a vertical force on the handholds to tilt the barrow or trolley while pushing it forward. The vertical force may be upward to counteract the backward moment created by the load about the single wheel of a barrow. However, in contrast, the force is usually downward when pushing a two-wheeled gas cylinder trolley or sack trolley because the center of gravity of the load normally lies in front of the pair of trolley wheels, thus creating a forward moment that must be counteracted by an opposing moment at the hands. In both these circumstances, the user pushing the barrow or trolley has some freedom to tilt the handles to a preferred height (as shown in Figure 12.10), which is different from the pushing of a four-wheeled cart, and tends to keep the handles at the same height regardless of load weight (Okunribido and Haslegrave 1999). It seems probable that the preference for handle height will be influenced by the optimum trunk and arm postures for exerting the pushing force (as with the

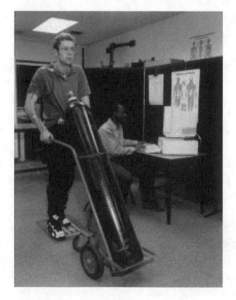

FIGURE 12.10
Pushing a gas cylinder trolley.

cart) but in addition (for the barrow) by the best arm posture to exert the significant vertical force to support the weight of the load or (for the trolley) the need for precise control of the balance to keep the load over the center of rotation of the trolley wheels and thus minimize the vertical force that must be exerted. Lateral forces needed for steering will be a third factor influencing posture and preferred handle height. Okunribido and Haslegrave (1999) studied the initial stages of moving a two-wheeled gas cylinder trolley and found that considerable wrist deviations occurred and high stresses were generated at the elbow as the handles were pulled downward and backward to tilt the trolley. The wrist and elbow posture altered at the transition point as the tilt was completed and a forward acceleration of the trolley had to be generated. The actual postures were strongly influenced by the height and orientation of the trolley handle.

The conclusion reached from the above studies of four-wheeled, two-wheeled, and single-wheeled load transporters is that handle position and design is a very important determinant of the loading on trunk, shoulder, and arm of the user. Because forces must be applied in three dimensions, the orientation and shape of the handle must afford a good interface in whichever posture of the hand grasp and wrist is most efficient, and this posture will alter in different phases of the motion. The orientation of the handle is particularly important for points at which force exertion is highest: pushing forward and decelerating the load when stopping for all three types of transporter, vertical support of the load in a barrow, and downward force to tilt a cylinder trolley ready to move off. It is also worth noting that well-designed handholds should allow some freedom to adjust the posture of the wrist and arm because this will change during the task, as when lifting or lowering a load through any distance or when steering a trolley around a corner.

There is some discussion over whether pushing or pulling is preferable (for example, when moving an industrial four-wheeled container such as a cage trolley), but there does not appear to be a clear answer to this. The height has implications for the placement of handholds, but other factors do need to be considered. Vision may be severely restricted when pushing a high-sided or heavily loaded container, but it is probably easier to push this with two hands and so be able to adopt a more symmetric and stable posture than to walk backward pulling it (which again creates a difficult and potentially hazardous visual situation) or to twist round and pull it with one hand (with consequential asymmetric musculoskeletal loading and possibly an increased risk of injury.

12.3 Evaluation Techniques

Given the large influence that environmental and task conditions have on posture and consequently on strength for a given force exertion, it cannot be expected that handbooks can give complete databases covering the design of all tasks. The NIOSH (1991) lifting equation provides a recommended weight limit (RWL) for lifting and carrying over relatively short distances, taking into account various risk factors influencing posture and the way that the load is held (Waters et al. 1994). Statistical regression models have been used to predict population strength from the basic characteristics of the population of interest, primarily gender, age, weight, and stature (Shoaf et al. 1997, Chaffin 1998). Another approach is the prediction of strength using biomechanical models, such as the University of Michigan's Three-Dimensional Static Strength Prediction Program 3DSSPP™, described in Chaffin and Erig (1991) and Chaffin et al. (1999). This can predict the percentage population strength capability for a given force exertion and posture. The model also provides a rating of the stability of the task by comparing the torque generated at the ankles with the counteracting moment created by ground reaction forces acting on the feet. Ayoub et al. (1998) have developed a model to predict human motion during lifting. Biomechanical models are also used to assess the risk of musculoskeletal injury. Modeling approaches are discussed in more detail in Chapter 14.

Simpler evaluation techniques that can be used for rapid screening in industry and during observation studies include checklists and posture classification tools, many of which have been mentioned in earlier chapters. ISO and CEN standards (ISO 2002, CEN 1998, 2002) define methods for assessing lifting and carrying tasks, and manual handling and force exertion during the operation of machinery. Most of these tools include some assessment of force exerted within the task (based on observation or the worker's subjective report) but, without direct measurement of the forces involved, this can only provide a rough estimate of the severity of the effects.

Summary

Postural strategies can be used to gain advantage in applying high forces and this is likely to be one of the components of skill in forceful tasks. The evidence provided from the many research studies serves to emphasize the complex nature of the use of the human musculoskeletal system in force exertion. Strength varies with task, posture, and environmental conditions. Task conditions involve many factors — including direction of force exertion, speed, layout (height, reach distance) and interface — each of which not only has a direct influence on strength but can also influence posture. The main conclusions that can be drawn are as follows:

- Posture has a major influence on strength capability, whether in terms of whole body posture or of the posture within the range of motion of a particular joint, because this determines the muscles that can be employed in the exertion and the mechanical advantage offered to these muscles as they exert force about a joint. Posture, or task layout, is one of the major limiting factors for strength exertion and conversely it is one of the most important determinants of risk of musculoskeletal injury.

- Even small constraints on free adoption of the most advantageous posture can significantly affect the strength that can be exerted. Thus, adequate space should be allowed in the layout of workstations, particularly for foot placement, headroom, and arm posture.

- There are extensive databases that can be used to assess population strength capabilities for particular tasks and postures. The most extensive are for isometric strength capability, but others are available for isokinetic force exertions and for specific lifting, carrying, pushing, and pulling tasks. However, there are still many gaps in the databases and strength cannot always be predicted for particular work tasks.

- The applicability of a particular database needs careful consideration — some identify maximal strength capability, others (based on psychophysical methodology) give the maximum capability that workers perceive to be acceptable for given work durations and repetition frequencies, while others indicate a risk of injury (based on muscular effort or biomechanical loading in the posture adopted).

- Environmental conditions can have a profound effect on both posture and strength. These include the condition of floor surfaces, space constraints due to restricted access or low ceilings, and the quality and orientation of the handholds provided.

- It is important that designers be aware of how constraints may reduce a person's force exertion capability below the levels that can be achieved in good working postures, and that this is likely to increase the risk of injury.

- Selection criteria and tests for jobs requiring strength must be designed to assess strength under conditions replicating the specific task situations.

References

Ayoub, M.M., Woldstad, J.C., Lin, C.J., and Bernard, T., 1998. A model to predict human motion during lifting, SAE Technical Paper 981303, Warrendale: Society of Automotive Engineers.

Barnes, R.M., 1968. *Motion and Time Study*, New York: John Wiley.

Burgess-Limerick, R. and Abernethy, B., 1997. Qualitatively different modes of manual lifting. *International Journal of Industrial Ergonomics*, 19(5), 413–417.

Burgess-Limerick, R. and Abernethy, B., 1998. Effect of load distance on self-selected manual lifting technique. *International Journal of Industrial Ergonomics*, 22, 4–5, 367–372.

Burgess-Limerick, R., Abernethy, B., Neal, R.J., and Kippers, V., 1995. Self-selected manual lifting technique: functional consequences of the inter-joint coordination. *Human Factors*, 37(2), 395–411.

CEN, 2002. *European Standard EN 1005-3 Safety of Machinery — Human Physical Performance*, Part 3: *Recommended Force Limits for Machinery Operation*, Brussels: European Committee for Standardization.

CEN, 2003. *European Standard EN 1005-2 Safety of Machinery — Human Physical Performance*, Part 2: *Manual Handling of Machinery and Component Parts of Machinery*, Brussels: European Committee for Standardization.

Chaffin, D.B., 1998. Prediction of population strengths, SAE Technical Paper 981307, Warrendale: Society of Automotive Engineers.

Chaffin, D.B. and Andersson, G.B.J., 1991. *Occupational Biomechanics*, 2nd ed., New York: John Wiley.

Chaffin, D.B. and Erig, M., 1991. Three-dimensional biomechanical static strength prediction model sensitivity to postural and anthropometric inaccuracies. *IIE Transactions*, 23(3), 215–227.

Chaffin, D.B., Gallay, L.S., Woolley, C.B., and Kuciemba, S.R., 1986. An evaluation of the effect of a training program on worker lifting postures. *International Journal of Industrial Ergonomics*, 1, 127–136.

Chaffin, D.B., Andres, R., and Garg, A., 1983. Volitional postures during maximal push-pull exertions in the sagittal plane. *Human Factors*, 25(5), 541–550.

Chaffin, D.B., Andersson, G.B.J., and Martin, B., 1999. *Occupational Biomechanics*, 3rd ed., New York: John Wiley.

Daams, B.J., 1993. Static force exertion in postures with different degrees of freedom. *Ergonomics*, 36(4), 397–406.

Dempster, W.T., 1955. Space requirements of the seated operator, WADC Technical Report 55-159, Wright-Patterson Air Force Base, OH: Aerospace Medical Research Laboratories.

Drury, C.G., 1975. Application of Fitt's law to foot-pedal design. *Human Factors*, 17(4), 368–373.

Elford, W., Straker, L., and Strauss, G., 2000. Patient handling with and without slings: an analysis of the risk of injury to the lumbar spine. *Applied Ergonomics*, 31(2), 185–200.

Ferguson, S.A., Gaudes-MacLaren, L.L., Marras, W.S., Waters, T.R., and Davis, K.G., 2002. Spinal loading when lifting from industrial storage bins. *Ergonomics*, 45(6), 399–414.

Fothergill, D.M., Grieve, D.W., and Pheasant, S.T., 1991, Human strength capabilities during one-handed maximum voluntary exertions in the fore and aft plane. *Ergonomics*, 34(5), 563–573.

Fothergill, D.M., Grieve, D.W., and Pheasant, S.T., 1992. The influence of some handle designs and handle height on the strength of the horizontal pulling action. *Ergonomics*, 35(2), 203–212.

Frings-Dresen, M.H.W., Kemper, H.C.G., Stassen, A.R.A., Crolla, I.F.A., and Markslag, A.M.T., 1995. The daily workload of refuse collectors working with three different collecting methods: a field study. *Ergonomics*, 38(10), 2045–2055.

Gagnon, M., Chehade, A., Kemp, F., and Lortie, M., 1987. Lumbosacral loads and selected muscle activity while turning patients in bed. *Ergonomics*, 30(7), 1013–1032.

Gagnon, M., Plamondon, A., and Gravel, D., 1993, Pivoting with the load — an alternative for protecting the back in asymmetric lifting. *Spine*, 18(11), 1515–1524.

Gallagher, S., 1991. Acceptable weights and physiological costs of performing combined manual handling tasks in restricted postures. *Ergonomics*, 34(7), 939–952.

Gallagher, S. and Unger, R.L., 1990. Lifting in four restricted lifting conditions. *Applied Ergonomics*, 21(3), 237–245.

Gallagher, S., Marras, W.S., and Bobick, T.G, 1988. Lifting in stooped and kneeling postures: effects on lifting capacity, metabolic costs, and electromyography of eight trunk muscles. *International Journal of Industrial Ergonomics*, 3(1), 65–76.

Grieve, D. and Pheasant, S., 1981. Naturally preferred directions for the exertion of maximal manual forces. *Ergonomics*, 24(9), 685–693.

Haslegrave, C.M., 1990. The role of arm and body posture in force exertion, in *Proceedings of the Human Factors Society 34th Annual Meeting*, Orlando, 2–5 October 1990, 771–775.

Haslegrave, C.M., 1994. What do we mean by a "working posture"? *Ergonomics*, 37(4), 781–799.

Haslegrave, C.M., Tracy, M.F., and Corlett, E.N., 1997a. Force exertion in awkward working postures — strength capability while twisting and working overhead. *Ergonomics*, 40(12), 1335–1362.

Haslegrave, C.M., Tracy, M.F., and Corlett, E.N., 1997b. Strength capability while kneeling. *Ergonomics*, 40(12), 1363–1379.

Hertzberg, H.T.E., 1972. Engineering anthropology, in *Human Engineering Guide to Equipment Design*, rev. ed., Van Cott, H.P. and Kinkade, R.G., Eds., New York: John Wiley, 467–584.

Hettinger, T. and Müller, E.A., 1953. Der Einfluss des Schuhgewichtes auf den Energieumsatz beim Gehen und Lastentragen [The influence of shoe weight on the energy cost in walking and load carrying]. *Arbeitsphysiologie*, 15, 33–40.

Hignett, S. and Richardson, B., 1995. Manual handling human loads in a hospital: an exploratory study to identify nurses' perceptions. *Applied Ergonomics*, 26(3), 221–226.

Hignett, S., Crumpton, E., Ruszala, S., Alexander, P., Fray, M., and Fletcher, B., 2003. *Evidence-Based Patient Handling. Tasks, Equipment and Interventions*, London: Routledge.

Holbein, M.A. and Chaffin, D.B., 1997. Stability limits in extreme postures: effects of load positioning, foot placement and strength. *Human Factors*, 39(3), 456–468.

HSE (Health and Safety Executive), 1996. Deer Farming. HSE Information Sheet AIS 7 (rev.), Sudbury: HSE Books.

HSE (Health and Safety Executive), 1999. Handling and Housing Cattle. HSE Information Sheet AIS 35, Sudbury: HSE Books.

HSE (Health and Safety Executive), 2000. Manual Handling Solutions for Farms. HSE Publication AS 23 (rev. 2), Sudbury: HSE Books.

Hultman, G., Nordin, M., and Örtengren, R., 1984. The influence of a preventive educational programme on trunk flexion in janitors. *Applied Ergonomics*, 15(2), 127–133.

Imrhan, S.N. and Ayoub, M.M., 1990. The arm configuration at the point of peak dynamic pull strength. *International Journal of Industrial Ergonomics*, 6(1), 9–15.

Imrhan, S.N. and Ramakhrishnan, U., 1992. The effects of arm elevation, direction of pull and speed of pull on isokinetic pull strength. *International Journal of Industrial Ergonomics*, 9(4), 265–273.

ISO, 2002. *ISO/FDIS 11228-1, Ergonomics – Manual Handling,* Part 1: *Lifting and Carrying,* Geneva: International Organization for Standardization.

Karwowski, W. and Mital, A., 1986. Isometric and isokinetic testing of lifting strength of males in teamwork. *Ergonomics,* 29(7), 869–878.

Karwowski, W. and Pongpatanasuegsa, N., 1988. Testing of isometric and isokinetic lifting strengths of untrained females in teamwork. *Ergonomics,* 31(3), 291–301.

Konz, S., 1990. *Work Design: Industrial Ergonomics,* 3rd ed., Worthington, OH: Publishing Horizons.

Kroemer, K.H.E., 1968. Push forces exerted in sixty-five common working positions, Technical Report USAF:AMRL-TR-68-143, Wright-Patterson Air Force Base, OH: Aerospace Medical Research Laboratory.

Kroemer, K.H.E., 1974. Horizontal push and pull forces exertable when standing in working positions on various surfaces. *Applied Ergonomics,* 5(2), 94–102.

Kroemer, K.H.E. and Robinson, D.E., 1971. Horizontal static forces exerted by men standing in common working postures on surfaces of various tractions, Technical Report AMRL-TR-70-114, Wright-Patterson Air Force Base, OH: Aerospace Medical Research Laboratory.

Kumar, S., 1991. Arm lift strength in work space. *Applied Ergonomics,* 22(5), 317–328.

Kumar, S. and Garand, D., 1992. Static and dynamic lifting strength at different reach distances in symmetrical and asymmetrical planes. *Ergonomics,* 35(7/8), 861–880.

Kumar, S., Chaffin, D.B., and Redfern, M., 1988. Static and dynamic lifting strengths of young males. *Journal of Biomechanics,* 21(1), 35–44.

Laubach, L.L., 1976a. Muscular strength of women and men: a comparative study, AMRL Report TR-75-32, Wright-Patterson Air Force Base, OH: Aerospace Medical Research Laboratory.

Laubach, L.L, 1976b. Muscular strength of women and men: a review of the literature. *Journal of Aviation, Space and Environmental Medicine,* May, 534–542.

Laubach, L.L, 1978. Human muscular strength, in Anthropometric Source Book, Vol. I: Anthropometry for Designers, Webb Associates, Eds., NASA Reference Publication 1024, Lyndon B. Johnson Space Center, Houston, TX: National Aeronautics and Space Administration.

Lavender, S.A., Chen, S-H., Li, Y.C., and Andersson, G.B.J., 1998. Trunk muscle use during pulling tasks — effects of a lifting belt and footing conditions. *Human Factors,* 40(1), 159–172.

Lavender, S.A., Conrad, K.M., Reichelt, P.A., Johnson, P.W., and Meyer, F.T., 2000. Biomechanical analyses of paramedics simulating frequently performed strenuous work tasks. *Applied Ergonomics,* 31(2), 167–177.

Lee, K.S., Chaffin, D.B., Waiker, A.M., and Chung, M.K., 1989. Lower back muscle forces in pushing and pulling. *Ergonomics,* 32(12), 1551–1563.

Lee, K.S., Chaffin, D.B., Herrin, G.D., and Waiker, A.M., 1991. Effect of handle height on low-back loading in cart pushing and pulling. *Applied Ergonomics,* 22(2), 117–123.

Li, S., Zhu, Z., and Adams, A.S., 1995. An exploratory study of arm-reach reaction time and eye-hand co-ordination. *Ergonomics,* 38(4), 637–650.

Marras, W.S. and Davis, K.G., 1998. Spine loading during asymmetric lifting using one versus two hands. *Ergonomics,* 41(6), 817–834.

Martin, J.B. and Chaffin, D.B., 1972. Biomechanical computerized simulation of human strength in sagittal plane activities. *AIIE Transactions,* 4, 19–28.

Mital, A., 1986. Effect of body posture and common hand tools on peak torque exertion capabilities. *Applied Ergonomics,* 17(2), 87–96.

Mital, A., and Faard, H.F., 1990. Effects of sitting and standing, reach distance, and arm orientation on isokinetic pull strengths in the horizontal plane. *International Journal of Industrial Ergonomics*, 6(3), 241–248.

Mital, A., Nicholson, A.S., and Ayoub, M.M., 1993. *A Guide to Manual Materials Handling*, London: Taylor & Francis.

Mital, A. and Sanghavi, N., 1986. Comparison of maximum volitional torque exertion capabilities of males and females using common hand tools. *Human Factors*, 38(3), 283–294.

National Back Pain Association/Royal College of Nursing, 1997. *The Guide to the Handling of Patients*, 4th ed., Teddington, Middlesex, U.K.: National Back Pain Association/Royal College of Nursing.

Nussbaum, M.A., Chaffin, D.B., Stump, B.S., Baker, G., and Foulke, J., 2000. Motion times, hand forces, and trunk kinematics when using material handling manipulators in short-distance transfers of moderate mass objects. *Applied Ergonomics*, 31(3), 227–237.

Okunribido, O.O. and Haslegrave, C.M., 1999. Effect of handle design for cylinder trolleys. *Applied Ergonomics*, 30(5), 407–419.

Peebles, L. and Norris, B., 1998. Adultdata: The Handbook of Adult Anthropometric and Strength Measurements — Data for Design Safety. DTI Publication URN 98/736, London: Department of Trade and Industry.

Peebles, L. and Norris, B., 2003. Filling "gaps" in strength data for design. *Applied Ergonomics*, 34(1), 73–88.

Pheasant, S.T. and Grieve, D.W., 1981. The principal features of maximal exertion in the sagittal plane. *Ergonomics*, 24(5), 327–338.

Pheasant, S.T., Grieve, D.W., Rubin, T., and Thompson, S.J., 1982. Vector representations of human strength in whole body exertion. *Applied Ergonomics*, 13(2), 139–144.

Pinder, A.D.J, Wilkinson, A.T., and Grieve, D.W., 1995. Omnidirectional assessment of one-handed manual strength at three handle heights. *Clinical Biomechanics*, 10(2), 59–66.

Resnick, M.L. and Chaffin, D.B., 1995. An ergonomic evaluation of handle height and load in maximal and submaximal cart pushing. *Applied Ergonomics*, 26(3), 173–178.

Resnick, M.L. and Chaffin, D.B., 1996. Kinematics, kinetics, and psychophysical perceptions in symmetric and twisting pushing and pulling tasks. *Human Factors*, 38(1), 114–129.

Rohmert, W., 1966. Maximum forces exerted by men in the zone of movement of the arms and legs [Maximalkräfte von Männern im Bewegungsraum der Arme und Beine], Forschungsberichte des Landes Nordrhein-Westfalen Research Report 1616, Koeln-Opladen: Westdeutscher Verlag. (English translation: Library Translation 1839, Farnborough: Royal Aircraft Establishment).

Rohmert, W. and Jenik, P., 1971. Isometric muscular strength in women, in *Frontiers of Fitness*, Shephard, R.J., Ed., Springfield, IL: Charles C Thomas, 79–97.

Rohmert, W., Mainzer, J., and Kahabka, G., 1987. Analyse biomechanischer und physiologischer Engpässe beim ausüben von Stellungskräften [Analysis of biomechanical and physiological bottlenecks when exerting static positional forces] *Zeitschrift für Arbeitswissenschaft*, 41(2), 114–120.

Sanchez, D. and Grieve, D.W., 1992. The measurement and prediction of isometric lifting strength in symmetrical and asymmetrical postures. *Ergonomics*, 35(1), 49–64.

Seminara, J.L. and Parsons, S.O., 1982. Nuclear power plant maintainability. *Applied Ergonomics*, 13(3), 177–189.

Shoaf, C., Genaidy, A., Karwowski, W., Waters, T., and Christensen, D., 1997. Comprehensive manual handling limits for lowering, pushing, pulling and carrying activities. *Ergonomics*, 40(11), 1183–1200.

Smith, J.L., Ayoub, M.M., and McDaniel, J.W., 1992. Manual materials handling capabilities in non-standard postures. *Ergonomics*, 35(7/8), 807–831.

Snook, S.H. and Ciriello, V.M., 1991. The design of manual handling tasks: revised tables of maximum acceptable weights and forces. *Ergonomics*, 34(9), 1197–1213.

Svensson, O.K., 1987. On Quantification of Muscular Load during Standing Work — A Biomechanical Study, Dissertation, Kinesiology Research Group, Karolinska Institute, Stockholm.

Takala, R.P. and Kukkonen, R., 1987. The handling of patients on a geriatric ward. *Applied Ergonomics*, 18(1), 17–22.

Tichauer, E.R., 1968. Potential of biomechanics for solving specific hazard problems, in *Proceedings of the ASEE 1968 Conference*, Park Ridge, IL: American Society of Safety Engineers, 149–187.

Tracy, M.F., 1995. Biomechanical methods in posture analysis, in *Evaluation of Human Work*, 2nd ed., Wilson, J.R. and Corlett, E.N., Eds., London: Taylor & Francis, 714–748.

Warwick, D., Novak, G., Schultz, A., and Berkson, M., 1980. Maximum voluntary strengths of male adults in some lifting, pushing and pulling activities. *Ergonomics*, 23(1), 49–54.

Waters, T.R., Putz-Anderson, V., and Garg, A., 1994. Applications Manual for the Revised NIOSH Lifting Equation. DHHS (NIOSH) Publication 94-110, Cincinnati, OH: National Institute for Occupational Safety and Health.

Williams, M. and Stutzman, L., 1959. Strength variations through the range of joint motion. *Physical Therapy Review*, 39(3), 145–152.

Woldstad, J.C., 1985. Human performance times in the use of manual hoist systems, in *Trends in Ergonomics/Human Factors II*, Eberts, R.E. and Eberts, C.G., Eds., Amsterdam: North-Holland, 599–608.

Woldstad, J.C., McMulkin, M.L., and Bussi, C.A., 1995. Forces applied to large handwheels. *Applied Ergonomics*, 26(1), 55–60.

13

Performance

Colin G. Drury and Victor L. Paquet

CONTENTS

Introduction

In the human factors community, performance is seen as one of the key objectives: human factors/ergonomics aims to improve human performance and well-being (Wilson 1995). Although much of the research on posture has focused on well-being, for example, the role of posture in causation of musculo-skeletal injuries, there exists a body of literature that has investigated postural effects on performance. A series of laboratory studies investigated the effect of gross body posture (lying, sitting, standing) on reaction time and found that standing reduced reaction times under certain conditions (Vercruyssen and Simonton 1994). A field study (Kim et al. 2001) found that postural and time-pressure measures together accounted for more than 50% of the variance in quality at different workstations on a production line.

Clearly, performance is highly relevant to human factors/ergonomics, and there is at least some evidence of a posture/performance relationship. In ergonomics practice the authors have encountered two "commonsense" views of the posture/performance relationship. One view holds that giving the workforce better posture should lead to higher performance, for one of a number of reasons. It could be due to a reduction in localized muscular discomfort, or perhaps to gratitude, or to a vague "Hawthorne effect" (i.e., the desire to please the management or experimenter whether the change is truly beneficial or not). Another view, primarily encountered in relation to inspection and vigilance tasks, suggests that giving workers better posture may have the effect of decreasing their alertness by encouraging comfortable dozing. The evidence that might support these possible viewpoints needs to be examined and suggestions made of possible mechanisms to explain the findings. This chapter first examines concepts and definitions of performance, then provides a model relating performance to posture, proceeds to review the relevant body of literature, and concludes with a summary of the important relationships.

13.1 Measuring Performance

For ergonomists, performance has acquired two relatively distinct meanings: physical performance and mental (or cognitive) performance. In the biomechanics and work physiology communities, performance has come to mean the achievement and maintenance of a level of physical output. Thus, the performance of a muscle can be defined by its maximum force production (maximum voluntary contraction, MVC) and/or its ability to maintain a given force over time. Performance as maintenance of a level of physical output is defined as endurance, and usually bears a known relationship to the fraction of MVC demanded (Rohmert 1973, Deeb and Drury 1990). Such long-duration exertion of muscular force can have direct relevance to posture, where stability must be maintained. Similarly, physical performance can include aerobic work, such as running, walking, or lifting and carrying objects. Here, performance can be maintained for long periods only at a fraction of the worker's maximum energy expenditure. In tasks requiring high muscular forces or heavy physiological loading, performance is directly limited by these mechanisms. Posture clearly affects the ability to exert forces (e.g., Ayoub and McDaniel 1974, Haslegrave et al. 1987) and also to perform repetitive heavy physical work (Snook 1978). These direct effects of posture on physical performance have been dealt with earlier in this book and are not covered again here.

Posture does not affect the second form of performance, mental or cognitive work, in the same way as it does heavy physical work. However, as the two examples mentioned earlier showed, there may be some important effects. In fact, studies of the effects of posture on performance are a subset of a research tradition examining the effects of physical work on performance (e.g., Mozrall and Drury 1996). To explore this further, performance needs to be defined more specifically.

Performance can be defined as achieving success in a task. The objective of any task is its successful completion, whether in manual materials handling, driving, marathon running, computer interaction, or process control. Success may be defined in different ways, but it typically consists of keeping speed and accuracy within known bounds. Speed is usually measured by task completion time (T) (for example, driving to work in time to start) while accuracy is usually measured in terms of error rate (E) (for example, crashing or breaking laws while driving to work). Thus, success implies:

$$T_{min} \leq \text{performance time} \leq T_{max} \tag{13.1}$$

$$E_{min} \leq \text{performance error rate} \leq E_{max} \tag{13.2}$$

As a practical matter, the lower limits (T_{min}, E_{min}) are typically zero as there is no penalty for too rapid or too accurate task completion. Within this definition of success, there is often an implied performance level ranking such that lower performance time and lower errors are better for reasons of productivity and quality. This even applies in overtly physical sporting events such as marathons (time) and soccer (inaccurate passes). Because task success is such a large driver of all activities, from survival to sports to work, performance needs to be measured. Performance is measured by the set of two constituent measures, speed and accuracy. This is subject to some restrictions on such well-being (W) factors as health and comfort, as, for example, task objectives must be achieved without significant levels of injury, physical overexertion, discomfort, or dissatisfaction that ultimately prevent the task from being completed. Thus,

$$\text{Performance} = [\text{speed, accuracy}] \tag{13.3}$$

subject to

$$W_{min} \leq \text{well-being} \leq W_{max} \tag{13.4}$$

where W_{min} and W_{max} are limiting values of well-being. As was noted for the lower limits of performance, the upper limit of W_{max} limit is typically not a practical limitation.

Well-being can also imply a longer-term performance measure, which may be termed task completion. If the task is part of a job that continues over extended time (weeks or months), then from both an individual's and an employer's viewpoint it is important for the same person to remain on the job long enough for task completion. If not, there are costs associated with labor turnover (e.g., compensation, hiring, or training) as well as with temporary absence from work (e.g., rescheduling, overtime for others, or output reduction). In civilian and military work, as well as in sports contexts, absenteeism and labor turnover represent a long-term performance measure that must be taken into account in any system. Thus, a third performance measure of task completion (C) can be added:

$$C_{min} \leq \text{task completion} \leq C_{max} \tag{13.5}$$

And the set of performance measures must be expanded to include:

$$\text{Performance} = [\text{speed, accuracy, task completion}] \tag{13.6}$$

subject to Equation 13.4:

$$W_{min} \leq \text{well-being} \leq W_{max}$$

There are occasions when we wish to maximize performance, as in competitive sports or emergency operations. Equally, there are tasks and occasions when a given level of performance must be satisfied but higher levels are not necessary. For example, when people walk to their workplace from the bus stop or parking lot, they are not usually concerned with minimizing transit time provided this time is below some acceptable limit. People can be satisfiers as much as optimizers (Simon 1988).

13.2 Speed–Accuracy Trade-Off in Performance

To some extent, performance can be traded for well-being, as in twisting while lifting to save time but increasing the chances of injury. However, the more common trade-off is between speed and accuracy, known as speed–accuracy trade-off (SATO) (e.g., Wickens 1992, Drury 1994). This can occur in performance of industrial tasks, as in defect detection as inspection speed is increased (Drury 1992) or in sports performance, for example, as reduced accuracy of passing in team sports under time pressure.

There are many quantitative models of SATO for different tasks. For example, random walk models of reaction time (Sperling and Dosher 1986) imply a reduced reaction time if the criterion of accuracy is relaxed. Other examples include Fitts' law (Fitts 1954), where a reduced time is required for low-accuracy movements compared to high-accuracy movements, or self-paced tracking (DeFazio et al. 1992), where cars can be driven more quickly as road width increases.

Because SATO exists, any measurement of task performance must take both speed and accuracy into account. If speed is measured without accuracy, or vice versa, then an incomplete and perhaps misleading picture of task performance results. In practice, this gives four ways to measure task performance:

1. *Fix speed, measure accuracy.* This is the typical measure for externally paced tracking tasks (e.g., Hess 1997) where the speed at which the target moves is externally determined. The performance measure is derived from the deviation between target location and follower location, for example, root-mean-square (RMS) error. Thus, in flying an aircraft manually, the RMS error in altitude would be an appropriate measure of one aspect of flying performance. Accuracy measures at fixed speeds are typical of paced tasks. An example referenced earlier (Kim et al. 2001) was a production line in a manufacturing plant run at constant speed. Quality, essentially an error measure, at this fixed speed, was measured.

2. *Fix accuracy, measure speed.* In self-paced tasks, compliance with a given level of accuracy is often assumed. For example, in control of discrete accurate movements (Gan and Hoffmann 1988) a target of fixed size must be reached in minimum elapsed time. The accuracy is defined by the Fitts' law index of difficulty and is related directly to speed of movement. Other examples include self-terminating search tasks (e.g., Drury and Hong 2000) where a target must be located in the minimum time to achieve task success.

3. *Ask for some defined combination of speed and accuracy.* The weighting between speed and accuracy can be defined by the context (e.g., manufacturing system) or be artificially engineered by the experimenter. The participant then chooses a performance strategy to match the rewards offered by the system. Thus, in many older uses of financial incentives at work (e.g., Barnes 1980) both fast times and low error rates had predefined financial rewards, or corresponding penalties for underachievement. Payoff matrices were the chosen incarnation of this idea in the "human operator as economic maximizer" vogue of the past generation (Sperling and Dosher 1986). Payoff matrices provide some guidance to the participant in the relative importance of speed and accuracy, simplifying the interpretation of mixed (speed, accuracy) outcomes for the experimenter.

4. *Allow each participant to define performance.* If experimenters, or managers, do not have prespecified relationships between the values of speed and accuracy, participants must invent their own criteria to make any meaningful performance possible. This may be realistic to some extent, although most employees are quite sensitive to the reward structures implied by the actions and reactions of their colleagues and managers. For example, well-publicized terrorist actions shift the implied reward structure of the aviation security system toward accuracy and away from speed. In more abstract settings, such as experiments, it may be most instructive to observe and

understand the trade-off strategies developed by participants in complex tasks. The strategy may be of greater interest than the performance itself as it allows us to understand why operators take a particular action, for example, in complex process control tasks (Moray et al. 1986).

There have been comparisons between some of these four measurement techniques. For example, Drury and Forsman (1996) used the first two in a visual search task and found comparable results for the SATO relationship.

The above discussion emphasizes only one set of complementary effects of speed and accuracy, those pertaining to what Norman and Bobrow (1979) termed *resource limited tasks*. There are other tasks, known as *data limited*, where no amount of additional time will improve performance. There are yet other tasks where multiple aspects of accuracy are possible, for example, signal detection tasks where missed signals covary with false alarms (McNichol 1972).

The main point to be made, however, is that performance in itself may be a complex construct to measure, and that inadequate measures can blur any conclusions one wishes to make regarding relationships between posture and performance.

13.3 Mechanisms and a Model of Posture and Performance

The relationship between posture and performance is only one manifestation of the more general relationship between physical work and task performance. If all types of physical work were the same and all tasks identical, then simple physical/cognitive relationships could reasonably be expected to emerge strongly from experimental and field studies. However, as Mozrall et al. (2000) found in their review and modeling, both physical work and mental tasks take many forms so that any consistent relationships emerge over much more restricted ranges of work and task. Too often, these relationships have been studied by physiologists with a somewhat naïve view of cognitive functioning, or by psychologists with restricted knowledge of the complexities of physical functioning.

Mozrall and Drury (1996) provided a framework for studying combined physical or mental tasks by, as noted earlier, extending Wickens' idea of multiple (mental) resources to provide a taxonomy of physically limiting human subsystems. In this more restricted review of posture and performance, limitations imposed by the postural muscles maintaining a required physical configuration (posture) for (perhaps) extended time periods are primarily dealt with by considering such postural work as taxing the ability of muscles to exert small and relatively static forces over time (i.e., static muscular endurance). The static force that can be exerted depends in a lawful way on the endurance time (e.g., Deeb and Drury 1990). Any model of the physical aspects of the task must include both the force to be exerted and a temporal description of the exertion. The force exerted should logically be expressed as a fraction of the maximum force available, while the temporal description needs to include durations of both exertions and rests between exertions. Some of these "rests" may arise from voluntary posture shifts that allow different muscle groups to take over posture maintenance, perhaps temporarily, so that primary postural muscles can experience some degree of recovery. A representation of the postural activity would then be given by the force/duration relationship for any muscle, as, for example:

$$\text{Postural Activity (muscle group } i) = F_i(t) \qquad (13.7)$$

A simple model of such a process might have the postural activity $F_i(t)$ of the N muscles affecting task performance:

$$\text{Performance} = f((F_1(t), F_2(t), \ldots F_N(t)) \qquad (13.8)$$

However, this would be too simplistic to represent performance, which is likely to be multidimensional as noted in Section 13.2. It would also ignore the potential for active feedback loops such as the voluntary posture shifts noted above, where the participant is not merely the mechanical exerter of a given set of postural forces, but an active determiner of the time course of these forces within limits imposed by the task and workplace. Just as an earlier generation expanded its simple cause-and-effect model of stress (e.g., Selye 1950) into a more active transactional model (e.g., Cox and Griffiths 1995), the notion of posture causing performance changes will probably need to be expanded into a more extended model.

One model to attempt this, while still a known oversimplification, is at least a start (Liao and Drury 2000). Figure 13.1 shows an update of their model in which the characteristics of the job, workplace, and individual impact task performance through postural mechanisms, as well as through other mechanisms that are independent of the individual's posture. The job characteristics include the sequence, frequency, and duration of tasks performed by the worker. The workplace characteristics include the physical characteristics of the environment (e.g., lighting, noise, temperature), as well as the layout characteristics of the tools and equipment available to the worker. Qualities of the individual include experience or skill level, motivation, strength, endurance, and, in some cases, body size. In this model, the combination of the job, workplace, and individual's characteristics together determine the work methods that are used.

Time on task and the work methods together define the resulting posture and the time variation of muscle activity from postures $F_i(t)$, although not perhaps in a direct manner. The forces exerted by the muscles over a period of time will result in the sensation of discomfort (e.g., Corlett and Bishop 1976) as well as loss of force production capability (Rohmert 1973). Either

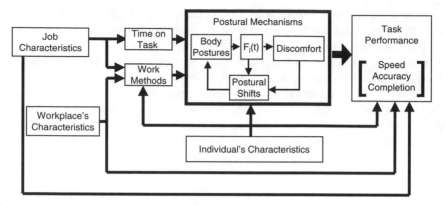

FIGURE 13.1
Model of the factors that affect task performance, which include, but are not limited to, postural mechanisms. $F_i(t)$ = pattern of muscle force exerted over time.

or both of these mechanisms can drive postural shifts, creating a postural feedback loop. The physical and anatomical characteristics of the individual (e.g., body segment sizes and weights, and the orientation of tendons with respect to the joint's center of rotation) will also impact these postural mechanisms directly. The outcomes of this whole system are then postulated to affect performance, although again perhaps not in a direct manner. Performance itself is a potentially complex phenomenon, as discussed above, so that postural stresses may affect some tasks (e.g., movement control) but not others (e.g., working memory). Task performance is also known to be affected by many other factors, both external and internal to the individual, in addition to any postural effects.

One way in which postures affect performance is through direct mechanical interference: a drawing task will probably be adversely affected if the drawing arm must also be used to aid postural stability. This mechanism through which posture affects task performance is related to whether or not the individual's body is positioned ideally at a given time during the task. For example, an individual's ability to detect a target may be impeded when the orientation of the eyes relative to the display is unfavorable, or the finger activation of controls may be slowed when the hands are oriented in an unfavorable position. A less obvious mechanism is that the demands on the postural system draw resources from the pool available for task completion. For example, the postural discomfort may be so distracting that tasks requiring complex cognitive manipulation (e.g., process control, inspection) degrade from the interruptions caused by this distraction. Subtle mechanisms may come into play, such as reduced task motivation in protest against workplace discomfort, or even increased motivation to complete the task early and hence relieve the discomfort. With sufficient resources, a person will be able to maintain the desired level of task performance during the task, even when the postures are unfavorable. As these resources are depleted with time, however, task performance may suffer in terms of speed and accuracy, and lack of resources may even, in extreme cases, prevent successful completion of the task.

All three of these postural mechanisms (direct mechanical interference, resource depletion, and reduced task motivation) may together affect task performance, although some experiments attempt to rule out one or more. For example, Mozrall et al. (2000) examined the effect of spatial restriction on posture by using a simulated visual inspection task. They ensured that their extreme physical space restrictions did not physically interfere with the task to rule out the influence of direct mechanical interference. As explained in more detail in Section 13.4.3, they found large effects of space restriction on posture, discomfort, and postural shifts but almost no significant effects on task performance, either speed or accuracy of inspection. The model in Figure 13.1 is consistent with such a finding, in that performance can be determined by many factors, of which posture is only one. Even extreme postural changes may not affect what is rather arbitrarily defined as task performance.

If the model in Figure 13.1 implies that posture would not always be expected to affect performance, it also implies that, if some of the intervening variables such as discomfort or postural shifts are neglected, posture may not be able to be understood well enough to determine any relationship to performance.

At this point it is appropriate to review the evidence in the literature for relationships between posture and the performance of tasks. Neither posture nor performance may have been defined or measured in a way consistent with the authors' views on the complexity of posture and performance. To expect a consistent relationship to emerge from the potentially disparate studies is perhaps still premature.

13.4 Studies of Posture and Performance

While few will argue that the relationship between working body posture and task performance is not an important human factors issue for the design of work systems, it is surprising how little research exists on the subject (Thomas et al. 1991). Much of the literature devoted to the evaluation of postures at work or in laboratory simulations of tasks emphasizes the effects postures have on an individual (e.g., muscular discomfort or illness, loss of balance, reduction in strength), with few studies evaluating the impact of these effects on task performance.

Some field studies provide information about the general relationships between working postures and productivity, but research that considers the effects of postures on individual performance is generally limited to the laboratory. A brief summary of some of the relevant fieldwork is first provided, followed by more detailed summaries of the laboratory research.

Because the measures of performance may vary according to the tasks that are performed, the relevant findings of the laboratory activities have been organized according to the types of tasks studied.

13.4.1   Field Studies

Field research related to posture and performance has attempted to study how productivity is affected by changes to tools, workstations, or workstation layout that are thought to be associated with changes in working postures. These studies have employed large groups of people and typically involved characterizing the relationships between the postures assumed by groups of workers and the overall performance for the group (e.g., relationship between changes in frequency of awkward postures within a department and changes in productivity for that department). These studies have rarely considered the work methods employed by individuals under different sets of work characteristics or the characteristics of the individuals, which would be needed to study the mechanisms by which postures affect task performance. In these studies, postures are manipulated indirectly with changes to the work environment (e.g., use of a new workstation or tool) and performance is usually evaluated with some measure of speed (e.g., productivity) or task completion (e.g., turnover or absenteeism). For example, Spilling et al. (1986) evaluated the impact of interventions (e.g., adjustable workstation, addition of armrest) designed to improve working posture with the goal of reducing labor turnover and sick leave in a population-based study at an assembly and wiring plant. They showed that these changes corresponded to improvements in recruitment costs, training, and sick leave payments. Similar benefits of workplace improvements on job performance have been reported by others (e.g., Hasslequist 1981).

Other studies have documented either the effects of body postures on muscular loading through, for example, the reporting of muscular discomfort experienced during the task, or have actually measured the types of and frequencies of postures employed during a task. For example, Corlett and Bishop (1976) studied spot welders' utilization of machines in terms of mean length of work period, and found that utilization of machines increased when perceived discomfort was reduced. In another study of welding, Heinsalmi (1986) used the Ovako Working posture Analyzing System (OWAS) (Karhu et al. 1977, 1981) to document awkward postures in welding tasks. They made improvements to the job that reduced the frequency of awkward postures and found that the improvements led to a 90% reduction in required task time. Wangenheim et al. (1986) measured the frequency of awkward trunk postures and productivity of concrete workers while using different concrete pouring machines. The machine associated with the lowest frequency of awkward low back posture resulted in the highest productivity in terms of volume of concrete poured per unit time.

The ability of a company to ensure that tasks are completed effectively and efficiently depends, at least in part, on employee injury, turnover, and absenteeism. Awkward postures and static muscular exertions are considered to be work-related risk factors for musculoskeletal disorders that clearly affect injury rate, turnover, and absenteeism. Although these outcomes are measures of task performance, the reader is referred to the introduction of this book and other chapters that deal specifically with the role of posture on the risk of development of musculoskeletal disorders for a discussion of the topic.

The field studies described in this section have high face validity, but it is impossible to determine exactly what role the change in working postures had on the outcomes, as well as which postural mechanisms affected performance. These studies rarely considered the work methods, time on task, and the individual characteristics of the employees simultaneously, which would be needed to study the mechanisms by which posture affects task performance. Additionally, performance has rarely been measured specifically in terms of speed and accuracy. Laboratory studies provide a few more insights about the mechanisms by which postures impact performance.

13.4.2 Reaction Time and Tracking Tasks

The studies described below have demonstrated that very large differences in body posture may have an effect on reaction time (RT) and tracking performance for some people (Figure 13.2). This could be the result of the postural mechanisms affecting task performance described earlier, which could include muscle fatigue, reduced arousal or motivation, or differences in the orientation of the display with respect to the eyes or the orientation of the controls with respect to the hands.

Several studies have provided some evidence of a relatively weak relationship between posture and reaction time. Vercruyssen et al. (1988) found that in a visual choice reaction time (VCRT) task, participants performed

FIGURE 13.2
There is evidence to demonstrate that reaction time differs across postures such as those shown, for some people under some task conditions of varying difficulty. (Modified from Vercruyssen and Simonton 1994.)

faster standing than sitting. Vercruyssen et al. (1989) reported an improvement in reaction time in standing rather than seated postures for males in a degraded stimulus condition, when they required participants to respond to the direction of an arrow (pointed either left or right) that was either degraded or of high image quality. Cann (1990a,b) tested the effects of posture (lying, sitting, and standing), age, gender, and task difficulty on reaction time and tapping performance, and found that task difficulty, age, and posture together affected task performance. Reaction time appeared faster for standing postures; however, older men benefited less from these postures, particularly when the task was more difficult. Simonton et al. (1991) found that reaction time for standing postures was slightly faster than for seated postures but that they resulted in a higher frequency of errors, which implies a speed–accuracy trade-off change rather than an overall improvement in performance. In other studies, significant associations between postures and reaction time have not been found. For example, Diggles and co-workers (cited in Vercruyssen and Simonton 1994) looked at the effects of inverted (i.e., upside down) vs. upright postures and did not find significant postural effects, although a slight (nonsignificant) decrease in simple reaction time was found for upright postures.

There is also some evidence to suggest that some seated postures have an impact on tracking performance. For example, Deaton and Hitchcock (1991) found differences in tracking performance measures for different seat pan recline angle conditions (27° and 67° recline).

The results of studies such as these may have important implications on the design of the systems in which tracking and reaction time is of critical importance (e.g., combat aircraft control), although the lack of strong posture effects independent of other factors does not provide conclusive evidence on the benefits of one posture over another. Postures cannot be recommended at this time that optimize reaction time and tracking performance outside of the laboratory. For example, while standing postures may improve reaction time for some people in some situations, it is much too early to apply the findings to the design of complex tasks such as driving (Figure 13.3).

13.4.3   Visual Inspection Tasks

Clearer postural effects on task performance have been found for visual inspection tasks. For example, Edwards et al. (1994) evaluated the impact of a prolonged reclined posture on performance in a target detection task performed by 12 male military personnel. Participants performed the task for the 70 min of the experiment in each of four seated posture conditions (upright, and reclined 25°, 45°, and 65°). Results showed that increasing the angle of recline was associated with increases in body discomfort, sleepiness, and stress, and with decreases in performance as measured by the number of targets detected.

FIGURE 13.3
In driving, tracking performance and reaction time are important, but there is little reason to believe that a standing driver will outperform a seated driver!

The results of a more general study by Mozrall et al. (2000) demonstrated weak but measurable differences in inspection performance for different postural conditions in short-term inspection tasks (Figure 13.4). In this study the effects of restricted spaces and postures on inspection task performance were evaluated. Portable walls were used to manipulate the amount of vertical workspace available for the task, and measurements were made of body posture holding times, heart and respiration rates, perceived workload with the NASA Task Load Index (NASA TLX) (Hart and Staveland 1988), body part discomfort (Corlett and Bishop 1976), and task performance measures of speed of task completion and accuracy. Groups of eight men and women participants performed two 12-min inspection tasks in one of nine possible workspace arrangements. Groups restricted vertically (i.e., low work heights in which participants typically worked while in trunk flexion) reported higher physical demands, experienced greater discomfort, and completed on average fewer trials than those who were not restricted in their working postures. The effect of posture on task completion speed, however, was not statistically significant. The results demonstrated that use of awkward working postures clearly has physiological and psychophysical effects, but that the impact on performance is relatively modest, at least in the short-term.

Bhatnager et al. (1985) studied a different inspection task that allowed the effects of posture on performance to be studied over a longer duration. For this study, four males were placed in one of three postural conditions that were manipulated by three different display heights, and inspected circuit boards for 3 h with two 5-min breaks per hour. Postures were videotaped, and positions of head, arms, back, and legs and reports of muscular discomfort were recorded. Inspection time, search time, and the frequency of two types of error (search errors and false alarms) were also measured. The "high display"

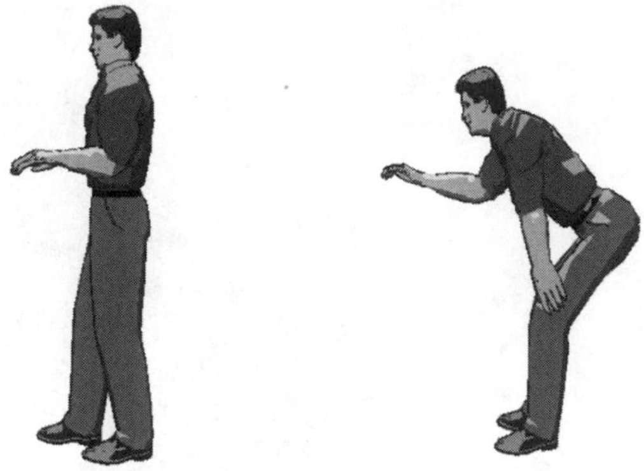

Measure	Control (Unrestricted)	Vertical Restriction
Posture Holding Time (sec)	56.5	34.0
Physical Demands (NASA TLX)	3.6	18.0
Change in Trunk Discomfort	0.6	3.3
Trials Completed	15.4	14.2
Search Hit Rate	0.95	0.96

FIGURE 13.4
Mean levels of reported physical workload, muscular discomfort, and task performance for two postural conditions in a 12-min inspection task. Vertically restricted postures resulted in lower posture holding times, higher physical workload scores, and greater increases in discomfort, with a reduction in the number of completed trials. Search hit rates were unaffected. (Modified from Reynolds 1994.)

condition was associated with the "worst" head-neck postures and worst task performance. As time-on-task increased during the tasks, participants tended to lean forward, change posture more frequently, report more discomfort, take more time inspecting circuit boards, and make more errors when inspecting the circuit boards.

13.4.4   Keyboarding Tasks

Much of the literature related to postures and performance has been focused on keyboarding tasks. Perhaps this is because the nature of these tasks allows for relatively easy postural assessment due to the relatively static body postures of the trunk and shoulders, and because quantifiable measures of task performance that include keystroke rate and typographical error rates can be obtained easily. Studies have often involved manipulating a workstation characteristic, such as keyboard height or orientation, and measuring upper extremity postures, muscular activation, or discomfort as well as keyboarding performance under controlled laboratory conditions (Figure 13.5). In studies such as these, it is implied that performance may be affected by the level of muscle fatigue of the upper extremities or shoulders experienced during the different experimental conditions. The postures assumed during different work conditions are thought to cause fatigue of

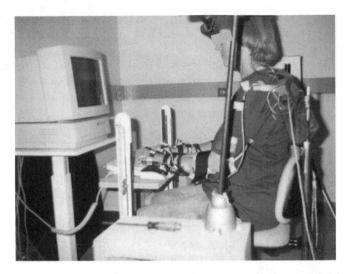

FIGURE 13.5
A research participant uses a wrist support while keyboarding with the eyes located at a fixed distance from the computer monitor. Wrist postures and muscle activity are monitored to evaluate postural stress, while typing error rates are recorded.

the forearm and hand muscles and to lead to reduced speed and accuracy of typing either directly (by impairing finger and wrist movements) or indirectly (by distraction). Differences in performance for different workstation conditions may also be the result of changes in the orientation of the hands and fingers relative to the keyboard. Shoulder muscle fatigue may also distract a keyboarder or may alter the upper arm posture, which in turn might affect the orientation of the hands relative to the keyboard.

In spite of the plausible postural mechanisms thought to affect performance, studies of posture and keyboarding demonstrate only weak effects, at best. While some studies such as Smith et al. (1998) provide some evidence that use of a wrist support appears to improve speed (keystrokes per minute), many more studies have failed to find relationships between postures and performance during keyboarding. Studies have shown that while changes in keyboard height affect wrist postures, keying performance often remains unchanged (e.g., Cushman 1984). Others have found that keyboard slope has minimal impact on keyboarding performance (e.g., Suther and McTyre 1982), while still others have shown minimal benefits to keying performance when a wrist support is used (Bendix and Jessen 1986). The lack of clear and measurable effects of posture on keyboarding performance may, in some cases, be related to the design of the earlier studies, which usually allowed discomfort and performance to vary as dependent variables. If, for example, a study was designed to require participants to maintain a particular level of discomfort across workstation conditions, differences in task performance would be more likely. Nevertheless, results of many studies suggest that individuals have the ability to adapt or cope with little or no performance change in spite of the changes in the temporal pattern of muscle activity and body orientation with respect to the display and keyboard when keyboarding tasks are performed.

Changes to computer workstations that affect working postures and muscular discomfort often have very modest effects on typing performance. A study by Porter et al. (1992) evaluated the effects of a tilting workstation on posture, muscular discomfort, and typing performance during a keyboarding task. They measured time for task completion and the numbers of errors made in a standardized typing test. Workstation design had a significant positive effect on neck and head flexion, ulnar deviation, and neck, buttocks, wrist, and hand discomfort, but did not have a significant effect on typing performance. Pustinger et al. (1985) evaluated the effects of an adjustable workstation on posture, task performance, and muscular discomfort, with participants performing a 3-h self-paced typing task at an adjustable workstation on one day and at a fixed workstation on another day. Categories of upper extremity, trunk, and lower extremity postures were recorded as they changed during the task, as well as the numbers of keystrokes, corrections, uncorrected errors, and backspaces for each of the sessions. The results demonstrated that an adjustable workstation was associated with fewer postural changes and less muscular discomfort than a non-adjustable workstation. However, there were no significant differences in keying performance between the two workstation types. Life and Pheasant (1984) evaluated how keyboard height and presence of a copy stand affected biomechanical stress, discomfort, and typing performance. Participants included 12 skilled female typists who worked at four different working heights, half of them using a copy stand while the others read materials that were placed on a flat table. Measurements included typing speed, inter-keystroke times, and muscular discomfort, as well as biomechanical evaluation of the postures assumed during the tasks. Despite an increase in estimated torque for the shoulder and an increase in overall discomfort for the higher keyboard conditions, performance was not affected. Hedge and Ng (1995) investigated the effects of a split keyboard with a central trackball on performance, posture, and comfort. In the study, 12 female typists participated and measurements included postures recorded with an electrogoniometer, discomfort, typing speed, errors, and keyboarding performance. In this study, no significant effects were found for wrist posture, discomfort, or typing performance.

For some of the more radical changes to keyboarding tasks, improvements in working postures were actually found to be associated with decrements in task performance. For example, Hedge and Shaw (1996) conducted a laboratory study in which 12 female typists performed tasks with a chair-mounted split keyboard (one half of the keyboard on the left armrest, the other on the right) and with a conventional keyboard to evaluate the effects of the chair-mounted split keyboard on body posture and performance. Wrist, shoulder, and neck postures were measured with electrogoniometry or video-based motion analysis, and speed and accuracy of typing were recorded. Results suggested that use of the chair-mounted split keyboard resulted in less stressful shoulder and back postures, but decreased typing speed without having an effect on accuracy. In a related study, Muss and Hedge (1999) evaluated the effect of a chair-mounted vertical split keyboard and forearm supports on wrist posture (measured with electrogoniometers), comfort, keyboarding speed, and accuracy. In this study, 12 experienced female typists performed tasks using this keyboard and a traditional keyboard. Better postures were found for the new keyboard but typing speed and accuracy were reduced. The reduction in performance in studies performed by Hedge and Shaw (1996) and Muss and Hedge (1999), however, is very likely due to a lack of experience or training with the new equipment, rather than to the effects of posture on performance.

Postures may begin to affect performance when working over fairly long time periods of continuous typing. Liao and Drury (2000) studied the effects of keyboard height and work duration of up to 2 h of typing on torso and upper extremity angles (measured in the sagittal plane), muscular discomfort of individual body areas, and keying performance (measured by mean keystroke rate, accuracy, and error correction frequency). Keyboard height had significant effects on all of the postural angles, and on body part discomfort of the right upper arm, but did not impact keying performance. Work interval did not have a significant effect on the magnitude of the postural angles over time but did have an effect on the frequency of upper body and overall postural shifts and on error rates. Discomfort and non-work-related postural shifts, such as head scratching, had weak associations with error rates.

13.5 Conclusions

Although postural mechanisms may affect speed, accuracy, and the full completion of tasks, task performance is also determined by characteristics of the job, workplace, and individual that are independent of posture. Many of the early studies suffered from limitations that may prevent a clear understanding of how postures affect task performance. In spite of the large number and complexity of factors that affect job performance, and in spite of the study limitations, the literature demonstrates that, in some cases, body postures sometimes have measurable effects on task performance.

Summary

In this chapter performance is defined in terms of speed, accuracy, and completion. Four methods of measuring task performance in terms of speed and accuracy are summarized, although only one method has been used extensively in past studies. A model is proposed in which the characteristics of the job, the workplace, and individual affect the postural mechanisms. These postural mechanisms, along with characteristics of the job, workplace, and individual that are independent of posture, are thought to affect task performance. The interactions between these characteristics and working postures on task performance are extremely complex, making it quite difficult to determine exactly how working posture influences performance. Although many studies have failed to find clear effects between working postures and task performance, this may be attributed to limitations in study design such as lack of control over job, workplace, or individual factors and short measurement periods. Some laboratory studies of reaction time, inspection, and keyboarding tasks of fairly long duration do show weak but measurable effects.

Large gaps in the understanding of the effects of posture on task performance still remain. Many tasks remain unstudied, while others have not been studied in enough detail to inform job design. It is still not yet possible to answer important design questions such as:

- Does seated posture in an automobile correlate with driving performance?
- Are there preferable postures for computerized inspection tasks, such as in the inspection of luggage in airports?
- Will body postures affect keyboarding performance over time periods of 6 or 8 h?
- What are the most important postural mechanisms that affect performance for different tasks? What are the effects on speed and accuracy?
- Can productivity and quality arguments be used to make the case for postural improvements? What are the types of errors that are most affected by postures?

Clearly, there is a need for additional research to study the effects of postures on task performance that will enable engineers to make informed decisions about appropriate the design of jobs and job tasks in the future.

References

Ayoub, M.M. and McDaniel, J.W., 1974. Effects of operator stance on pushing and pulling tasks. *AIIE Transactions*, 6, 185–195.

Barnes, R., 1980. *Motion and Time Study Design and Measurement of Work*, New York: John Wiley & Sons.

Bendix, T. and Jessen, F., 1986. Wrist support during typing — a controlled electromyographic study, *Applied Ergonomics*, 17(3), 162–168.

Bhatnager, V., Drury, C.G., and Schiro, S.G., 1985. Posture, postural discomfort and performance. *Human Factors*, 27(2), 189–199.

Cann, M.T., 1990a. Age and Speed of Behavior: Effects of Gender, Posture-Induced Arousal, and Task Loading, Master's thesis in human factors, Los Angeles, CA: University of Southern California.

Cann, M.T., 1990b. Causes and correlates of age-related cognitive slowing: effects of task loading and CNS arousal, in *Proceedings of the Human Factors Society 34th Annual Meeting*, Santa Monica, CA: Human Factors Society, 149–153.

Corlett, E.N. and Bishop, R.P., 1976. A technique for assessing postural discomfort. *Ergonomics*, 19(2), 175–182.

Cox, T. and Griffiths, A., 1995. The nature and measurement of work stress: theory and practice, in *Evaluation of Human Work*, 2nd ed., Wilson, J.R. and Corlett, E.N., Eds., London, Taylor & Francis, 783–803.

Cushman, W.H., 1984. Data entry performance and operator preferences for various keyboard heights, in *Ergonomics and Health in Modern Offices*, Grandjean, E., Ed., London: Taylor & Francis, 495–504.

Deaton, J.E. and Hitchcock, E., 1991. Reclined seating in advanced crewstations: human performance considerations, in *Proceedings of the Human Factors and Ergonomics Society 35th Annual Meeting*, Santa Monica, CA: Human Factors and Ergonomics Society, 132–136.

Deeb, J.M. and Drury, C.G., 1990. A methodology for muscular isometric endurance and fatigue research, *International Journal Industrial Ergonomics*, 6, 235–260.

Defazio, K., Wittman, D., and Drury, C.G., 1992. Effective vehicle width in self-paced tracking. *Applied Ergonomics*, 23(6), 382–386.

Drury, C.G., 1992. Performance: do speed and accuracy trade off? in *Proceedings of the 28th Annual Conference of the Ergonomics Society of Australia*, London: Taylor & Francis, 11–22.

Drury, C. G., 1994. The speed–accuracy trade-off in industry. *Ergonomics*, 37, 747–763.

Drury, C.G. and Forsman, D.R., 1996. Measurement of the speed accuracy operating characteristic for visual search. *Ergonomics*, 39(1), 41–45.

Drury, C.G. and Hong, S.-K., 2000. Generalizing from single target search to multiple target search. *Theoretical Issues in Ergonomics Science*, 1(4), 303–314.

Edwards, R.J., Streets, D.F., and Bond, G., 1994. The effects of back angle on target detection, in *Proceedings of the Human Factors and Ergonomics Society 38th Annual Meeting*, Santa Monica, CA: Human Factors and Ergonomics Society, 1252–1255.

Fitts, P.M., 1954. The information capacity of the human motor system in controlling the amplitude of movement. *Journal of Experimental Psychology*, 47, 381–391.

Gan, K.C. and Hoffmann, E.R., 1988. Geometrical conditions for ballistic and visually-controlled movements. *Ergonomics*, 32, 829–840.

Hart, S.G. and Staveland, L.E., 1988. Development of NASA TLX: results and empirical and theoretical research, in *Human Mental Workload*, Hancock, P.A. and Meshkati, N., Eds., Amsterdam: Elsevier Science, 139–183.

Haslegrave, C.M., Tracy, M., and Corlett, E.N., 1987. Industrial maintenance tasks involving overhead working, in *Contemporary Ergonomics 1987*, Megaw, E.D., Ed., London: Taylor & Francis, 197–202.

Hasslequist, R.J., 1981. Increasing manufacturing productivity using human factors principles, in *Proceedings of the Human Factors Society 25th Annual Meeting*, Santa Monica, CA: Human Factors Society, 204–206.

Hedge, A. and Ng, L., 1995. Effects of a fixed-angle, split keyboard with center trackball on performance, posture and comfort compared with a conventional keyboard and mouse, in *Proceedings of the Human Factors and Ergonomics Society 43rd Annual Meeting,* Santa Monica, CA: Human Factors and Ergonomics Society, 957.

Hedge, A. and Shaw, G., 1996. Effects of a chair-mounted split keyboard on performance, posture and comfort, in *Proceedings of the Human Factor and Ergonomics Society 40th Annual Meeting,* Santa Monica, CA: Human Factors and Ergonomics Society, 624–638.

Heinsalmi, P., 1986. Method to measure working posture loads at working sites (OWAS), in *The Ergonomics of Working Postures,* Corlett, N., Wilson, J., and Manenica, I., Eds., London: Taylor & Francis, 100–108.

Hess, R., 1997. Feedback control models — manual control and tracking, in *Handbook of Human Factors,* Salvendy, G., Ed., New York: John Wiley & Sons, 1249–1294.

Karhu, O., Kansi, P., and Kuorinka, I., 1977. Correcting working postures in industry: a practical method for analysis, *Applied Ergonomics,* 8, 199–201.

Karhu, O., Harkonen, R., Sorvali, P., and Vepsalainen, P., 1981. Observing working postures in industry: examples of OWAS application. *Applied Ergonomics,* 12(1), 13–17.

Kim, S.-W., Drury, C.G., and Lin, L., 2001. Ergonomics and quality in paced assembly lines. *Human Factors and Ergonomics in Manufacturing,* 11(4), 1–6.

Liao, M.-H. and Drury, C.G., 2000. Posture, discomfort and performance in a VDT task. *Ergonomics,* 43(3), 345–359.

Life, M.A. and Pheasant, S.T., 1984. An integrated approach to the study of posture in keyboard operation. *Applied Ergonomics,* 15(2), 83–90.

McNichol, D., 1972. *A Primer of Signal Detection Theory,* Sydney, Australia: Allen & Unwin.

Moray, N., Lootseen, P., and Pajak, J., 1986. The acquisition of process control skills. *IEEE Transactions,* SMC-16, 497–505.

Mozrall, J.R. and Drury, C.G., 1996. Effects of physical exertion on task performance in modern manufacturing: a taxonomy, a review and a model. *Ergonomics,* 39(10), 1179–1213.

Mozrall, J.R., Drury, C.G., Sharit, J., and Cerny, F., 2000. The effects of whole-body restriction on inspection performance. *Ergonomics,* 43(11), 1805–1823.

Muss, T.M. and Hedge, A., 1999. Effects of a vertical split-keyboard on posture, comfort and performance, in *Proceedings of the Human Factors and Ergonomics Society 43rd Annual Meeting,* Santa Monica, CA: Human Factors and Ergonomics Society, 496–500.

Norman, D.A. and Bobrow, D.J., 1979. On data limited and resource limited processes. *Cognitive Psychology,* 5, 44–64.

Porter, J.M., Gyi, D.E., and Robertson, J., 1992. An evaluation of a tilting computer desk, in *Contemporary Ergonomics 1992,* Lovesey, E.J., Ed., London: Taylor & Francis, 54–59.

Pustinger, C., Dainoff, M.J., and Smith, M., 1985. VDT workstation adjustability: effects on worker posture, productivity, and health complaints, in *Trends in Ergonomics/Human Factors II,* Eberts, R.E. and Eberts, C.G., Eds., Amsterdam: North-Holland/Elsevier Science, 445–451.

Reynolds, J.L., 1994. The Effects of Whole-Body Restriction on Inspection Performance, Dissertation, Buffalo, NY: State University of New York at Buffalo.

Rohmert, W., 1973. Problems in determining rest allowances. *Applied Ergonomics*, 4, 91–95.

Selye, H., 1950. *Stress*, Montreal: Acta Incorporated.

Simon, H.A., 1988. Rationality as process and as product of thought, in *Decision Making: Descriptive, Normative and Prescriptive Interactions*, Bell, D.E., Raiffa, H., and Tvershy, A., Eds., New York: Cambridge University Press, 58–77

Simonton, K., Vercruyssen, M., and Kashizume, K., 1991. Effects of posture on reaction time: influence of gender and practice, in *Proceedings of the Human Factors and Ergonomics Society 38th Annual Meeting*, Santa Monica, CA: Human Factors and Ergonomics Society, 768–771.

Smith, M.J., Karsh, B.-T., Conway, F.T., Cohen, W.J., James, C.A., Morgan, J.J., Sanders, K., and Zehel, D.J., 1998. Effects of a split keyboard design and wrist rest on performance, posture and comfort. *Human Factors*, 40(2), 324–336.

Snook, S., 1978. The design of manual handling tasks. *Ergonomics*, 21, 963–985.

Sperling, G. and Dosher, B.A., 1986. Strategy and optimization in human information processing, in *Handbook of Perception and Human Performance*, Vol. 1, Boff, K.R., Kaufman, L., and Thomas, J.P., Eds., New York: John Wiley & Sons, 2-12–2-65.

Spilling, S., Eitrheim, J., and Aarås, A., 1986. Cost–benefit analysis of work environment investment at STK's telephone plant at Kongsvinger, in *The Ergonomics of Working Postures*, Corlett, N., Wilson, J., and Manenica, I., Eds., London: Taylor & Francis, 380–397.

Suther, T.W. and McTyre, J.H., 1982. Effect on operator performance at thin profile keyboard slopes of 5°, 10°, 15° and 25°, in *Proceedings of the Human Factors Society 26th Annual Meeting*, Santa Monica, CA: Human Factors Society, 430–434.

Thomas, R.E., Jr., Congleton, J.J., Huchingson, R.D., Whiteley, J.R., and Rodrigues, C.C., 1991. An investigation of relationships between driver comfort, performance and automobile seat type during short term driving tasks. *International Journal of Industrial Ergonomics*, 8, 103–114.

Vercruyssen, M. and Simonton, K., 1994. Effects of posture on mental performance: we think faster on our feet than on our seat, in *Hard Facts about Soft Machines — The Ergonomics of Seating*, Lueder, R. and Noro, K., Eds., London: Taylor & Francis, 119–131.

Vercruyssen, M., Cann, M.T., McDowd, J.M., Birren, J.E., Carlton, B.L., and Burton, J., 1988. Effects of age, gender, activation, and practice on attention (preparatory states) and stages of information processing, in *Proceedings of the Human Factors Society 32nd Annual Meeting*, Santa Monica, CA: Human Factors Society, 203–207.

Vercruyssen, M., Cann, M.T., and Hancock, P.A., 1989. Gender differences in postural activation effects on cognition, in *Proceedings of the Human Factors Society 33rd Annual Meeting*, Santa Monica, CA: Human Factors Society, 896–900.

Wangenheim, M., Samuelson, B., and Wos, H., 1986. ARBAN — a force ergonomic analysis method, in *The Ergonomics of Working Postures*, Corlett, N., Wilson, J., and Manenica, I., Eds., London: Taylor & Francis, 243–255.

Wickens, C.D., 1992. The speed–accuracy operating characteristic, in *Engineering and Human Performance*, New York: HarperCollins, 312–322.

Wilson, J., 1995. A framework and a context for ergonomics methodology, in *Evaluation of Human Work*, Wilson, J.R. and Corlett, E.N., Eds., London: Taylor & Francis, 1–40.

14

Digital Human Models for Ergonomic Design and Engineering

CONTENTS

0-415-27908-9/04/$0.00+$1.50
© 2004 by CRC Press LLC

Introduction

Don B. Chaffin

The preceding chapters have emphasized how much we now know about human anthropometry, biomechanics, physiology, psychophysics, postures, and movement control. Design guidelines and human performance models have been described. This chapter delineates the current status of how these data and models are being implemented within computer-aided design and engineering systems to enhance the use and safety of consumer products and manufacturing processes.

Digital human models (DHMs) of various types have existed since the 1960s. Some of these are described in Peacock and Karwowski (1993), Raschke et al. (2001), and Chaffin (2001). Many of these models allow the user not only to graphically depict a digital human form, sometimes referred to as a hominoid or avatar, but also to insert this digital human form into a complex virtual environment, such as a computer-rendered interior of a vehicle or manufacturing work cell. Once such a virtual marriage has taken place, the user can then simulate many different types of situations related to how certain design changes in a proposed environment might affect human performance or health risk for a specified population of people in such an environment.

Motivation for Interest in Digital Human Models

Why are companies and individuals buying licenses to use a variety of commercial digital human modeling software? As is many times the case, several reasons seem evident and these are briefly discussed below:

First and perhaps foremost is the relatively large amount of cost and time consumed in designing, producing, and testing hardware prototype designs of a new product or work process. This is especially true when the effectiveness of a proposed design depends on a tight coupling of a human operator and new hardware. An example would be when designing the interior of a vehicle or a manual workstation. In the process of finalizing a design for these scenarios there may be three or more different prototypes built. Each of these will be sequentially tested by panels of people who supposedly represent the intended user population. A hypothetical cost of such a traditional serial process is depicted in Figure 14.1. In contrast, many companies believe that by spending more in computer-aided engineering (with digital

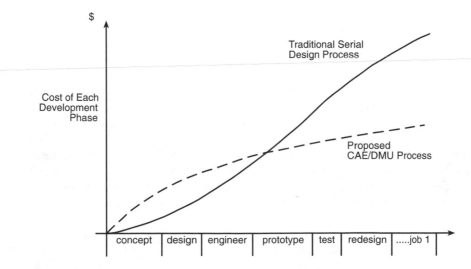

FIGURE 14.1
Example comparison between the typical accrued costs (solid line) associated with a traditional design process (with several physical prototypes necessary for sequential testing) and the use of contemporary computer-aided engineering (CAE) with digital mockup (DMU) evaluations (dashed line).

mockups) early in the design process, one can reduce downstream prototype building and testing costs, as shown by the dashed line in Figure 14.1. An example of this shift in design effort was seen in the recent Boeing 777 aircraft that used extensive mathematical modeling and simulation, which avoided four or more prototype planes being built and tested to assure a proposed aircraft could meet its performance and safety specifications under a large variety of operating conditions.

A second compelling reason for using DHMs in design is that the known variability in population human performance is high, but is often not obvious to a product or workplace designer. For example, the muscular strength capability of a population of healthy people might vary by 10:1 (Chaffin et al. 1999). Worse yet, there is little correlation between a person's size, which is a visible attribute, and his or her strength. It also is well documented that even small variations in the postures one chooses to use during an exertion can result in large differences in the resulting strength, as discussed in Chapter 12. Thus, being able to predict with a model the interrelationship between population strengths and postures can be of great importance in a particular design. Likewise, the variability in many other human attributes present similar challenges to a designer, and hence there is a need to have easy-to-use models of population performance capabilities to guide decisions affecting a variety of people's interactions with the designed environment.

Another reason that DHMs are becoming popular is that desktop and laptop computers are now powerful enough to handle the sometimes large memory and high speed data processing requirements of complex analytical and graphical human models. The fact is that early human prediction models were running on the first PCs in the early 1980s, but it has only been in the last 5 years or so that a typical designer could have several different human performance models running simultaneously with full-color three-dimensional rendering of an avatar performing a task in a graphically rich environment.

The last reason for an increasing interest in digital human modeling is probably related to the fact that graduates of design and engineering schools are increasingly being encouraged to take courses in human psychology, physiology, and statistics, along with their normal analytical engineering and design courses. By using the powerful new digital human design software applications, these cross-trained people can combine their engineering/design skills with human performance and ergonomic skills, and thus are able to render designs effectively that consider population attributes and limitations without the need for extensive follow-up prototype building and testing.

Digital Human Performance Prediction Modeling Is Different from More Common Human Analysis Modeling

There is no doubt that the majority of contemporary ergonomics practice is still concerned with analyzing existing tasks. Most often, such analyses are meant to determine if a significant number of people either would be harmed by performing a specified task or would be too slow or error prone to meet the system design goals. What this means in practice is that ergonomics is most often utilized after a group of people has become convinced in a work or product marketing process that a potential, problematic human interface situation exists. To analyze such a potential problem, often people can be videotaped and forces can be measured while the task of concern is performed. In this context the video and force data provide excellent inputs to models of human population capabilities, but the results pertain only to the way the task is being performed by the selected group of users. In this sense, human models are used to access population limits, with posture, movement, and force inputs from a sample of people, much like a quality control process is used to check a machine to determine if it is producing acceptable parts. Although this use of human models is very important, it would be much better in a proactive sense if a designer of the product or workplace had access to the human models before the product or process was prototyped, so that unacceptable or hazardous performance requirements could be avoided early in the design process.

As described in the earlier chapters, some human performance capabilities can be accurately predicted by existing models, and more robust models will be developed in the near future. Strength modeling is an excellent example of this trend. Population strength models and data have existed since the middle 1960s (Chaffin 1969), but these require a user to input accurate postural data, which often can only come from either photographs or videos of people performing a task of interest, or from a highly skilled user manually manipulating a computerized hominoid to simulate the task in question. To facilitate the latter approach, some digital human strength models have graphical user interfaces with inverse kinematics that make it easier to input correct postures before analyzing these for population strength capabilities (Chaffin 1998). More recently, video frame grabber technology has been incorporated into DHMs to allow a user to stream a video of a person directly into an analysis program to obtain postural data (Gleicher 2001). Even with these advances, however, the results are still dependent on how well the person being videotaped portrays the postures that would be used by other people, and hence existing models are often not very good for predicting alternative movements that different people might use to perform a task of interest in the future. In this context many existing DHMs provide sophisticated analysis tools for reactive ergonomics practice, but they are not very good proactive design tools.

Attributes of Digital Human Models When Used for Design

The following are some desirable attributes to allow a DHM to be used for proactive design simulations:

1. A DHM, when used as a design tool, must enable the user to simulate quickly a large variety of input scenarios, or "what if" exercises. In this context, both speed of accessing and executing many different human performance models contained within the digital human simulation software are important.

2. A design-oriented model must be able to simultaneously analyze many different population attributes, such as people's size, shape, strength, flexibility, etc., to assure the designer using it that not just one but several human performance capabilities are being considered.

3. The precision of each particular human performance prediction model may not be as important in an early design application as when used for analysis of an already existing situation. In other words, the design model should be able to alert the user that some human performance function may be overly stressed by a proposed design. As a particular design is then refined in subsequent simulations, further in-depth analysis can be performed to evaluate the importance of the suspected problem more precisely.

4. Because most people who are practicing designers today are not well trained in ergonomics, a digital human design-oriented model must have good graphic user interfaces (GUIs) that will allow a nonskilled ergonomics specialist to easily understand and manipulate the I/O (input/output) of a DHM without extensive training in human performance or ergonomics.

5. Because a designer will want to study people performing a proposed task in an appropriate graphical environment, (e.g., if driving is the task of interest, a realistic interior of the car and even a driving scene must be rendered), the DHM application must interface easily with other graphical databases to allow for subsequent insertion of the DHM into the proposed computer-rendered environment.

6. Finally, it is not enough to analyze the capabilities of only a few different people in a design. The designer may wish to consider a very wide disparity in population groups to assure that a proposed design is applicable to different user groups. This is in contrast to when a DHM is used to analyze a specific job or product, wherein only a limited group of people who are currently involved in operating the existing system needs to be considered.

So what are some human attributes that are listed as important by designers? According to one survey of practicing designers, conducted for the Society of Automotive Engineers (SAE) G-13 Standards Committee by Nelson in 1998, the following are some attributes that should be included in future DHMs:

1. Anthropometrics of a variety of population groups
2. Population strengths and endurance capabilities

3. Realistic human motions
4. Motion movement times and complex reaction times
5. Maximum volitional reach capabilities and sight lines

Clearly this is not an exhaustive list, but it does indicate that currently many designers would be pleased if future DHMs assisted them in better understanding how their proposed designs could affect several different manual and psychomotor aspects of human performance with diverse populations. What follows are descriptions of some of the more popular commercial DHMs available today. These are quite sophisticated models, and are evolving. It is hoped that the reader will appreciate the scope and capabilities of these models, as well as their limitations as they relate to posture and movement control and simulation.

14.1 The Jack Human Simulation Tool

Ulrich Raschke

Introduction

In the introduction to this chapter, some of the motivations for using high-end DHMs were described, along with functional capabilities that might be desired for these models to be useful in any engineering process. This section describes the Jack human simulation model, which is now supported as part of UGS-PLM Solutions. It was born from research into real-time manipulation of complex kinematics systems (Badler et al. 1993), and has since evolved into a commercial product incorporating a great deal of published human modeling data.

The tool can be used by engineers to ask questions of their designs regarding how well the design accommodates the range of human sizes that might be interacting with it, and how a proposed design might affect human performance in terms of comfort, efficiency, and injury risk prediction. These analyses can be animated and exported as movies, providing graphic content for management review, training, and service materials. This overview provides some of the fundamental technologies comprising the Jack human figure model.

14.1.1 Kinematic Representation in Jack

For physical ergonomics investigations in digital environments, human models need to mirror the structure, shape, and size of people in sufficient detail to allow the figures to realistically assume the static postures observed of actual individuals performing similar tasks. Such human form models typically consist of an underlying kinematic linkage system that closely parallels our own skeletal structure, and an attached geometric shell that duplicates our surface shape.

The Jack human figure is an articulated, linked system that is similar to our own skeletal makeup. The joints of the skeleton obey physiological range of motion restrictions, keeping the user from posturing the Jack avatar in poses that are not typically achievable by the population. Particular attention is paid to the construct of the skeleton, to make sure that it represents reality as much a possible. This is critical, as it has been shown that the biomechanical models used for subsequent analyses of human performance are sensitive to the avatar's posture. The Jack figure has a fully articulated spine below the neck of 17 segments, fully articulated hands, and a sophisticated shoulder complex. Manually adjusting this large number of joints individually (68 of them in Jack) to define a posture would be intractable for a user, so degree of freedom reduction methods, called "behaviors," have been implemented. These behaviors are kinematic models created from observations of how people often move. For example, most of us cannot move individual vertebrae of the spine. The skeleton is held together with ligaments and tendons that couple the movement of skeletal bones. These motion couplings can be modeled to dramatically reduce the number of degrees of freedom the user needs to adjust, making Jack easier to posture than would be the case if all 68 links had to be set in a particular posture. For example, the shoulder complex, consisting of the sternoclavicular, acromioclavicular, and glenohumeral joints, is modeled to move as a group when the upper arm is adjusted, moving the elevation and fore–aft position of the shoulder as the arm moves through its range of motion as depicted in Figure 14.2. Behavior posture prediction models exist in Jack for walking, spine bending, shoulder posturing, head tracking, and balance.

14.1.2   Anthropometry in Jack

The internal skeletal structure and surface topography of a DHM influences both the qualitative and quantitative use of the figures. As an engineering tool, the accuracy of the internal link structure affects the dimensional measures made between the environment and the human figure, such as viewpoint and reach. Similarly, the ability of the figure to acquire a physiologic surface shape directly adds to the perception of reality when viewing a simulation. While both of these aspects are important, to date effort has concentrated on improving scaling of the link lengths. This bias is in part a result of the large amount of traditional, one-dimensional anthropometric data available (i.e., stature, sitting height, shoulder breadth, etc.), in contrast to the largely non-existent three-dimensional surface contour data. Second, a driving factor of human modeling in visualization environments has been

FIGURE 14.2
The Jack human figures used in the analysis of an automotive interior. The Jack functionality can be used to look at specific issues such as reach and visibility for a particular user demographic group. The SAE J-Standard reach and head clearance zones have also been incorporated to aid design and benchmarking tasks.

to produce a system that works in near real time (Badler et al. 1993). The complexity of the figure surface description presents a computational burden on real-time simulation performance, so a balance is sought in which there is sufficient surface detail for visual reality without undue computational load.

A variety of anthropometric databases can be used in Jack to represent the dimensions of a population. Primarily, the ANSUR 1988 U.S. Army database is used (Gordon et al. 1988). This very detailed database includes more than 120 measures per person from a collection of about 1700 males and 2200 females of the U.S. Army. Multivariate statistical models of the proportions of individuals can be derived from these data. While it is collected on Army personnel, it included clerical and support personnel, and the dimensions have been estimated to be within a few percent of the civilian population as a whole (Roebuck 1995). Often there are proprietary databases or country-specific anthropometric databases that are desired. Unfortunately, these are almost always only summary statistics, containing mean and standard deviation values of a subset of dimensions, which do not provide sufficient information to specify the many dimensions or proportions necessary to create a comprehensive human figure model. Jack includes advanced anthropometric interfaces to allow users to create figures scaled from these incomplete data, while drawing on statistical models from more complete data to provide missing proportionality information.

In response to the recent availability of three-dimensional body scan data through the CAESAR (Civilian American and European Surface Anthropometry Resource) project, Jack also is able to utilize these data, morphing the figure surface polygons to match the surface topography defined by these scans. Although these latter data and their application to design are in their infancy, they are expected to enhance the visual look of the human figures and allow for additional analyses, such as more comprehensive accommodation and clothing studies.

14.1.3   Posturing and Motion in Jack

As mentioned earlier, the Jack human figure has the concept of postural behaviors to help collapse the many degrees of freedom of the figure into a few that adhere closely to the parlance of the human factors community, such as shoulder abduction, adduction, and humeral rotation, or torso flexion, axial rotation, and lateral bending. Even with a substantial reduction in degrees of freedom that are provided by coupled joint motions, there still are far too many degrees of freedom remaining to allow rapid and accurate posturing by a user.

To address this, Jack uses inverse kinematics (IK) to help specify joint kinematics based on a desired end-effector position. IK operates on a linked chain of segments, for example, the torso, shoulder, arm, forearm, and wrist, which provides, when given the location of the distal segment (e.g., hand), all of the joint postures along this chain based on a chosen optimization criterion. For Jack, the optimization criteria include that the joints do not separate and that the joint angles remain within their physiological range of motion. The IK also operates through the coupled joint behavior definitions, so that the empirical rules defining the motion of these complexes is preserved. Using IK, the practitioner is able to grab Jack's hand in the three-dimensional visualization environment, and manipulate its position in real time, while the rest of the figure modifies its posture (e.g., torso, shoulder, arm) to satisfy the requested hand position, as depicted in Figure 14.3.

14.1.4   Motion in Jack

Although static posturing is often sufficient to analyze many ergonomic issues, such as reach, vision, clearance, and joint loading, there are often times when figure motion in the form of animation is important. Examples include simulated training material, managerial presentations, and analyses that depend on observations of a person performing an entire task cycle, such as when assessing the efficiency a workplace layout. The Jack animation system provides a variety of methods to create simulations, from operations that interpolate the joint locations between two postures, such as is commonly used in robotic interpolation, to constraint-based methods operating through use of the IK system.

FIGURE 14.3
Serviceability analysis of a design can be performed prior to a prototype build. Here the Jack human figure is used to analyze the maintenance access to an electrical box inside an aircraft nose cone by placing the hand in the required position and allowing the IK and coupled joint motion behaviors to posture the rest of the body.

14.1.5 Virtual Reality in Jack

Often a designer might ask how a person will move to perform an operation, or ask if there is sufficient clearance for a person to grasp a part within the confines of surrounding parts. Situations that require nontypical, complex motions currently cannot be answered adequately with the movement prediction algorithms available. Although there are numerous promising efforts under way to model natural human movement, currently the use of immersive virtual reality (VR) technology provides the best solution to create complex human movement sequences rapidly. For these reasons, immersive VR is increasingly used in both design and manufacturing applications. The Jack system supports a variety of immersive hardware, including gloves, head-mounted displays, and whole-body motion trackers. As will be described later, many of the human performance models available with Jack can work in real time while the figure postures are being manipulated. Motions can be captured and then played back for human performance analysis or presentation purposes.

14.1.6 Performance Models with Jack

One of the major application areas of Jack is in the analysis of manufacturing workplaces, where issues of assembleability, work cell layout, work cell adjustability, and worker risk of exertion injury can be evaluated. A wide variety of human performance tools are available in Jack that provide assessments of worker risk of injury, strength capability, fatigue potential, and task

TABLE 14.1

List of Human Performance Tools Available in the Jack Human Simulation System

Performance Model	Data Source	Input Parameters
NIOSH lifting equation	Waters et al. (1993)	Posture and lift begin and end, object weight, hand coupling
Low back injury risk assessment	Chaffin et al. (1999)	Joint torques, postures
Strength assessment	University of Michigan 3D static strength equations	Anthropometry, gender, body posture, hand loads
Psychophysical strength tools	Ciriello and Snook (1991)	Task description, hand coupling, gender
Fatigue analysis	Rohmert (1973a,b), Laurig (1973)	Anthropometry, gender, body posture, hand loads, cycle time
Metabolic energy expenditure	Garg et al. (1978)	Task descriptions, gender, load description
Rapid upper limb assessment (RULA)	McAtamney and Corlett (1993)	Posture assessment, muscle use, force description
Ovako (OWAS) working posture	Karhu et al. (1977)	Posture assessment
Comfort	Variety of sources including Henry Dreyfuss Associates (1993), Rebiffé (1966), and Krist (1994)	Posture assessment

timing (Table 14.1). To facilitate the use of these tools, many have been integrated in such a way that they can run in the background, allowing the designer to concentrate on the design and only be flagged by situations that may be potentially injurious to the worker.

14.1.7   Low Back Injury Risk Assessment in Jack

Low back injury is estimated to cost U.S. industry tens of billions of dollars annually through compensation claims, lost workdays, reduced productivity, and retraining needs. The low back analysis tool in Jack builds on the Jack skeleton and posturing capabilities, adding a biomechanical model to estimate the forces and moments at the joints, and a sophisticated muscle recruitment model that estimates the activity of the torso muscles in response to these forces and moments (Raschke et al. 1996). These internal muscle contributions to the overall spinal forces can be an order of magnitude larger than the applied loads. NIOSH (1981) has recommended guidelines against which the predicted compression forces can be compared, and job design decisions made. The implementation of this tool in Jack works in real time as the Jack figure is manipulated with an alert to the user if a high back-stress level is predicted.

14.1.8   Population Strength Assessment in Jack

Strength assessments are a typical human performance analysis, regardless if the application involves manual handling tasks, serviceability investigations, or product operation. Questions of strength can be posed in a variety of ways. Designers may want to know the maximum operating force for a lever, dial, or wheel, such that their target demographic population will have the strength to operate it. Asked in a slightly different way, the engineer may create a job design, and might ask what percentage of the population would be expected to have the strength to perform the required tasks within the job. Jack provides a variety of strength tools, both based on psychophysical methods (Ciriello and Snook 1991) and on maximum voluntary exertion empirical data models (Chaffin 1998).

14.1.9 Fatigue Assessment in Jack

Jack also has tools that predict possible fatigue of the worker, making sure that there is sufficient rest time in the work cycle to avoid worker fatigue during the workday. Although there are only sparse data available for fatigue assessment, the implementation in Jack draws on the University of Michigan three-dimensional strength data to identify the level of muscle group strength required by a loading situation, and then uses this with empirical endurance models to estimate the recovery time required by this exertion.

Summary of Jack Functions

The Jack human simulation tool is used to analyze physical human factors issues for many different aspects of product and population engineering. This overview presents some of the fundamental concepts upon which the Jack figure is constructed and manipulated. In addition some of the available human performance tools are described. However, the value of the Jack DHM comes not only from the wide range of incorporated human factors data and models, which continue to be augmented at a rapid pace by others, but also by the integration of Jack with other software products used for the design and manufacture of contemporary products and processes. This allows the assessment of human factors issues to be pushed "upstream" in the product engineering process, making more people aware of the human factors impact their design decisions will have at a stage in the design process when necessary changes can be implemented at a minimum cost.

14.2 The SAFEWORK Human Simulation Tool

Mark Morrissey

Introduction

SAFEWORK is the brand name for the DHM developed by the Safework, Inc. Business Unit of Dassault Systemes. SAFEWORK structures multiple

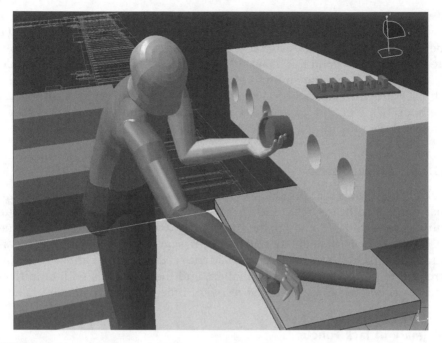

FIGURE 14.4
SAFEWORK human model simulating a manufacturing assembly task.

human modeling systems to facilitate detailed investigation into human-centered design issues. It is intended to provide very accurate simulation of people from many different populations, and their physical interactions with the environment, to ensure they can perform naturally in a workplace tailored to their tasks, as depicted in Figure 14.4.

14.2.1   Enterprise-Wide Human Modeling in SAFEWORK

In the same way that CATIA and other such computer-aided design systems facilitate the design of digital geometrical object models, SAFEWORK provides a digital geometrical representation of humans and allows a designer to evaluate humans in terms of the products they must use and tasks they must perform, as shown in Figure 14.5. However, humans are the most complex system components to be considered due to the diversity of size, the number of body segments, the degrees of freedom in movements, the complex limitations of these movements, the muscle forces that people can produce, and the various behaviors that have to be modeled.

Any software solution is only as strong as the foundations and assumptions on which it is built. For SAFEWORK, these foundations consist of skeletal definition, anthropometry, posture, and movement.

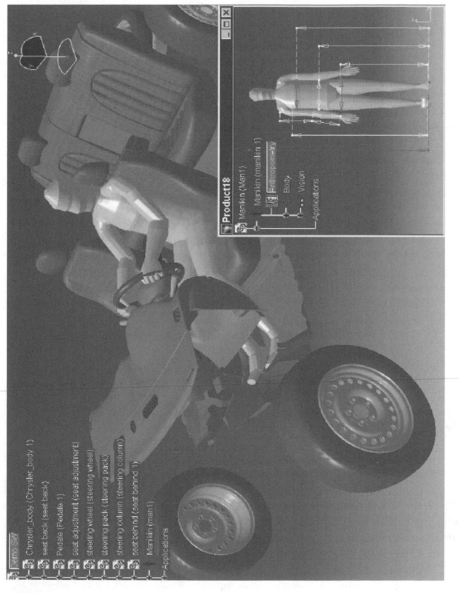

FIGURE 14.5
SAFEWORK manikin integrated into a CATIA digital mockup.

14.2.2   Anthropometry in SAFEWORK

The importance of anthropometry in design is to provide a highly scalable graphic human form in which, for example, a global automotive vehicle manufacturer can optimize its designs for a target audience consisting of millions of potential consumers across a worldwide marketplace.

Detailed anthropometry surveys can record more than 100 variables that become extremely useful input variables in the development of a human form model. The effectiveness and validity of human modeling analysis tools, however, are directly correlated with the accuracy with which the human models represent the population they are simulating. As such, the SAFEWORK solution defines the human body in terms of 104 anthropometric variables, including a fully articulated spine, shoulder, and hand models. This attention to detail is intended to minimize erroneous assumptions often associated with the definition of anthropometric variables used in human models, and to ensure that the multifactored variations in "population" size are represented in life-cycle design applications.

For underlying cost–benefit trade-off reasons, a traditional general rule of thumb for designers has been to accommodate 90% of the population (from 5th to 95th percentile). This concept, in itself, is multidimensional, in that it involves the analysis of multiple anthropometry variables simultaneously. Unfortunately, anthropometry variables are often analyzed individually in a univariate manner. Confusion regarding the appropriate application of anthropometric data is well documented (Zehner et al. 1993). Brandenburgh (1999) noted that MIL-STD-1472D states:

> Design limits shall be based upon a range from the 5th percentile female to the 95th percentile male values for critical body dimensions.... [U]se of a design range from the 5th to 95th percentile values will theoretically provide coverage for 90 percent of the user population for that dimension.

However, MIL-STD-1472 qualifies this statement where more than one variable is to be considered, by stating:

> Where two or more dimensions are used simultaneously as design parameters, appropriate multivariate data and techniques should be utilized.

MIL-HDBK-759c concedes that the univariate approach is inadequate in scenarios where two or more anthropometry variables are used simultaneously as critical design parameters. This standard further indicates:

> Extreme caution should be used when two or more dimensions are simultaneously used as criteria for design. Percentile values are not additive between different dimensions.... For example, it is incorrect to assume that the combination of the 5th percentile values will describe the dimensions of a "5th percentile man."

Brandenburgh's intent is clear. When two or more dimensions are required as design parameters, a multivariate approach should be employed. The multivariate approach used by SAFEWORK generates a population of manikins that statistically represent the identified target audience. Each manikin in the population possesses a distinct set of anthropometric relationships that must be analyzed in unison. The exact number of "boundary" manikins increases with the number of critical design criteria. Thus, the user specifies "critical design variables" upon which a special boundary manikin algorithm unique to SAFEWORK automatically adjusts all other manikin anthropometric variables, based on the statistical correlations of individual variables. Resultant manikins are then "created" using links, ellipses, lines, and flat and gouraud shading. Gender and morphological profile can be specified including seven somatotype choices ranging from ectomorph to mesomorph.

In addition to six default size manikins produced in a simulation, SAFEWORK can create, by means of an anthropometry module known as the "Human Measurement Editor," any human size or shape from published population data. Users can manually define any of the 104 anthropometric variables on the manikin by inputting desired measurements in terms of percentile values or unit measurements. SAFEWORK also affords users the capacity to define the mean and standard deviation for each variable. The multivariate algorithm then generates a manikin corresponding to the most probable human being in the target population.

SAFEWORK allows users to access up to 94 recently completed male and female population surveys containing data relating to anthropometric variables, including their standard deviations, percentile values, and correlations. Functional anthropometry, by its very definition, is domain and application specific. As such, SAFEWORK's Human Measurement Editor presents anthropometry data derived from both standing and seated reference postures. In addition, SAFEWORK permits the intuitive construction of user-defined, or proprietary, anthropometric databases for truly global human modeling requirements.

14.2.3 Realism of Skeletal Structure and Movement in SAFEWORK

Ensuring that a human model moves and behaves in a realistic, task-oriented fashion is the next "foundation" element required for validity of a human modeling solution. SAFEWORK defines the human body with 100 links and segments representative of the skeletal structure, with 148 degrees of freedom ensuring realistic joint movement capability, which, when combined with fully articulated spine, shoulder, and hand models, provides the user with exceptional flexibility for specific design problems, such as depicted in Figure 14.6.

The SAFEWORK human model can be manipulated using a number of posturing techniques including direct kinematics, IK, custom movement libraries, low-level simulation primitives, or VR motion capture technology.

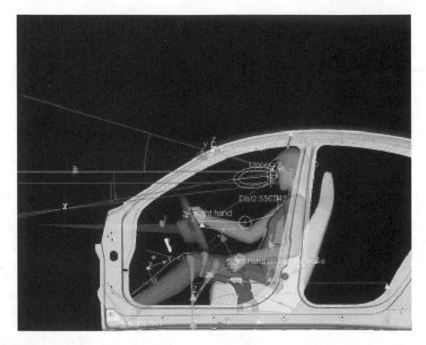

FIGURE 14.6
SAFEWORK manikin vision analysis window using a posture and scaling algorithm.

IK is employed by SAFEWORK to define the final position of an end-effector at the end of a kinematics chain (for example, hands on steering wheel). IK techniques require the system to "inverse" all joint positions to determine what joint angles are required to reach the desired postural goal. SAFEWORK possesses seven default IK handles to control manikin motion and predict postures. In addition, the user can define up to 20 end-effectors as scenario-specific constraints. The manikin's 148 degrees of freedom take into account joint limits, and support a coupled range of motion for enhanced realism of movement.

IK chains can also be extended to include tools, clothing, objects, or accessories that are manipulated by the human model. For example, a power tool attached to the hand of the manikin can serve as the end-effector for the IK calculations, and would drive the movement of the manikin. Simultaneous multiple end-effectors enable the designer to describe a task as a series of goals and geometrical constraints.

In contrast, direct kinematics can be employed to accurately fine-tune manikin postures by manipulating individual segments in each available degree of freedom. For example, changing the degree of forearm flexion would result obviously in a new hand position, without affecting the position of the upper arm segment in space.

Both direct kinematics and IK play an important role in determining the predicted posture of the manikin. All manipulation and posturing techniques need to ensure that the resulting posture lies within the functional limitations of the human model segments. The challenge is to try to make sure that the resulting posture will be as natural as possible. To that effect, SAFEWORK permits all movements to be defined not just by absolute physical limits, but also by "preferred angles," which may reference joint comfort angle data (range of angles between segments where the least discomfort is observed).

Range of motion data representing task-specific functional limitations can also be applied to each segment so that each available degree of freedom (flexion/extension, rotation, abduction/adduction, etc.) conforms to best practice, as outlined by general human factors principles or task-specific criteria. The concept of "population accommodation" does not relate uniquely to anthropometry data. In the same way that anthropometry data can be normalized around a mean, flexibility should also be subject to accommodation analysis. For example, some demographic elements of the population are more flexible than others and, as such, will enable a certain percentage of the population to perform certain tasks that others could not. SAFEWORK permits users to utilize available population statistical data that can be used in conjunction with coupled range-of-motion limitations to analyze the range of motion of a given segment, not just in an isolated sense, but also according to the position of neighboring segments.

14.2.4 Analysis Tools in SAFEWORK

The ability to define, create, and manipulate human models that represent the appropriate target audience is, in itself, merely establishing the foundation for "adding value" to the design process. SAFEWORK possesses a range of analysis tools for evaluating task-specific human factors criteria. For example, gaining an understanding of what an operator or maintenance person could "see" in a task environment is a fundamental element of a human factors analysis. The SAFEWORK Vision Module shown in Figure 14.7, derived from the NASA 3000 standard (1989), contains an accurate vision behavior model to replicate the realistic movement of the human eyes, so that "what the manikin sees, the designer sees." Four types of vision simulation are provided: binocular, ambinocular, monocular left, and monocular right (stereoscopic viewing with advanced depth perception is available in the Virtual Reality Feature). Visual characteristics are displayed as peripheral cones, central cones, blind spot cones, and central spot cones that permit the user to gain an insight into the manikin's view.

14.2.5 Human Modeling Data Interoperability in SAFEWORK

The SAFEWORK architecture is based on the concept of libraries. Manikin-oriented libraries, containing variables, such as angular limitations, comfort angles, maximum force exertion, and other preferred variables are provided, as well as global posture libraries and local posture libraries (grasp, pinch,

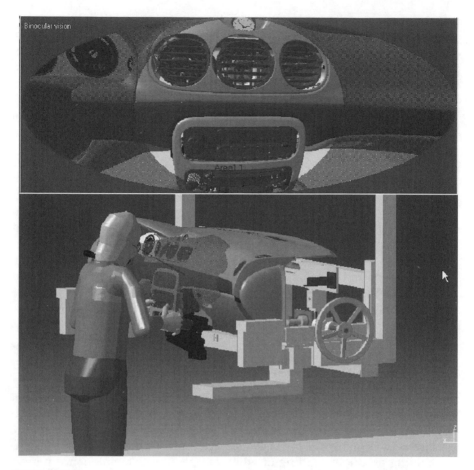

FIGURE 14.7
SAFEWORK manikin vision analysis window allowing the designer to see the operator's view along a projected sight line.

grip, hook, etc.). Libraries of clothing (part of the Clothing Module) can be used to indicate the functional limitations of the manikins are also provided. No single source of data in SAFEWORK is "hard coded" so a user can edit, modify, or create new data sets as appropriate.

Summary of SAFEWORK Functions

The concept of human scalability provided by SAFEWORK is an important factor in the widespread adoption of digital human modeling technology within large organizations. SAFEWORK also has pioneered the "authoring/reviewing" paradigm within the design and manufacturing communities, which permits human factors engineers to "author or produce intellectual collateral" in their designs, which can then be shared, reviewed, and utilized by other entities within the extended enterprise. For example, this sharing can be done within the CATIA V5 architecture, wherein an underlying SAFE-WORK human model kernel is used for a wide range of manufacturing, assembly, packaging, and product design applications. Although each of these applications requires a specific set of features/functions, a common underlying file format means that human modeling data can be seamlessly shared among all key stakeholders. This ability to share SAFEWORK simulations with other designers is critical to building the knowledge base needed for human-centered design simulation across large organizations.

14.3   The RAMSIS and ANTHROPOS Human Simulation Tools

Andreas Seidl

Introduction

In the past, human templates or simple DHMs were used to represent defined percentiles of the population in static postures. Today, complex vehicles are designed solely in computer-aided design systems with an effective reduction of design time, or in VR systems to reduce the use of hardware mockups.

To overcome the deficiencies of using conventional templates or written ergonomic design guidelines and standards for vehicles, automotive companies; namely, Audi, BMW, Ford, DaimlerChrysler, Opel, Porsche, and VW, as well as the seat manufacturers Keiper Recaro and Johnson Controls initiated the development of a specialized DHM for vehicle design. This DHM, named RAMSIS, is meant to provide efficient design of interiors of cars, trucks, and airplanes within existing computer-aided design systems (Seidl 1997).

VR developers would have enlivened their virtual scenes with these human figure models long before now if such models had been available to them with high graphic quality, and with biomechanical intelligence. These developers and users of VR techniques require an efficient tool with an adaptable interface within their VR systems. For this purpose the DHM ANTHROPOS was developed. What follows is a brief description of both RAMSIS and ANTHROPOS TECHNOLOGIES.

14.3.1   Anthropometrical Realization in RAMSIS

At the beginning of RAMSIS development it was realized that the amount and types of anthropometric data were insufficient for the complete definition of a comprehensive three-dimensional human figure model. At present anthropometrical data are obtained by two- and three-dimensional body

FIGURE 14.8
Overlay technology of measured postures with RAMSIS DHM to fit postures and anthropometric data.

scanners of test subjects in a number of select postures. The data images are read into the computer, where they are overlaid with RAMSIS. The simulated length, thickness, and circumference of each human body element are then varied until the scanned data are completely congruent with the corresponding digital human form (Figure 14.8).

The heart of the tool RAMSIS is a three-dimensional human form model with its archives containing data regarding postures and seated comfort, as well as an anthropometric database. While the appearance of a human being is completely described by its body surface, the mobility is largely determined by the skeleton. In a similar way, RAMSIS is represented on two levels: an internal and an external modeling level. The internal RAMSIS model plays the role of a "human skeleton" and is the basis for the definition of RAMSIS kinematics. The external RAMSIS model represents the body surface. In contrast to most existing human models, the body surface of RAMSIS is not modeled by rigid, geometrically simple objects (prismatic bodies or ellipsoids) but rather by use of a set of posture-dependent control points. These control points (about 1200 in the standard model) are attached to the internal RAMSIS model. The attachment is not static, but varies in accordance with the joint positions.

The statistical results of the process of overlaying measured postures onto a digital surface are realized in the anthropometric database module RAMSIS/BodyBuilder. Using a classification scheme it is possible to describe the human population in a realistic way. From 90 real physique types obtained from scanned populations of people, the user chooses statistically defined population groups by providing key measurements of height, proportion, and corpulence. The database of BodyBuilder provides the size of a defined population segment by borderline typologies. This functionality is combined with standard anthropometric databases from Germany, France, USA/Canada, Japan/Korea, South America, and Mexico.

14.3.2 Research of Posture and Movement Simulation for Cockpit Design with RAMSIS

RAMSIS includes a three-dimensional posture and movement simulation. Postures of different analysis tasks of test subjects were measured by cameras located at arbitrary positions, as shown in Figure 14.8. In addition to posture measurement with the pedals, steering wheel, and seat placed in various locations, typical functions such as reach and entering and exiting the car have been included in the test range. Based on the distribution of postures with respect to various body dimensions and various tasks, a multidimensional "postural function" has been developed for each joint (Seidl 1994).

The measurement of postural discomfort has been performed with the aid of a vehicle mockup. The position of the controls may be varied in this mockup to such a large extent that the dimensions of nearly any vehicle — from a sports car to a small truck — can be simulated. In addition, the mockup has been extended to include simulations of driving views and acoustics. For the measurement of discomfort, the test subjects were requested to maintain different postures. A standardized questionnaire was used for evaluating the feeling of discomfort of the test subjects. The questionnaire data for each test subject were compiled with the corresponding recorded postures. Thus, it was possible to calculate regression coefficients, which when applied to postures of the RAMSIS model can then predict the expected postural discomfort of a given seated position.

In practice, the designer describes the task to be fulfilled by RAMSIS by interactively defining complex model restrictions, for example, hands on the steering wheel, feet on the pedals, drive looking backward, including a chosen viewing posture, as shown in Figure 14.9. These restrictions can be stored and reused with other human simulations. Additional constraints, such as avoiding penetration of body parts, also are provided.

14.3.3 Modeling of Vehicle Seating Spinal Postures and Belt Factors with RAMSIS

Accurately predicting driver posture in a new vehicle/seat design allows for efficient planning and verification of what an occupant can reach, see, and access while in a specific posture. Additionally, proper occupant posture prediction allows for better determination of safety restraint locations and compliance with regulations. Based on the Michigan State University JOHN model, which provides a biomechanical simulation of the lower torso posture, experiments were conducted to examine the change of postures due to seat and interior package factors. This research provided a biomechanical simulation of the torso posture and the postural effects due to current seat and interior package factors (Gutowski et al. 2001). For example, many new seat designs include aggressive side bolsters and/or lumbar support, which did not exist when prior standards and occupant prediction methods were developed. The results are integrated into the posture prediction model of

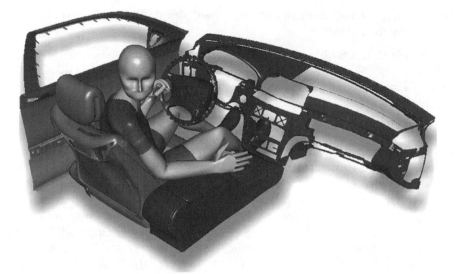

FIGURE 14.9
Complex model postural restrictions and the simulation of complicated drivers' postures are possible in RAMSIS.

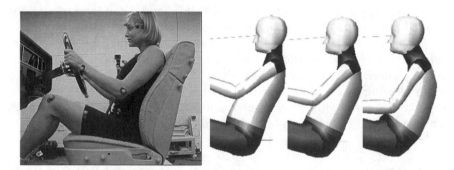

FIGURE 14.10
Measurements in different car seats and simulation of curvature of lumbar spine by kinetics coupling of pelvis and thorax in RAMSIS.

the RAMSIS program to give a more detailed prognosis of the spine curvature, and thus can be used to refine the model–seat interactions, as shown in Figure 14.10.

Seat belt efficacy to prevent injuries in a collision is directly related to how well the seat belt design matches occupant body dimensions. Research by Transport Canada developed a Belt Fit Test Device (BTD) to forecast potential occupant injuries resulting from discrepancies between seat belt designs and occupant sizes. The BTD consists of torso and lap shells corresponding to the shape and surface of a 50th percentile Canadian adult male, attached to the SAE 3-D H-point machine. Scales at the clavicles, sternum, and lap areas

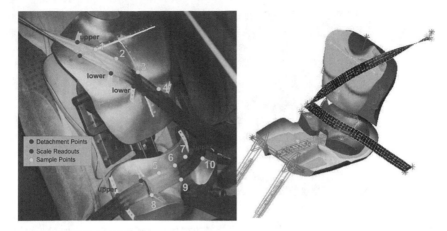

FIGURE 14.11
The belt fit test device (left) and the realization of the device with the belt course simulation within RAMSIS.

measure the location of the seat belt, where it touches the BTD surface, and indicate areas that are less likely to cause serious bodily harm to internal organs and soft tissue from belt intrusion during a collision. As a supplementary proposal the Joint Working Group on Abdominal Injury Reduction (JWG-AIR) suggested creation of an "electronic" version of the BTD (eBTD). The digital eBTD module is integrated with RAMSIS and allows vehicle manufacturers to use computer-aided design data to evaluate seat belt designs before a vehicle is produced. It positions the computer-aided design data representation of the physical BTD in a three-dimensional vehicle interior, including vehicle geometry, for a given driver (Pruett et al. 2001). Users specify the location and types of anchor points for various seat belt configurations. The software module then simulates the predicted routing over the eBTD, and measures the belt position over the clavicle, sternum, and lap scales, with respect to belt width and contact to the surface, as shown in Figure 14.11.

14.3.4 Additional Functions in RAMSIS

Similar to other DHMs, a variety of functions are integrated into RAMSIS. A reach analysis tool makes it possible to determine a predicted surface envelope of reachable limits for any chain of body elements set in various seated postures. These surfaces are actually calculated, taking into consideration the kinematics of the model.

A line-of-sight simulation allows the user to sit inside the model and "look" at the proposed design of the workplace through the eyes of the manikin. The user can switch from the left eye to the right eye, or a combined eye view with a "shadow" function for covered objects. A mirror simulation

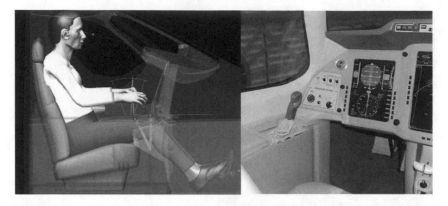

FIGURE 14.12
Design of seat position, armrest, throttle, and pedal design of an airplane cockpit using RAMSIS and the result of the analysis.

allows a simple method of detecting hidden or partially hidden objects. As shown in Figure 14.12, complex cockpit analysis and design are possible with these ergonomic tools in RAMSIS.

In addition, some specialized automotive technologies are integrated in RAMSIS. For example, a parametric package designer or the most important checking procedures of national and international car regulations are integrated. This is realized by implementing related SAE standards in RAMSIS, like the Driver's Eye Location (SAE J914), the head position (SAE J1052), the driver selected seat position (SAE J1517), and the accommodation tool reference (SAE J1516).

14.3.5   The Digital Human Model ANTHROPOS

Users of VR techniques need an autonomous DHM that has the ability to avoid collisions and has time-related loading and stress prediction of human joints at each phase of movement being simulated. Additionally, the visual quality of a DHM in a VR environment has to meet high standards, because VR systems are often used as sales mockups or demonstrators. ANTHROPOS was developed to provide realistic visualization of people performing a variety of whole-body tasks.

14.3.6   Design of ANTHROPOS

The ANTHROPOS graphic models consist of 90 parts of the body and a corresponding number of joints, some of which have five axes. The surface is constructed internally from 3200 skin points and as many as 40,000 reference points, which are also used for recognition of collisions. The resulting

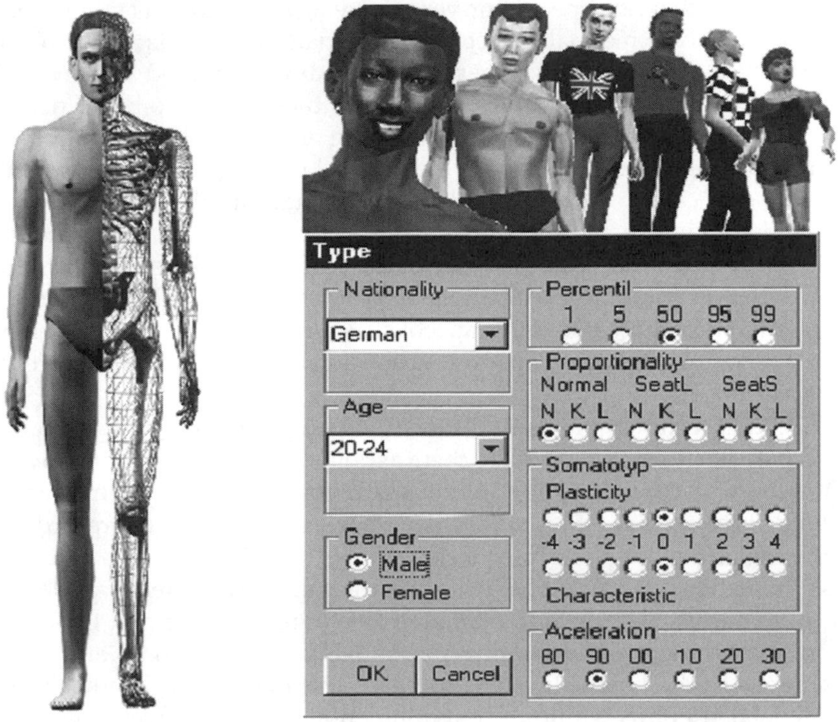

FIGURE 14.13
Model structure of ANTHROPOS — flexible skeleton and skin definition with anthropometric database and clothing generator.

human figure can be displayed with various graphic qualities, from wire frame to shading, depending on the computer-aided design system being used. The movement intelligence of the models is integrated in the internal support and movement apparatus (skeleton) including a 24-part spine and five-finger hands. By deforming the outer skin or clothing layers in the model, the skin or clothes adapt to the skeleton movement. One also can select the reference points that are to make contact with the environmental graphic. The models are lightly clothed, as depicted in Figure 14.13.

14.3.7 Working Posture Animation and Simulation with ANTHROPOS

The primary movement simulation process in ANTHROPOS is called Auto Animation. This algorithm recognizes the various movement limits of the joints, and the movement dependencies of pelvis and thigh when sitting, as well as the movement in the shoulder region when lifting the arms. Body postures believed to be injurious to health cannot be generated with auto animation. To generate specific or very awkward postures, as still occur at many workplaces, manual direct postural manipulation is used, during which each joint can be moved separately. With this method, however, the mobility of the spine, depending on age and fitness, can also be influenced. All settings are made dynamically by entering angles or with potentiometer settings. Standard body postures (bending, kneeling, crouching, sitting, climbing, etc.) can be generated. Also, pelvis rotation with fixed foot positions and walking to a distant goal are possible. When the floor and seat height are given, along with the angle of the seat and backrest, the optimum foot position is computed and the model assumes this sitting position. In parametric animation the user can construct relationships to a reference point, and mutual relationships to touch points and objects (lever, steering wheel, car door).

In addition to touch points and touch planes, restriction surfaces can be defined; ANTHROPOS recognizes them as collision surfaces relating to parts of the body with the help of its numerous skin points. It positions the manikin to avoid these. If restriction angles have been entered, then it records the magnitude of the variation from these values.

For the hands, gripping postures with the hand straightened and slack, and in an adjustable gripping diameter, forefinger straight and bent, with dynamic movement of single fingers are available. Rotational and lateral movements of the kinematical chains can be carried out in free space, but also to defined points within the environmental graphic. Direct movements to the chosen point and motion capture gathering of several points are possible.

Using Newtonian mechanics, the kinematic chains move in different ways, depending on age, to goal points where they remain fixed until further notice. During these movements, the joint at the end of the chain (hip) tests whether and how the following joint (knee) has already been moved, and computes the biomechanical correct limiting angles to be derived from it. To specify the position of the hands and feet end target or goal, exact touch planes can be specified. As shown in Figure 14.14 the combination of the different animation algorithms of ANTHROPOS allows the analysis of a complex lifting movement.

ANTHROPOS automatically tests the physical loading of the joints caused by an animation (movement space used in percent, joint point resistance, torque, and normal force) and compares them with an alternative movement specified by the user. The values can be displayed alphanumerically and graphically. Using the Burandt method (Burandt 1978), load limits for lifting and carrying are computed with the person standing, facing the load. Sex and age, body height and fitness, load weight and frequency, as well as distance lifted, are included in the calculation (Seidl 2000). Depending on a person's anthropometry and sitting posture, ANTHROPOS computes the static and dynamic leg forces.

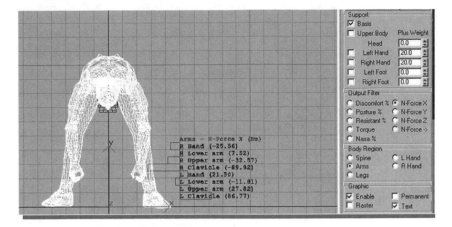

FIGURE 14.14
Simulation of a complex lifting posture with integrated load and body balance calculation within ANTHROPOS.

14.3.8   Additional Ergonomic Functions in ANTHROPOS

To recognize reach capabilities, not only after single animations, a reachability module for hands and feet is available with which all defined goal points are tested simultaneously, and (if necessary with offset values for tools) are noted as "reachable, only just reachable, or unreachable." The reachability curves are displayed in defined planes.

ANTHROPOS can provide sight line analysis in various ways. With the "eye" switched on, it shows its graphic environment in a separate window on each animation as a person would see it with a given head position. Angle of sight and distance seen can be specified by the user. The various sight functions, however, also include projection of points on defined planes (road, house front, etc.). Objects that hinder sight (steering wheel in front of instrument panel) are projected as shadows on the visual environment, and anything seen in a rearview mirror to be defined with curvature factors can be displayed on the mirror. In Figure 14.15 the animation algorithm combined with sight, load, and reachability functions allow efficient ergonomic analysis of a papercutter machine (Lippmann and Roessler 1998).

Summary of RAMSIS and ANTHROPOS Functions

The RAMSIS/BodyBuilder has provided a flexible system for creating human forms from a variety of anthropometric databases. These can be inserted efficiently into a vehicle interior. A software module for lap and shoulder belt fit testing has been perfected. These have been used with the ANTHROPOS software to creating a visually appealing and useful human

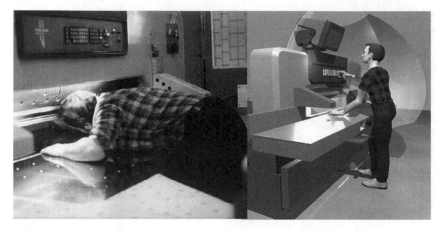

FIGURE 14.15
Analysis and improvement of a paper cutter with ANTHROPOS. (Left) The old design wherein a worker had to look into the machine; (Right) the new design after improvement with a video monitor.

figure. This figure is built on a human skeletal framework, and thus motion animations can be created using a variety of motion input command structures. Once realistic motion or postures are created, then line of sight and other ergonomic analyses can be performed.

14.4   SAMMIE: A Computer-Aided Ergonomics Design Tool

J. Mark Porter, Russell Marshall, Martin Freer, and Keith Case

Introduction

SAMMIE (System for Aiding Man Machine Interaction Evaluation) is a computer-aided human modeling, ergonomics design, and evaluation system. Since its conception in the late 1960s, SAMMIE has been continuously employed in research, and as a consultancy tool (Bonney et al. 1979; Case et al. 1980, 1990a,b; Porter et al. 1991, 1993, 1996, 1999). This section details the functionality of the SAMMIE system, and the SAMMIE team's approach to supporting the use of computer-aided ergonomics throughout the product development process. Some of the latest work is introduced that concerns the novel approach to supporting such methodologies as "design for all" through the creation of a database of individuals in terms of their anthropometry and functional capabilities, and the support for such data through SAMMIE and the new tool HADRIAN.

14.4.1 Functionality with SAMMIE

SAMMIE's primary goal is facilitated through the provision of the following:

- A fully articulated human model capable of representing people of the required gender, age, and nationality
- A knowledge base of maximal and comfort limits for the major joints of the body to represent realistic human reach capability
- The ability to model/import graphical models from other computer-aided design systems of the product or environment concepts, together with the ability to simulate model functionality, such as ranges of adjustment, control limits, and the structural and functional relationships between model elements
- The ability to assess the kinematic interaction between the human model and the product or environment in terms of fit, reach, vision, and detection of surface collisions
- The ability to assess concept designs and subsequent modifications to ensure good ergonomics before physical mockups and user trials are required

The SAMMIE human model is a fully articulated manikin capable of utilizing standard published anthropometric data or custom data obtained or taken by the user. The system then provides the flexibility to modify the human model's size through the percentile range from below 1st to above 99th percentile for the whole body, or for individual limb dimensions. In addition, the technique of somatotyping is used to control the shape of the avatar allowing the user to represent the degree of endomorphy (fatness), mesomorphy (muscularity), and ectomorphy (thinness) as described in Sheldon (1940).

SAMMIE has its own equipment/workplace modeling capability, in addition to supporting the importation of graphic data from the user's preferred computer-aided design system. Assessment focuses on whether or not the human models can work efficiently with the product or in the environment, and can adopt "comfortable" postures (Figure 14.16). The human model may be "driven" through direct manipulation of individual limb positions, but also through a library of postures and automated reach and vision checks. Additionally, lifting and materials-handling risk evaluation is supported by linking to the NIOSH lifting equation and RULA risk assessment tools for given postures. (NIOSH 1981, McAtamney and Corlett 1993).

Finally, the SAMMIE system provides a macro command language that allows processes to be automated and assessments to be chained together.

14.4.2 Development and Application of SAMMIE in Achieving Design for All

Recent work has focused on the concept of design for all or what some professionals refer to as "universal design." While young and able people

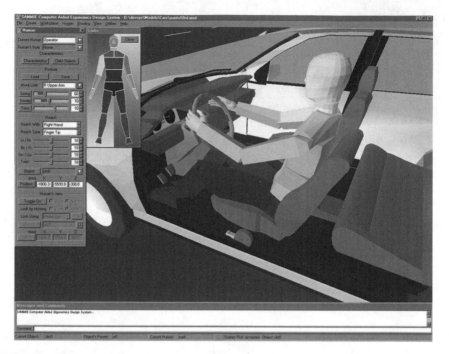

FIGURE 14.16
A development version of the SAMMIE system on the Windows NT platform.

are often considered to be able to "adapt" to a poor design, there is typically an associated human cost. For example, a poor posture that has to be maintained for prolonged periods will result in a high incidence of musculoskeletal complaints and possibly sickness absence. If important displays are not clearly visible or controls are difficult to operate, then safety will be compromised. People who are older or disabled have less opportunity to adapt to a poor design. In many cases, they are effectively "designed out" and cannot use the product or service. The design for all philosophy aims to reduce, if not eliminate, this problem.

Attempting to design for all, including people who are older or who have disabilities, exposes a number of limitations of current anthropometric and biomechanics databases. It is believed that there is a need for a new approach to effectively support designers when attempting to design for all, be it in the workplace, at home, or in public areas. The main limitations of current anthropometric and biomechanics databases for this purpose include (1) their mode and format of presentation, (2) their lack of support for investigating multivariate issues, and (3) the lack of holistic information including specific task and environmental factors (Porter et al. 2002).

In essence, designing for all requires access to a large library of publications to compile information on the physical size and abilities of people of all ages. Current anthropometric and biomechanics databases and guidelines present information typically as univariate percentiles with a separate table of numbers for each variable, such as eye height, arm reach, or hand grip strength. These percentile tables are prepared for either a healthy population, often aged 19 to 65 years, or for specific populations, such as people who are older and with specific disabilities. Sadly, most of these databases do not promote the need for multivariate analysis.

Statistical methods exist that can be used by specialists to conduct multivariate analysis, such as principal component analysis and Monte Carlo simulation. Both are complex and often lack face validity. Although many designers have doubts about the validity of combining different percentile body segments based on statistical calculations, the fact that there are no actual faces that can be put to these anonymous statistical creations is a bigger problem. Designers need to have empathy with the people they are designing for — they find it difficult to design for statistical calculations. Empathy comes from seeing people and getting to know and understand their needs and desires.

The data also need to be task and environment specific. For example, an assessment of an oven design will require the user to hold the hot baking tray in two hands using oven gloves, not just with a simple one-arm reach, as sometimes presented in existing guidelines. In addition, it is likely that users will have developed some "coping strategies," which help them to carry out the various tasks in the kitchen despite certain impairments. It would be most beneficial to record these and be able to pass this knowledge to the designer.

The approach to supporting design for all in SAMMIE includes two main elements. First is the creation of a novel computer database of "individuals," so that multivariate analysis can be conducted on a wide range of people of all ages, abilities, shapes, and sizes. The traditional creation of tables of percentiles for each body dimension effectively dismembers individuals — it becomes impossible to recreate the original individuals who were measured in the survey. Our approach is to preserve important information for each individual as a complete data set (Oliver et al. 2002). This literally enables us to "put faces" to the data (see Figure 14.17, which shows some of the individuals in our database), and makes multivariate analysis more straightforward, at least conceptually. To promote design for all, the designer needs to identify which of these computerized individuals are currently designed out by a proposed prototype design, together with the reasons why, so that subsequent design modifications can be made that will include these people.

The database initially comprises 100 individuals, including a large proportion that is older and/or disabled. This sample, while not intended to be representative of the whole population, provides a useful measure of the extent of variation in physical characteristics and capabilities and provides a preliminary database for the development and validation of the predictive tool. Design relevant information concerning each individual's task behavior (including coping strategies) and environmental issues have been recorded using test rigs to simulate typical activities of daily living that are known to be problematic for people who are older or disabled (Oliver et al. 2001).

FIGURE 14.17
Designers can now put faces to data with the SAMMIE computerized database of individuals.

The second element is the development of a computer-based tool to support the use of the database in design situations. Thus, the database of individuals also is supported by an integrated ergonomics analysis tool.

HADRIAN (Human Anthropometric Data Requirements Investigation and Analysis) is a computer-aided design tool that integrates the database of individuals, including their anthropometry, mobility/capability, disability, coping strategies, and a wealth of background data with a simple task analysis tool.

HADRIAN has been developed to work in conjunction with the SAMMIE system. Together these systems provide the capability to investigate data on individuals in addition to allowing task analysis and virtual fitting trials to be carried out on a design without the need for prototypes and user trials. However, it is not the intention to replace physical models and user trials, but rather to complement them.

The two systems HADRIAN and SAMMIE provide the designer with the ability to accomplish the following:

- Model a product/environment, or import a model generated on another computer-aided design system
- Select a target user base, which should be the whole database when designing for all

- Create a task description with as much or as little data as wished (e.g., viewing distances, which hand to use, etc.)
- Run the task analysis with the chosen user base
- Inspect the results of the analysis including:
  - Estimation of the percentage of the individuals in the database who are accommodated by the product/workstation/environment, which informs the designer of the extent to which design for all has been achieved, and is a useful metric for comparing alternative designs
  - Identification of those individuals who were designed out because they failed certain parts of the analysis, which should promote an understanding of why the failure occurred and lead to the development of design improvements
- Modify the design/task parameters and rerun the analysis, which promotes iterative design and evaluation until the design solution has been optimized

14.4.3 Task Analysis in SAMMIE

HADRIAN's task analysis features are aimed at providing designers with a simple and flexible mechanism for constructing a task description for the use of, or interaction with, their chosen product or environment design. Although most of the actual tools for performing individual elements of a task analysis are part of SAMMIE's inherent functionality, HADRIAN attempts to simplify their use and remove the overhead of driving the system, while concentrating on the application of sound ergonomics principles (Figure 14.18).

To construct a task description the designer first loads the graphic model to be assessed, from which the system extracts the interactive objects; those elements that will be sat on, reached to, viewed, activated, etc. The designer then decides what the user is to do by selecting the type of task element (e.g., reach) and then selecting the object to be reached for (e.g., keys). While the system provides users with ability to enter as much detail as they wish, it does not require it. Information that may be supplied includes which hand should be used, the duration of the reach, the importance of this task element, any maximum viewing distances, and orientation information for objects (this way up). Any information that the system needs to perform the analysis that is not explicitly specified by the designer will be set to a default that is task specific. Thus, the system may decide to use the nearest hand to perform a reach, but this may be overridden if the individual being assessed has a limited capability with that particular hand or has specified a preference to only use a particular hand for a particular type of task.

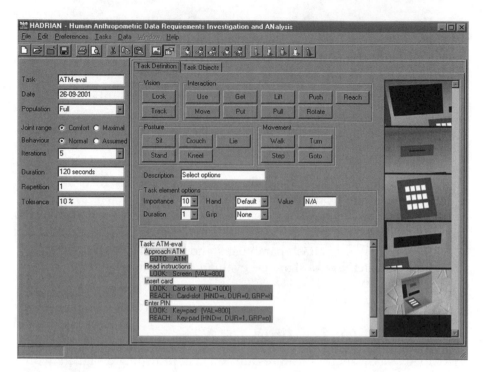

FIGURE 14.18
A screenshot from the prototype HADRIAN task analysis tool showing the task input selection screen in center and graphic displays of the environment on the right for an ATM machine evaluation.

The techniques behind the analysis have been developed to be as robust as possible, allowing for the multivariate nature of the analysis. The system also employs a framework that overlies the task description in an attempt to more accurately represent a dynamic process (performing the task) from a sequence of static task elements (reach x, view y, etc.) (Marshall et al. 2002). Such features have little or no impact on the designer, but lead to a much more flexible and realistic analysis.

14.4.4   Feedback and Result Reporting from SAMMIE

One of the most important aspects of the HADRIAN tool with SAMMIE concerns the results obtained from an analysis. Again, the designer is able to configure exactly how the tool behaves, and thus is able to customize the level and format of the feedback obtained. At one extreme, the system can perform the analysis without any user intervention, logging results, making assumptions where required, and skipping any failures. The final results are then presented when the analysis is complete for the designer to examine. The other extreme allows the designer to be involved in any decision-making processes where the system has to resolve some issue. Such issues may

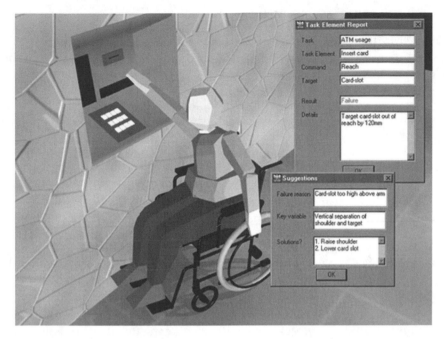

FIGURE 14.19
A HADRIAN task element failure report and suggestion dialogue.

include what to do in the event of a failed task element or an inconsistent task definition, such as explicitly specifying a hand for a reach task when the hand is already holding an item.

The flexibility of being able to intervene during an evaluation allows the designer to refine the task analysis during early runs and then run more autonomous analyses when the process is more robust. Alternatively, this facility allows the designer to run through the analysis in a more step-by-step approach to understand the issues faced by a particular user at every stage and actively think about how all aspects of the design can affect its usability. Figure 14.19 demonstrates one stage of a simple case study analysis performed on an automated teller machine (ATM). The user has failed on a particular task element and the system has been set to report all failures immediately and report suggestions. Although HADRIAN is not an intelligent design system, it can highlight some of the key variables that are involved in the failure and direct the designer's attention to the fundamental reasons for the problem.

A large range of results may be examined to determine who has successfully completed the task analysis, and potentially more importantly, who has failed, or been designed out, and why. A particular statistic presented is the percentage accommodated by the design. Although this is a very complex metric, it provides a powerful indicator of the usability of the design when compared against alternative design concepts.

Summary of SAMMIE Functions

The SAMMIE system has been introduced as a computer-aided ergonomics design tool. It provides those involved in the design process with a highly visual environment in which to assess the interaction of a design with its potential users. To aid this process SAMMIE incorporates functionality to help in representation of human physical form and capability together with assessment tools to perform virtual fitting trials. Recent developments have concentrated on addressing the issues of supporting designers, who may have little experience in ergonomics methods or may lack access to appropriate data. In addition, shortcomings of traditional data sources have been highlighted, especially when applied to groups that fall outside of the able bodied 19- to 65-year-old population.

Our approach to these concerns has been to develop our own database of individuals including a much richer set of data, not only to support multivariate analysis, but also to aid designers in familiarizing themselves with their target users and their behaviors/capabilities. Furthermore HADRIAN, a partner tool to SAMMIE, has been developed to focus the designer's efforts on the use of ergonomics principles. While still in development of these approaches, clearly the ease of use of the system also encourages application early on in the design process, when its effects can effectively and efficiently influence the design.

14.5   Boeing Human Modeling System

Steve Rice

Introduction

The Boeing Human Modeling System (BHMS) was developed starting in 1987 as a tool for human factors engineers to analyze proprietary product data for human fit, reach, and vision in aircraft cockpit design, using three-dimensional computer-aided design data. It has evolved based on user requirements to include a manikin with a 24-link flexible spine. More than 100 input measures can be accessed with the capability to analyze new designs, maintenance, and assembly scenarios for various populations of individuals, while supporting a variety of computer-aided design formats.

Through the years, many analysis features have been incorporated into BHMS, such as the capability to sweep three-dimensional volumes of body segment motions through space in order to define human motion paths and three-dimensional engineering "stay-out" volumes. This capability allows a manikin's arm/hand/tool to define a required three-dimensional volume for a population while performing such operations as maintenance tasks

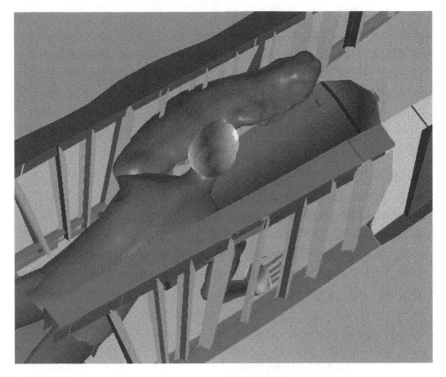

FIGURE 14.20
Swept volume of left arm and hand tool in a ratchet motion.

with various hand tools. Figure 14.20 depicts the swept volume generated from a ratcheting motion when using a socket wrench to remove a bolt during this BHMS maintenance task simulation.

14.5.1   Torso Modeling in BHMS

Another advance in human modeling analysis within BHMS was the development of the full 24-link spine and spine motion algorithms. The spine can be driven through its range of motion via a forward kinematics interface where the motion of the spine is split into groups: (1) head/neck, (2) lumbar, (3) torso, and (4) full spine. Figure 14.21 depicts the spine with attached skin enfleshment representation. Control points on the geometry representing the manikin torso/shoulder/neck skin enfleshment allow the surface to stretch as a normal torso would as the spine moves. The motion and reach range of the manikin was greatly extended by adding the spine motion algorithms to BHMS.

14.5.2   Predicting Reach Envelopes in BHMS

The ability to create "population union" reach envelopes with BHMS allows engineers to determine predicted maximum reach for all manikins in the

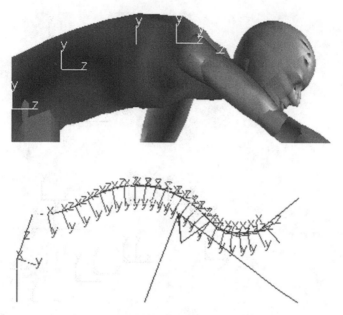

FIGURE 14.21
BHMS 24 link spine and enfleshment with control points used to shape the spinal curvature.

FIGURE 14.22
Reach envelope generated by BHMS.

"design to" population. This feature combines the maximum reach of any number of manikins, and then creates a reach envelope that will accommodate all those in the targeted design group. Figure 14.22 depicts a reach envelope shown in wire frame display mode.

FIGURE 14.23
Tool use with special fixed axis constraints for left hand with right hand attached to the tool handle.

14.5.3   Hand Tool Modeling in BHMS

Another analysis feature of BHMS is the ability to simulate the use of special tools, such as the speed wrench shown in Figure 14.23. This feature allows the full range of animation of the tool by keeping the left hand attached to the tool in a fixed location while the tool motion drives the inverse kinematics for the right hand, thus rendering a real-time evaluation of manikin-tool use.

Summary of Boeing Human Modeling System Functions

These types of analysis tools allow users to obtain critical human factors information early in the design phase of products, thus reducing cost and ensuring reach and visual access. It is believed that these types of human modeling analyses help to build products that are safer to use and more cost-effective to own.

Summary of Digital Human Modeling

Don B. Chaffin

This chapter has attempted to describe and illustrate the evolution of digital human modeling to improve the use of ergonomics knowledge in the early design of products and production processes. Five contemporary modeling activities have been presented. These represent a good cross section of the major developments in this field today.

What should be clear in reviewing the functionality of these models is that they offer a designer a means to literally visualize how a person might operate or service a commercial product (e.g., car, airplane, ATM, etc.) or perform a manual task in a production facility. The developers of these systems have made improvements in overcoming one major problem in the past, and that is to allow users to more easily import the graphic data needed to describe a particular geometric environment in which their human avatar will exist.

Another recent advance in most of these systems is their ability to consider the large, multidimensional variability that we humans represent. Several different statistical and empirical approaches to this important anthropometrics problem are described. It is in this area that the DHMs begin to show their true value over the two-dimensional manikin or template-fitting procedures traditionally used by ergonomists when attempting to determine what percentile of the population can be accommodated by a particular design. Only through the ability of these new DHMs, with their access to large arrays of population data, along with their ability to quickly perform complex statistical manipulations, can people of all sizes and shapes be considered in future designs.

The ability of the present models to capture and represent three-dimensional human size and shape has meant that maximum reach, clearance, and line-of-sight analyses are much more representative of select population capabilities. In addition, the newer models are now providing a user with ergonomic analysis tools to predict discomfort levels when certain postures are required, or load lifting limits for a worker population. Some of these ergonomic analysis tools are being integrated into the digital models in ways that make them easy to access. How well these tools are being used by designers with varying degrees of ergonomics knowledge is an interesting question, however. It would appear that at present the use and interpretation of the ergonomics predictions are still complicated enough that professional ergonomists are often necessary, thus making it more difficult to influence early design conceptualization unless a trained ergonomist is part of the design team.

Another important emerging attribute of these human models is their ability to use either inverse kinematics or motion capture data to predict the postures that people use when performing a task. At present these methods can be used to create realistic static postures and some simple motions for a variety of common tasks (e.g., walking, kneeling, stooping, etc.), thus avoiding the time-consuming and difficult job of inputting specific body segment positions for each task. The shear complexity of human motion behaviors and how these vary with a person's size, age, and gender, as discussed in earlier chapters of this text, however, makes this one of the most formidable problems for DHM developers to overcome in the future. For the present, modelers who need to present realistic human simulations of complex tasks often rely on motions captured from diverse people performing the tasks in a prototype or mockup of a new design. In the future such studies may be needed only to verify a final design concept. The hope is that future DHMs will include valid motion prediction models for a large array of motions, which then can be assembled to allow a truly dynamic visualization and ergonomic analysis of a task early in the design process.

In this latter regard, creating realistic dynamic human animations is not needed just to demonstrate to potential customers and executives how a new concept interfaces with a defined group of people, but having realistic motion prediction is the only means to truly understand the many different limitations that might affect a person's capabilities to perform a task. For example, consider body balance while performing a reaching motion. It is not difficult to compute the static balance limitation that a person might realize when reaching out to move or apply force to an object. It is much more difficult to predict the way balance might affect the actual capability when the person jerks on an object. Such dynamic motions are often used when people cannot perform a task in a slow and deliberate fashion. It allows them to compensate for lack of static balance or strength, thus extending their capabilities, but it also puts people at increased risk of falling or overstressing a particular musculoskeletal component. These are complicated human behaviors that are just now being studied and incorporated into DHMs.

Last, it is clear that human perceptual and cognitive capabilities are not being included in the present suite of DHMs. It is one thing for a model to indicate how a person would have to position the head to see an object, it is another thing to predict how much additional time it might require for a person to scan and identify a visual target, and it is still another thing to predict how much time it might take a person to perceive and respond correctly to a complex important visual message. When DHMs include accurate predictions of motion times, response times, and complex decision times, as well as probabilities of wrong or alternative motion responses, then designers will have a much more powerful tool than exists at present. As described in earlier chapters of this book, many relevant human performance models exist and others are being developed. It should not be too long before some of these are accessible within DHMs. Then, no doubt we will debate how valid their predictions are for different types of human perceptual and cognitive interface problems, just as we currently debate how valid the physical models are today.

One thing is certain: digital human modeling and simulation is just beginning to gain the support needed to have a real impact on design. Its importance will only increase, as will controversies about the validity of a particular model's output. It is hoped that this chapter serves to illustrate the potential and the limitations in this exciting aspect of ergonomics.

Acknowledgments

The development of HADRIAN (refer to Section 14.4) has been funded by the U.K. Engineering and Physical Sciences Research Council (EPSRC) through its EQUAL Initiative.

The contributors to this chapter wish to acknowledge and thank Pat Terrell, at the Center for Ergonomics at the University of Michigan, for her organizing and proofreading of the materials presented in this chapter.

References

Badler, N.I., Phillips, C.B., and Webber, B.L., 1993. *Simulating Humans. Computer Graphics Animation and Control,* New York: Oxford University Press.

Bonney, M.C., Blunsdon, C., Case, K., and Porter, J.M., 1979. Man-machine interaction in work systems. *International Journal of Production Research,* 17(6), 619–629.

Brandenburgh, R.L., 1999. Getting a digital foot in the door: how virtual human modeling is changing the way human factors influences cockpit design. *Flight Deck International,* September.

Burandt, U., 1978. *Ergonomie für Design und Entwicklung,* Cologne: Verlag Otto Schmidt KG.

Case, K., Porter, J.M., Bonney, M.C., and Levis, J.A., 1980. Design of mirror systems for commercial vehicles. *Applied Ergonomics,* 11(4), 199–206.

Case, K., Bonney, M.C., Porter, J.M., and Freer, M.T., 1990a. Applications of the SAMMIE CAD System in workplace design, in *Work Design in Practice,* Haslegrave, C.M., Wilson, J.R., Corlett, E.N., and Manenica, I., Eds., London: Taylor & Francis, 119–127.

Case, K., Porter, J.M., and Bonney, M.C., 1990b. SAMMIE: a man and workplace modelling system. In *Computer-Aided Ergonomics: A Researcher's Guide,* Karwowski, W., Genaidy, A.M., and Asfour, S.S., Eds., London: Taylor & Francis, 31–56.

Chaffin, D.B., 1969. A computerized biomechanical model: development and use in studying gross body actions. *Journal of Biomechanics,* 2, 429–441.

Chaffin, D.B., 1998. Prediction of population strengths, in *Digital Human Modeling for Design and Engineering. Proceedings of the Conference,* Dayton, OH, April 28–29.

Chaffin, D.B., 2001. *Digital Human Modeling for Vehicle and Workplace Design,* Warrendale, PA: Society of Automotive Engineers.

Chaffin, D.B., Andersson, G.B.J., and Martin, B.J., 1999. *Occupational Biomechanics,* 3rd ed., New York: John Wiley & Sons.

Ciriello, V.M. and Snook, S.H., 1991. The design of manual handling tasks: revised tables of maximum acceptable weights and forces. *Ergonomics,* 34, 1197–1213.

Garg, A., Chaffin, D.B., and Herrin, G.D., 1978. Prediction of metabolic rates for manual materials handling jobs. *American Industrial Hygiene Association Journal,* 39(8), 661–674.

Gleicher, M., 2001. Animation from observation: motion capture and motion editing. *Computer Graphics*, 33(4), 51–54.

Gordon, C.C., Bradtmiller, B., Churchill, T., Clauser, C.E., McConville, J.T., Tebbetts, I.O., and Walker, R.A., 1988. 1988 Anthropometric Survey of U.S. Army Personnel: Methods and Summary Statistics, Technical Report NATICK/TR-89/044, AD A225 094.

Gutowski, P., Bush, T.R., Hubbard, R., and Balzulat, J., 2001. Influence of automotive seat and package factors on posture and applicability to design models, Paper DHM 2001-01-2091, in *Proceedings of the SAE Digital Human Modeling Conference 2001*, June 26–28, Arlington, VA.

Henry Dreyfuss Associates, 1993. *The Measure of Man and Woman: Human Factors in Design*, New York: Whitney Library of Design/Watson-Guptill.

Karhu, O., Kansi, P., and Kuorina, I., 1977. Correcting working postures in industry: a practical method for analysis. *Applied Ergonomics*, 8, 199–201.

Krist, R., 1994. *Modellierung des Sitzkomforts: Eine experimentelle Studie*, Weiden: Schuch.

Laurig, W., 1973. Suitability of physiological indicators of strain for assessment of active light work. *Applied Ergonomics* (cited in Rohmert 1973b).

Lippmann, L. and Roessler, A., 1998. Virtual Human Models in Product Design, presented at IEEE YUFORIC Germany 98, June 16–18, 1998 Stuttgart, Germany, 53–61.

Marshall, R., Case, K., Oliver, R., Gyi, D.E., and Porter, J.M., 2002. Collection of Design Data from Older and Disabled People. The XVI Annual International Occupational Ergonomics & Safety Conference 2002, Toronto, pp. 1–5, CD-ROM International Society for Occupational Ergonomics & Safety (038-0L~1.PDF).

McAtamney, L. and Corlett, E.N., 1993. RULA: a survey method for the investigation of work-related upper limb disorders. *Applied Ergonomics*, 24(2), 91–99.

National Aeronautics and Space Administration (NASA), 1989. NASA–STD–3000: Man–System Integration Standards, Vol. 1, chap. 4, Washington, D.C.: National Aeronautics and Space Administration.

NIOSH, 1981. Work Practices Guide for Manual Lifting. NIOSH Technical Report 81-122, Cincinnati, OH: U.S. Department of Health and Human Services, National Institute for Occupational Safety and Health.

Oliver, R., Gyi, D., Porter, J.M., Marshall, R., and Case, K., 2001. A survey of the design needs of older and disabled people, in *Contemporary Ergonomics 2001*, *Proceedings of the Ergonomics Society Annual Conference*, Hanson, A., Ed., London: Taylor & Francis, 365–370.

Oliver, R.E., Marshall, R., Gyi, D.E., Porter, J.M., and Case, K., 2002. Collection of design data from older and disabled people, in *Proceedings of the XVI International Annual Occupational Ergonomics and Safety Conference*, Toronto, Canada, June 9–12.

Peacock, B. and Karwowski, W., 1993. *Automotive Ergonomics*, Washington, D.C.: Taylor & Francis.

Porter, J.M., Almeida, G.M., Freer, M.T., and Case, K., 1991. The design of supermarket workstations to reduce the incidence of musculoskeletal discomfort, in *Designing for Everyone and Everybody*, Queinnec, Y. and Daniellou, F., Eds., London: Taylor & Francis.

Porter, J.M., Case, K., Freer, M.T., and Bonney, M.C., 1993. Computer-aided ergonomics design of automobiles, in *Automotive Ergonomics*, Peacock, B. and Karwowski, W., Eds., London: Taylor & Francis, 43–78.

Porter, J.M., Freer, M.T., Case, K., and Bonney, M.C., 1996. Computer aided ergonomics and workspace design, in *Evaluation of Human Work: A Practical Ergonomics Methodology*, 2nd ed., Wilson, J.R. and Corlett, E.N., Eds., London: Taylor & Francis, 574–620.

Porter, J.M., Case, K., and Freer, M.T., 1999. Computer aided design and human models, in *The Occupational Ergonomics Handbook*, Karwowski, W. and Marras, W.S., Eds., Boca Raton, FL: CRC Press, 479–500.

Porter, J.M., Case, K., Gyi, D.E., Marshall, R., and Oliver, R.E., 2002. How can we "design for all" if we do not know who is designed out and why? The XVI Annual International Occupational Ergonomics & Safety Conference 2002, Toronto, pp. 1–5, CD-ROM International Society for Occupational Ergonomics & Safety (037-PO~1.PDF).

Pruett, C.J., Brown, C.M., and Balzulat, J., 2001. Development of an Electronic Belt Fit Test Device (eBTD) for Digitally Certifying Seat Belt Fit Compliance, Paper 01DHM-1, presented at SAE Digital Human Modeling for Design and Engineering Conference, June 26–28.

Raschke, U., Martin, B.J., and Chaffin, D.B., 1996. Distributed moment histogram: a neurophysiology based method of agonist and antagonist trunk muscle activity prediction. *Journal of Biomechanics*, 29, 1587–1596.

Raschke, U., Schutte, L., and Chaffin, D.B., 2001. Simulating humans: ergonomic analysis in digital environments, in *Handbook of Industrial Engineering: Technology and Operations Management*, Salvendy, G., Ed., New York: John Wiley & Sons.

Rebiffé, R., 1966. An ergonomic study of the arrangement of the driving position in motor cars. *Journal of the Institute of Mechanical Engineers*, London: Symposium.

Roebuck, J.A., Jr., 1995. *Anthropometric Methods: Designing to Fit the Human Body*, Santa Monica, CA: Human Factors and Ergonomics Society.

Rohmert, W., 1973a. Problems in determining rest allowances. Part 1: Use of modern methods to evaluate stress and strain in static muscular work. *Applied Ergonomics*, 4(2), 91–95.

Rohmert, W., 1973b. Problems in determining rest allowances. Part 2: Determining rest allowance in different human tasks. *Applied Ergonomics*, 4(2), 158–162.

Seidl, A., 1994. RAMSIS — a newly developed CAD tool for ergonomic analysis vehicles and its 3-dimensional system of analysis of postures and movements, in *Proceedings of the 12th Triennial Congress of the International Ergonomics Association*, Toronto.

Seidl, A., 1997. RAMSIS — a new CAD tool for ergonomic analysis of vehicles developed for the German automotive industry, SAE Technical Paper Series 970088, presented at International Congress & Exposition, Detroit, MI, February 24–27, 1997.

Seidl, A., 2000. The ergonomic tool ANTHROPOS in virtual reality – requirements, methods and realization, in *Proceedings of the IEA Congress 2000*, San Diego, CA.

Sheldon, W.H., 1940. *The Varieties of Human Physique*, New York: Harper & Bros.

Snook, S.H. and Ciriello, V.M., 1991. The design of manual handling tasks: revised tables of maximum acceptable weights and forces. *Ergonomics*, 34(9), 1197–1213.

Waters, T.R., Putz-Anderson, V., Garg, A., and Fine, L.J., 1993. Revised NIOSH equation for the design and evaluation of manual lifting tasks. *Ergonomics*, 36(7), 749–776.

Zehner, G.F., Meindl, R.S., and Hudson, J.A., 1993. A multivariate anthropometric method for crew station design. Technical Report AL-TR-1992-0164, Wright Patterson Air Force Base, OH: Crew Systems Directorate, Human Engineering Division, Armstrong Laboratory.

Index